Teratogenic Mechanisms

Advances in the Study of Birth Defects

VOLUME 1

Teratogenic Mechanisms

EDITED BY

T. V. N. Persaud

MTPPRESS LIMITED
International Medical Publishers

Published by
MTP Press Limited
Falcon House
Lancaster, England

ISBN 978-94-011-5912-8   ISBN 978-94-011-5910-4 (eBook)
DOI 10.1007/978-94-011-5910-4

Copyright © 1979 MTP Press Limited

British Library Cataloguing in Publication Data

Advances in the study of birth defects
   Vol. 1: Teratogenic mechanisms
   1. Abnormalities, Human
   2. Abnormalities (Animals)
   I. Persaud, T V N
   616'.043   QM691

Contents

List of Contributors

F. BECK
Department of Anatomy
School of Medicine
University of Leicester
Leicester, England

F. G. BIDDLE
Faculty of Medicine
Basic Sciences
Memorial University of Newfoundland
St. John's, Newfoundland, Canada
A1B 3V6

L. DENCKER
Department of Toxicology
University of Uppsala
Box 573
S-751 23 Uppsala, Sweden

ELIZABETH M. DEUCHAR
Department of Biological Sciences
University of Exeter
Hatherly Laboratories
Prince of Wales Road
Exeter EX4 4PS
Devon, England

E. M. JOHNSON
Daniel Baugh Institute
Thomas Jefferson University
1020 Locust Street
Philadelphia, Pennsylvania 19107, USA

S. KAPLAN
Department of Anatomy
The Medical College of Wisconsin
8701 W. Watertown Plank Road
Milwaukee, Wisconsin 53226, USA

L. A. KENNEDY
Department of Anatomy
Faculties of Medicine and Dentistry
University of Manitoba
Winnipeg, Manitoba, Canada R3E 0W3

W. LANDAUER
The Galton Laboratory
Department of Human Genetics and
Biometry
University College London
Wolfson House, 4 Stephenson Way
London NW1 2HE, England

W. G. McBRIDE
Director, Foundation
The Women's Hospital
Crown Street
Sydney, N.S.W. 2010, Australia

A. ORNOY
Department of Anatomy & Embryology
Hebrew University
Hadassah Medical School
Jerusalem, Israel

T. V. N. PERSAUD
Department of Anatomy
Faculties of Medicine and Dentistry
University of Manitoba
730 William Avenue
Winnipeg, Manitoba, Canada R3E 0W3

BERTHE SALZGEBER
Institut d'Embryologie du College de
France et du C.N.R.S.
49bis, Avenue de la Belle-Gabrielle
94130 Nogent-sur-Marne, France

R. M. SHAH
Faculty of Dentistry
The University of British Columbia
Vancouver, British Columbia, Canada
V6T 1W5

J. A. THLIVERIS
Department of Anatomy
Faculties of Medicine and Dentistry
University of Manitoba
730 William Avenue
Winnipeg, Manitoba, Canada R3E 0W3

Preface

The study of birth defects has assumed an importance even greater now than in the past because mortality rates attributed to congenital anomalies have declined far less than those for other causes of death, such as infectious and nutritional diseases. It is estimated that as many as 50% of all pregnancies terminate as miscarriages. In the majority of cases this is the result of faulty development. Major congenital malformations are found in at least 2% of all liveborn infants, and 22% of all stillbirths and infant deaths are associated with severe congenital anomalies.

Teratological studies of an experimental nature are neither ethical nor justifiable in humans. Numerous investigations have been carried out in laboratory animals and other experimental models in order to improve our understanding of abnormal intra-uterine development. In less than two decades the field of experimental teratology has advanced phenomenally. As a result of the wide range of information that is now accumulating, it has become possible to obtain an insight into the causes, mechanisms and prevention of birth defects. However, considerable work will be needed before these problems can be resolved.

This book brings together some of the more recent and important research findings related to the mechanisms and pathogenesis of abnormal development. It is not only a documentation of the latest experimental work, but it also points out future directions that seem productive and challenging.

I am most grateful to the distinguished panel of contributors. Their enthusiasm and cooperation have made this volume possible. My sincere thanks are due to the publishers, especially Mr D. G. T. Bloomer, Managing Director, MTP Press Ltd., for their encouragement and for extending to me every kindness. Finally, I am much indebted to my secretary, Mrs Barbara Clune, who has lightened the burden of editing this book.

This volume is dedicated to the memory of Professor Walter Landauer, a pioneer in the field of experimental teratology, who died during the preparation of this book.

Winnipeg, Canada   T. V. N. Persaud
February, 1979

1
Embryonic-fetal localization of drugs and nutrients

L. DENCKER

INTRODUCTION

Research in teratology has been mostly concerned with amassing data on the fetal effects of administered substances in experimental animals. The mechanisms of damage are however seldom recognized. Little is known about the relation between the placental passage of a particular substance and its tendency to induce fetal damage. Such data may be collected by excising fetal organs and then doing chemical or impulse counting analysis depending on the type of experiment. It can also be done by conventional autoradiography techniques after tissue fixation in different media. This approach is however not generally applicable to the wide range of soluble substances of current interest in the teratological literature. A technique which has contributed to this field of research is whole-body autoradiography – first presented by Ullberg[1] and successively developed in his laboratory[2]. This technique has been used mostly during late gestation and has facilitated the localization of drugs even in small fetal organs[3–5]. Reviews of important observations have been presented[6,7]. This technique – somewhat modified especially for the sectioning of excised whole uteri – has lately been used for studies on drug accumulation in the embryos at the early organogenetic period. This chapter summarizes results from these studies.

METHODS
Experimental schedule

In our group mice are most often used, but rats, hamsters or even as large specimens as whole pregnant macaca monkeys (5 kg) with a total length of 40 cm may also be sectioned intact[5]. The animals are first mated during the night (next day called day 1) and are then, on the selected gestational days, administered the radioactively labelled drug to be studied (usually intravenously). They are then anaesthetized after a predetermined time. In a late stage of pregnancy, the animals are most often used intact[2,6]. In early preg-

1

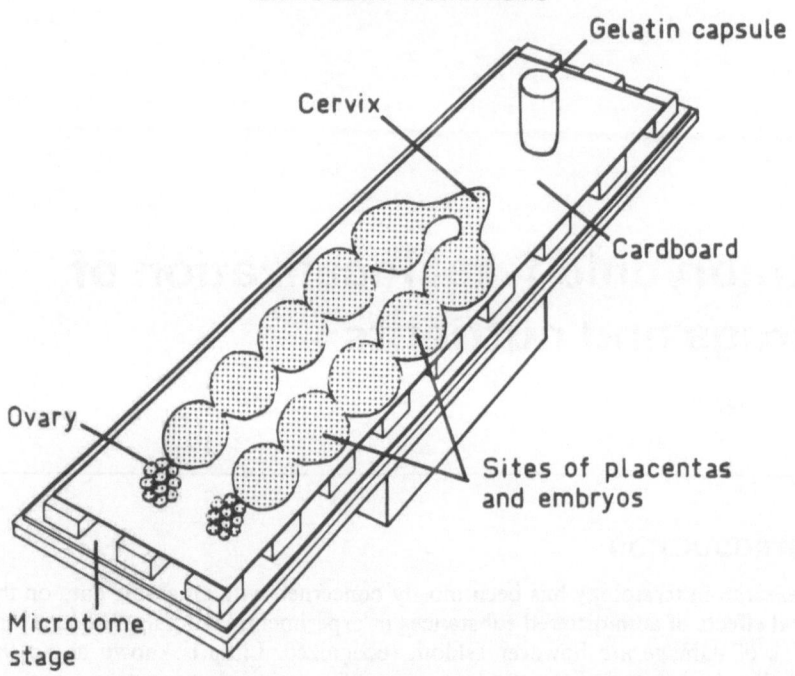

Figure 1.1 Schematic drawing of arrangement of rodent uteri for simultaneous sectioning through all embryos. Gelatin capsule used for maternal serum

nancy, however, it is better to remove the whole uterus and to arrange it in a horizontal plane[8] (Figure 1.1). It is also worthwhile to place beside the uterus a maternal organ with a known accumulation of the pertinent drug. The uterus, or the whole animal, is circumscribed by a frame and surrounded by an aqueous gel of carboxymethyl cellulose (CMC). It is then immediately frozen by immersion in a mixture of hexane cooled with dry ice. Maternal serum after centrifugation of the blood is placed in a capsule of gelatin previously mounted in the CMC together with the uterus (Figure 1.1). The specimen is then sectioned in a cryostat microtome (Palmstiernas Mekaniska Verkstad, Stockholm) at −15 °C. Scotch tape (No. 810, Minnesota Mining and Manufacturing Co.) is placed on the cut surface and 10–20μm thick sections are cut attached onto the tape. They are freeze-dried and then apposed against X-ray films (e.g. Industrex C, Kodak) under light pressure. Together with the section an isotope staircase may be pressed against the film in order to get a more quantitative evaluation of the tissue concentration[9]. After the exposure time − which varies in length depending on the isotope, dosage, section thickness, and film used − the latter is developed and the section is stained.

Parallel with the autoradiography, sometimes scintillation counting experiments are carried out. The maternal serum concentration is measured together with the total fetal concentration plus the concentration in the placenta, the maternal liver, kidney and brain and some other relevant organ(s). If of significance, fetal organs like the liver or brain are investigated separately.

2

Experimental schedule comments

The use of sections of whole pregnant mice may be advantageous, especially in late pregnancy. It is then often of interest to compare the drug concentration of a fetal organ with the corresponding organ of the mother. During the earlier stages, it is better to section the excised uterus. The section quality will be improved as sectioning through the maternal skeleton is avoided. All embryos may be sectioned simultaneously, and the risks for missing interesting embryonic structures are minimized. The 'maternal reference' is still present in the uterine tissues, the separated serum and some maternal organ(s) of interest.

One of the objectives of the studies has been to compare the embryonic-fetal uptake and distribution as a function of gestational age. Therefore, one fixed time interval between drug administration and autopsy has generally been used for each gestational day studied, and the interval has been chosen on the basis of previous data on the elimination of the specific drug from the mother. Briefly, one has to choose a time point after the maternal serum concentration peak, when the fetal to maternal concentration ratio has largely been stabilized. More than an eight hour interval is not usually chosen, as longer time intervals would result in a diluting effect on the fetal concentration due to fetal growth. This effect would in all probability be essentially more pronounced in early than in late pregnancy. For the confirmation of possible long term retention of drugs in the conceptus, some longer intervals have also been used and occasionally even a few shorter time intervals.

Drugs used in our investigations are usually labelled with carbon-14 or tritium and sometimes with iodine-125 or sulphur-35. A radioactive element, e.g. a heavy metal, may of course be used 'as is'. When labelled compounds are used, one has to consider the possibility of drug metabolism. The autoradiographic investigation therefore may be complemented by appropriate separation methods.

RESULTS AND DISCUSSION

So far only a few compounds have been investigated in respect of their kinetics in the feto-placental unit during the early organogenetic period. Those which have been studied were chosen mainly because they are more or less discussed in the teratological literature. This has favoured the use of substances with a wide range of chemical properties, and great variations in the embryonic distribution have been found. This concerns the rate of placental transfer as well as the distribution pattern within the embryos and their variations with developmental age.

Heavy metals

Heavy metals have received considerable attention as possible health hazards for the public and for known occupational health problems. They have been considered from the point of view of fetal damage as well. From reports by Ferm and his co-workers[10] and others, we know that cadmium and inorganic

3

mercury for example may cause severe malformations in rodents when given during early pregnancy. This information was of interest to us because the metals mentioned have been shown not to pass to the fetus in late gestation but instead to accumulate in the placenta[9, 11]. This might indicate the interesting possibility of a placental site of action in the induction of teratogenic effects.

We therefore found it appropriate to study the autoradiographic localization of cadmium, the more fetotoxic of the two metals, during the early organogenetic period[8, 12]. The majority of embryonic tissues did not show up in the autoradiograms, but in certain spots, recognized as different areas of the primitive gut (embryonic endoderm), a high accumulation was found (Figure 1.2). Upon cadmium administration (to mice and hamsters) up to the time of closure of the vitelline duct, cadmium accumulated in the gut wall, where it was retained for several days. After this closure, however, cadmium did not reach the embryo in significant amounts. The same pattern of uptake in the primitive gut was found for inorganic mercury[8], although lower.

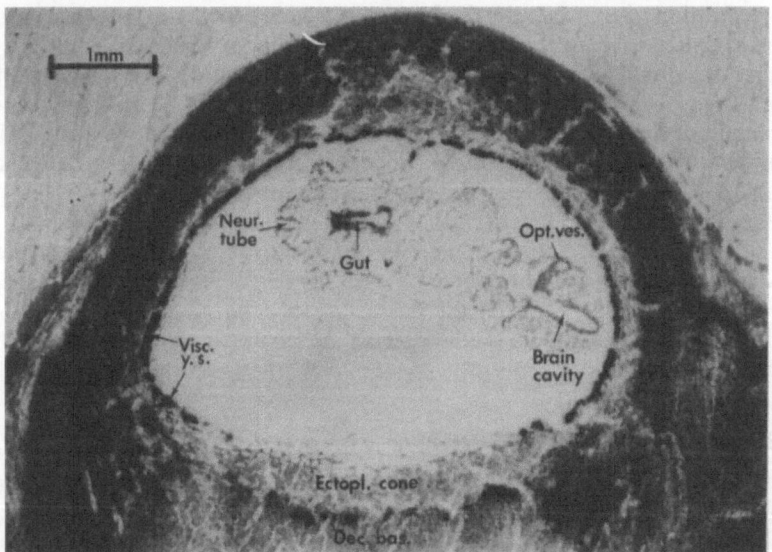

Figure 1.2 Photomicrograph of section with superimposed autoradiogram (black grains). Hamster uterus 24 h after i.v. injection of ^{109}CdCl$_2$ (day 8). Notice high uptake in endoderm of embryonic gut and visceral yolk sac placenta (visc.y.s.). No uptake in other embryonic tissues

Trypan blue

Upon evaluation of the published data concerning trypan blue, it was found that the embryonic susceptibility to this teratogen both in mice[13] and rats[14] chronologically largely coincided with the availability of the gut for drug accumulation via the vitelline duct. Experimentally it was possible to show that trypan blue also accumulated in the gut (mouse, rat) but not in other embryonic structures[15].

4

The hypothesis put forward by Beck and Lloyd, claiming that teratogenic effects may be provoked by disruption of the normal yolk sac placental function (using trypan blue as a model substance)[16] has received strong support in the literature. The basic concepts of this mechanism may well be valid not only for trypan blue but also for heavy metals, since they accumulate readily in the yolk sac as well as in the chorioallantoic placenta. Alternative hypotheses may however be discussed.

Looking into the literature of inductive processes in embryogenesis, one mostly finds models for the interaction between the ectodermal and meso-dermal tissues. With the accumulation of teratogens in the embryonic endo-derm, however, knowledge about endodermal influences on the behaviour of the other germ layers would be of value. To examplify this, it has been claimed that cadmium disturbs the development of the facial mesoderm in hamsters (causing cleft palate)[17] and trypan blue is known to cause a high percentage of exencephaly in mice[13]. The close contact of the foregut to the mesoderm and neuroectoderm of the head fold suggests a possible interaction in this region. Furthermore, attention to a developmental disturbance in the organs of endodermal origin seems to be indicated.

Zinc

The embryonic-fetal uptake and distribution of zinc is of interest for several reasons. First of all because of the striking and rapid onset of teratogenic effects after introduction of zinc deficient diets to experimental animals[18, 19]. Secondly, there seems to be a geographic relation between a clinically mani-fest zinc deficiency and CNS malformations (Egypt, Iran)[20, 21]. Thirdly, zinc inhibits many of the toxic effects of cadmium, among which are the

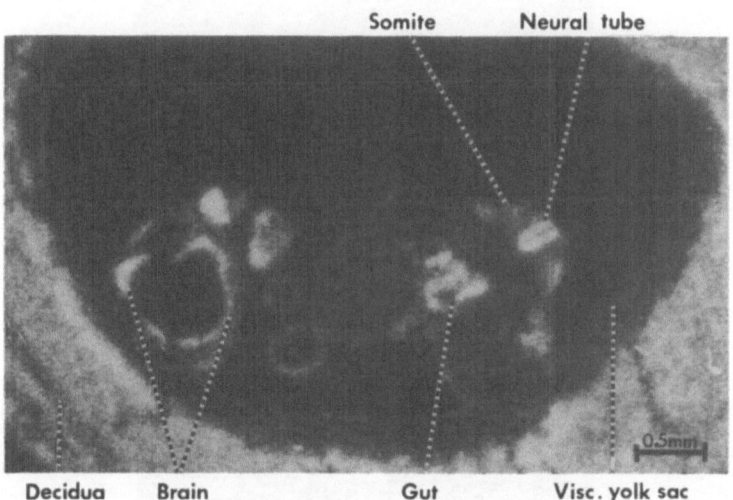

Figure 1.3 Autoradiogram showing distribution of ^{65}Zn in hamster embryo 24 h after i.v. injection of ^{65}ZnCl$_2$ to the mother (day 8). High concentration (white areas) in neuroepithelium (brain and neural tube) and gut and adjacently in tissues developed from the endoderm. Low concentration in visceral yolk sac endoderm

teratogenic ones[22]. Thus a comparison of the distribution of zinc and cadmium in the embryo may be of interest.

In contrast to cadmium, zinc[8] accumulated in all the embryonic structures of mice and hamsters, especially in the neuroepithelium but also in the endodermal structures (Figure 1.3). On the other hand, there was no apparent placental accumulation. Later in gestation, high concentrations were seen in the fetal liver. In the brain, some cell layers showed a higher concentration than others. Generally, a correspondence was found between the zinc and thymidine distribution. This may be explained by the participation of zinc in the DNA-polymerase[23] or in other enzyme systems functioning in growing cells[24]. The incorporation of thymidine into the neuroepithelial cells has been shown to decrease in rat embryos when the dams were kept on a diet deficient in zinc[25].

No apparent conclusions could be drawn from this study which would explain the inhibition by zinc of the toxic effects of cadmium. The endodermal cells (cf. Figures 1.2 and 1.3) may be the site of interference (in specific enzyme systems) between the two metals, but this may take place outside the embryo as well. Regardless, the capacity of the placenta to distinguish between two metals with such physicochemical similarities is fascinating.

Salicylic acid

When salicylic acid is given to pregnant mice, it causes fetal disturbances at different stages of gestation[26]. In line with this, the fetal concentration of

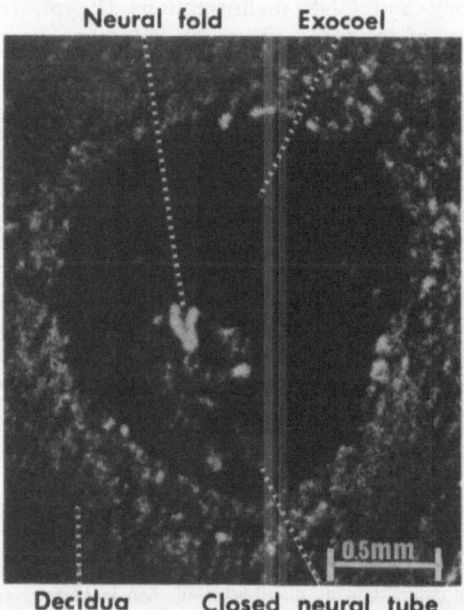

Figure 1.4 Autoradiogram of mouse uterus 4 h after i.v. injection of [^{14}C]salicylic acid to the mother (day 9). Highest concentration in the neural fold

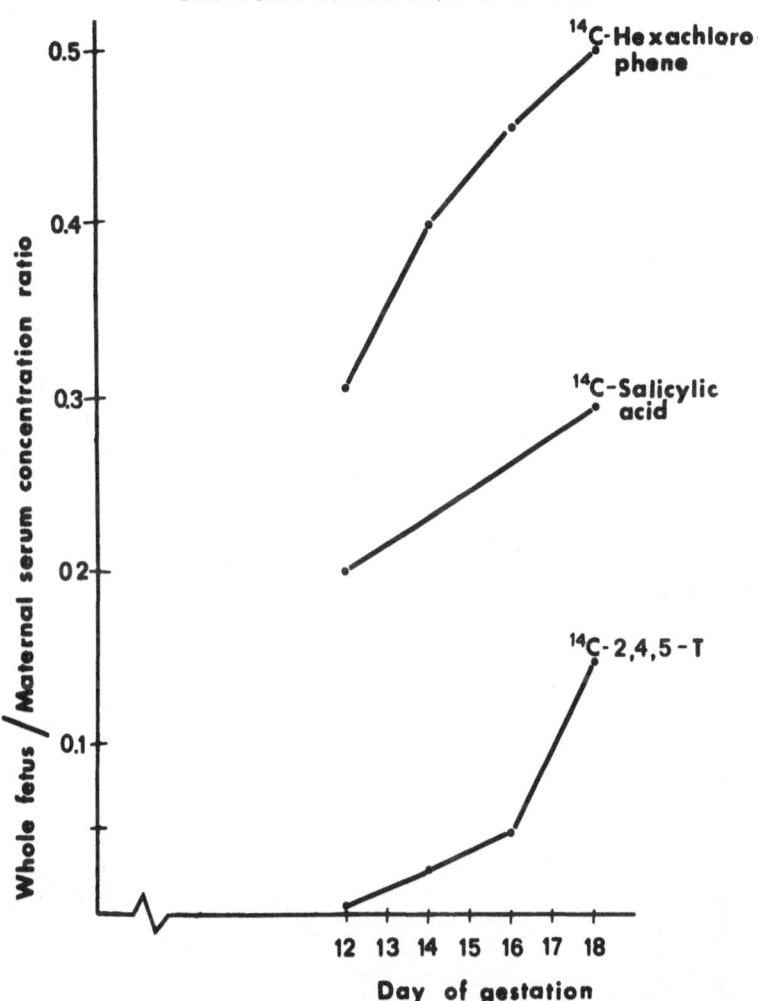

Figure 1.5 Gestational age-related variations in radioactivity concentration ratio between homogenized mouse fetuses (per gram tissue) and maternal serum (per ml) 4 h after i.v. injection at each indicated day. The ratio for all substances studied is higher in late than in mid-gestation. This is most pronounced for [14C]2,4,5-trichlorophenoxyacetic acid ([14C]2,4,5-T), which shows a 20-fold increase from day 12 to day 18. (Each value represents the mean of three fetuses or more from each of three mothers)

salicylic acid seems to be high throughout gestation[8, 27] (Figures 1.4 and 1.5). A finding judged to be of great significance, and which will be further considered in this paper, was the change in the distribution pattern which could be followed throughout the whole of fetal development[8]. The fetal pattern in late gestation was characterized by a relatively high concentration in the blood and fluids (body cavities and amniotic fluids). While at this stage a developed blood–brain barrier was present, the embryo showed a high concentration in the neuroepithelium but a lower one in the mesodermal structures and in the fluids (Figure 1.4). This means that a quantitative evaluation of the average fetal concentration is of limited value when the risks to the individual organs are considered.

2,4,5-trichlorophenoxyacetic acid (2,4,5-T)

2,4,5-T[8], another relatively low molecular organic acid, was transported rather easily to the late fetus. The activity in the body fluids was fairly high. There was a developed blood–brain barrier. In early gestation, however, 2,4,5-T could not be detected in the different embryonic tissues by means of autoradiography. On day 12 (mouse), when traces of activity could be detected in the embryos, the whole fetus/maternal serum concentration ratio was still very low (0.007) (Figure 1.5). Similar results have been found for the biphenyl molecule[28]. No rational explanation has been put forward for these gestational age-related differences between for example salicylic acid and hexachlorophene (see below) on the one hand, and 2,4,5-T and biphenyl on the other. An interesting observation is that 2,4,5-T causes malformations (cleft palate[29] and renal anomalies[30]) predominantly in the later part of the organogenetic period, i.e. when significant amounts of the substance is transferred to the fetus.

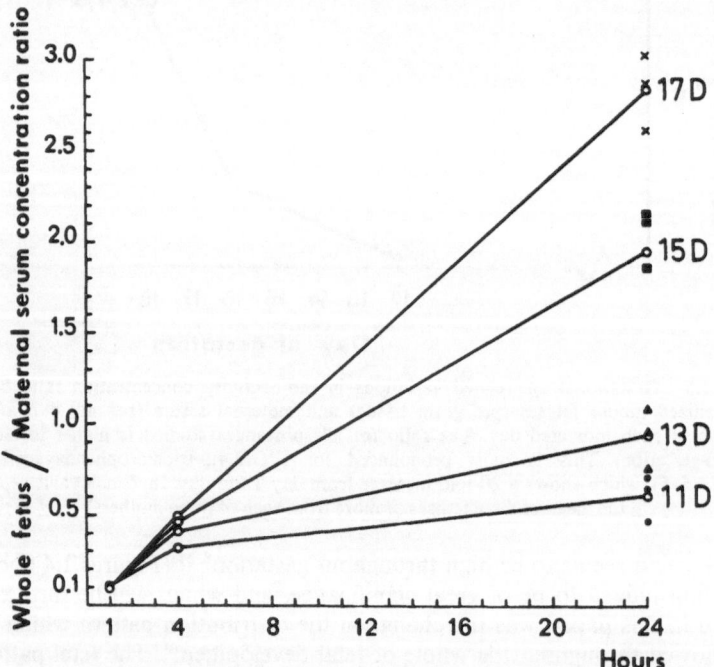

Figure 1.6 Gestational age-related variations in radioactivity concentration ratio (as in Figure 1.5) at different times after injection of [^{14}C]hexachlorophene to mice. The later in gestation, the more pronounced is the retention in fetal tissues. (Mean of three fetuses from each of three mothers. At 24 h the mean within each mother is indicated)

Hexachlorophene

Hexachlorophene[31] was highly transferred to the fetus (Figure 1.5). The most conspicuous feature was the long retention time in the fetus as compared to the mother. Even when hexachlorophene was injected into the mother prior to mating, it was transferred to the fetus and was found there in higher concentration than in the maternal tissues. Among the numerous drugs studied in our laboratory, hexachlorophene is the one which has shown the longest fetal retention time compared to the mother (nutrients excluded). The rate of maternal to fetal transfer as well as the fetal retention increased with the stage

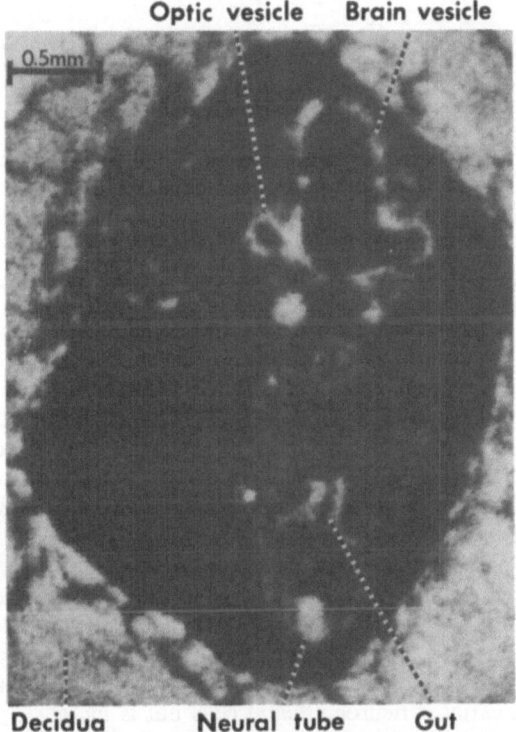

Figure 1.7 Autoradiogram showing distribution of radioactivity in mouse embryo 6 h after i.m. injection of [14C]hexachlorophene to the mother (day 10). Observe high uptake especially in brain and optic vesicles and neural tube

of gestation (Figure 1.6). During organogenesis, the uptake in the neuro-epithelium was fairly high (Figure 1.7). This may be related to the eye and CNS malformations found by Kimmel et al.[32].

Phenothiazine derivatives

Chlorpromazine, tricyclic antidepressants and chloroquine have been extensively studied in our department mainly because of their affinity to

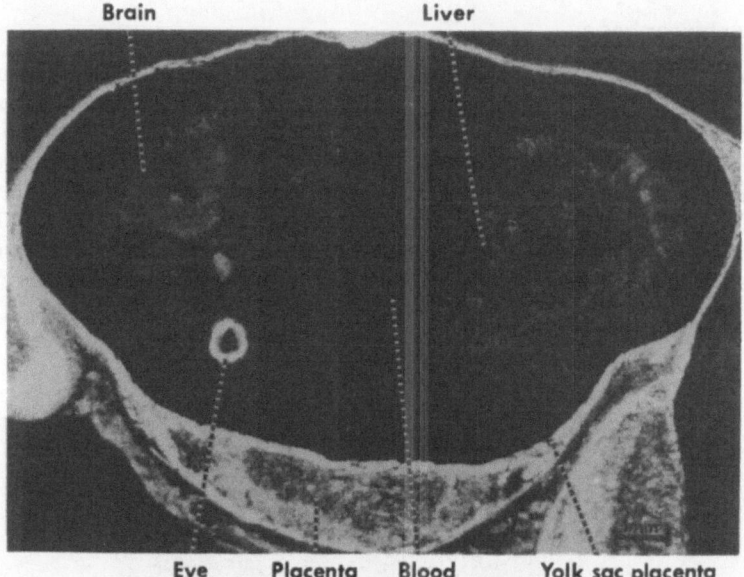

Figure 1.8 Fetal distribution pattern of radioactivity 4 h after i.v. injection of [^{35}S]chlor-promazine to pregnant mouse (day 14). Note the accumulation in pigment epithelium of eye. The brain concentration is comparatively high

melanin[33, 34]. These drugs have been shown to accumulate in the fetal eye (Figure 1.8) from the earliest stage at which melanin can be observed, and also in certain structures of the inner ear (chloroquine)[5]. As chlorpromazine has been used prenatally to induce postnatal changes in neurotransmitter activity as well as in behavioural performance experimentally (rats)[35, 36], it was decided to study its uptake in the fetal brain throughout gestation. Its transplacental passage and brain uptake were low during the early embryogenetic period, while the whole fetal as well as the brain uptake increased with gestation. Thus at day 14 the fetal brain showed a considerable concentration (Figure 1.8) compared with the serum. Consequently, this psychoactive drug behaves contrary to the drugs previously described in that it has a low affinity for the undifferentiated neuroepithelial cells but is increasingly accumulated with maturation. The fetal brain concentration was far lower than the maternal one also in late gestation. The results fit well with the finding that chlorpromazine affected the offspring if given just prior to parturition[36].

The quaternary phenothiazine derivative Aprobit® (N-hydroxyethylpromethazine) showed a very low embryonic uptake in early gestation. The transport to the late stage fetus was lower than that of the tertiary chlorpromazine. The placenta however accumulated this drug. The fetal brain had a low concentration but it was high in the choroid plexus, indicating the existence of either a barrier to the brain or a transport mechanism out of the brain.

Nicotine

Nicotine is still another drug with a melanin affinity[37]. It has been shown to accumulate in the bronchi of the mature fetal lung as well[38]. The previously described pattern of a high brain concentration in early gestation (Figure 1.9)

Optic vesicle   Brain vesicle   Decidua

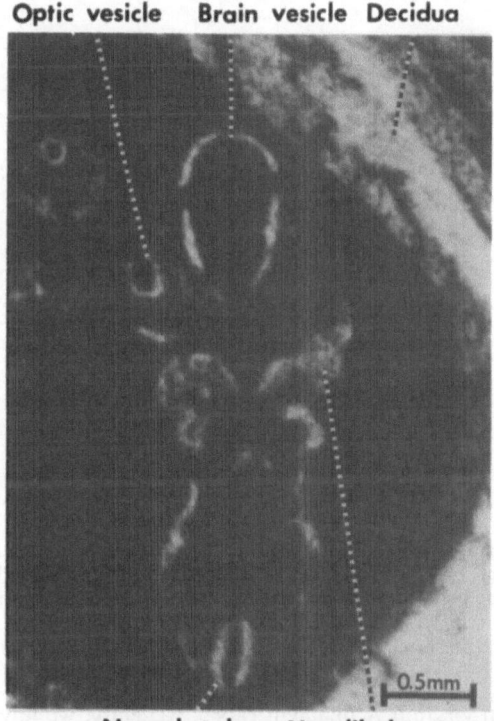

Neural tube   Mandibular area

Figure 1.9 Embryonic radioactivity distribution pattern 4 h after i.v. injection of [^{14}C]nicotine to mouse (day 10). Notice accumulation especially in the brain and optic vesicles and neural tube, but also in mandibular area

and a lower concentration in late gestation was repeated. It may be mentioned that – at least in late gestation – most of the radioactivity in the fetus after the injection of labelled nicotine represents cotinine or more water soluble metabolites[38].

Vitamins

Vitamin A is one of the vitamins which has received great attention in teratological research ever since the discovery by Hale (1937)[39] that vitamin A deficiency causes severe malformations. Hypervitaminosis A also causes malformations[40,41]. Retinoic acid, which has recently been shown to be formed in the body from retinol[42], is an even more potent teratogen than retinol itself[43]. Retinoic acid may, on the other hand, maintain embryogenesis

Eye   Brain   Visc.yolk sac Placenta   Placenta   Eye   Brain

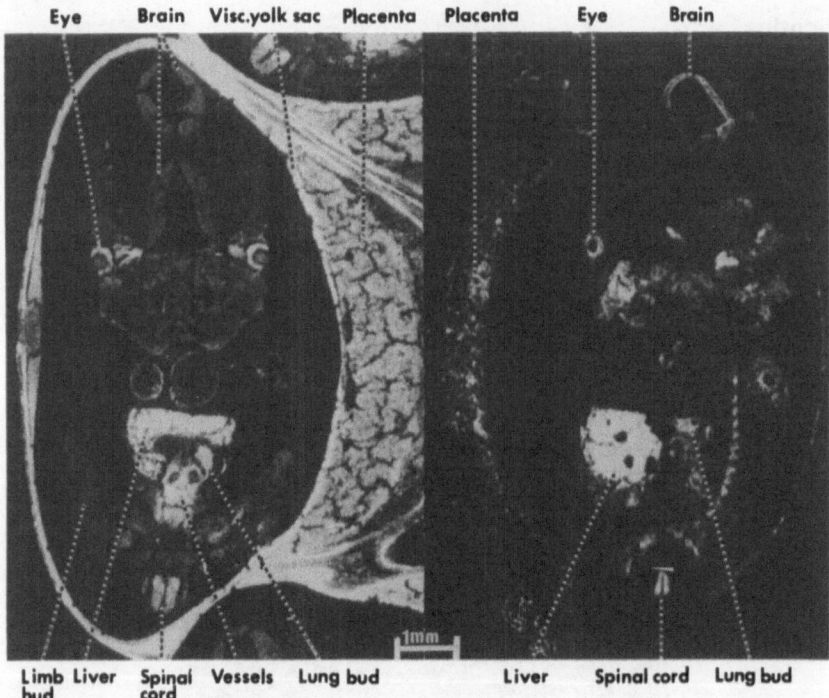

Limb Liver Spinal Vessels Lung bud   Liver   Spinal cord Lung bud
bud   cord

Figure 1.10 Comparison of mouse fetal distribution pattern 4 h after maternal injection of 15-[^{14}C]retinyl acetate (left) and [^{14}C]thymidine (right) (day 13). Similarly high uptake is seen in the eye, liver, lung buds, and spinal cord. For both substances a high uptake was seen in certain areas of the ependymal lining of the brain, heart and great vessels (not seen in these pictures). Then similarities indicate a role for vitamin A in connection with cellular division and/or growth in the embryos

(up to day 14) in vitamin A deficient rats[44]. Recently, we have compared the distribution pattern of these two highly interesting vitamins in mice at different stages of gestation. As was the case with the mother, in the embryo-fetus there were also important differences between the two related compounds. In the primitive streak to the first somites stage, vitamin A accumulated in what was identified as the extraembryonic endoderm, and a weaker uptake was seen in the embryo proper. During the organogenetic period, sometimes a very high uptake was seen in the neuroepithelial structures. In the endodermal derivatives, limb buds, mandibular arches, etc., a medium uptake was found. The distribution pattern at the different stages rather well equalled that of thymidine. At day 13 and 14 an apparent increase in accumulation was seen in the liver, lung buds, heart and great vessels, and in the retina (Figure 1.10). These sites of accumulation again correspond to the thymidine pattern (Figure 1,10). In late gestation, day 18, there was no apparent correspondence with the thymidine distribution pattern. The liver showed a high concentration, like the pleura and intestinal contents.

Retinoic acid showed an extremely high affinity for the neuroepithelial structures from day 9 up to day 13 (Figures 1.11 and 1.12) – more than any

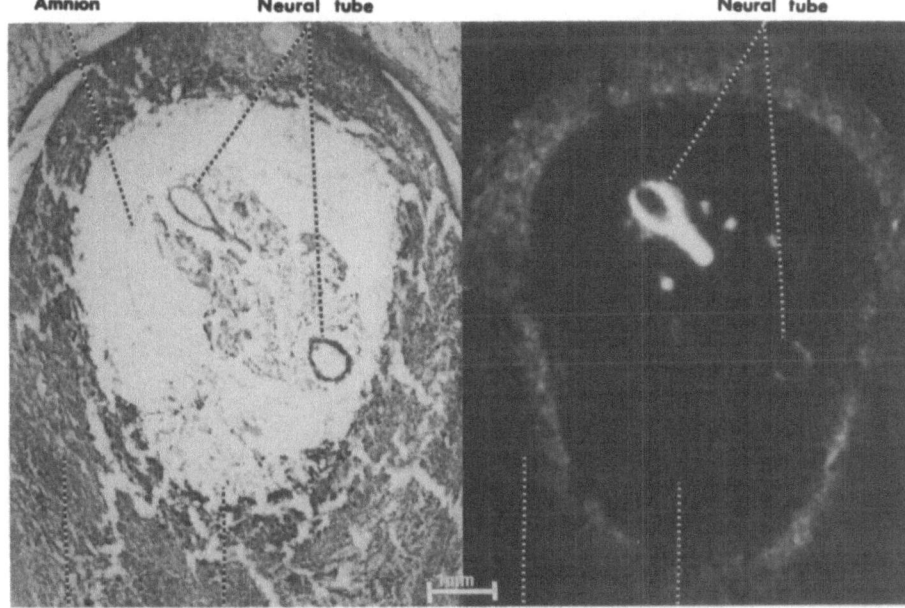

Figure 1.11   Section (left) and autoradiogram (right) showing a mouse embryo (day 10) 4 h after i.v. injection of 15-[¹⁴C]retinoic acid to the mother. Notice selective accumulation in neuroepithelial cells. but not in all parts of the neural tube

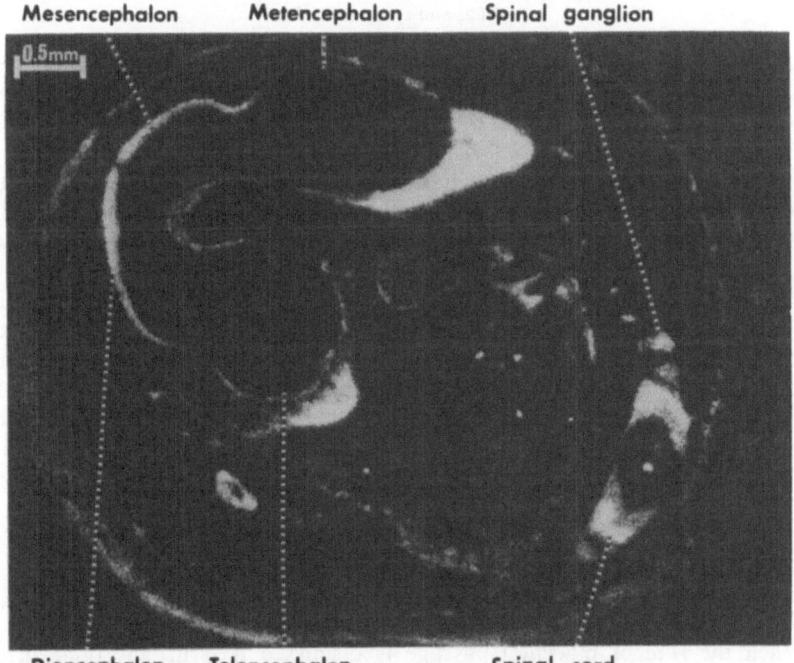

Figure 1.12   Autoradiogram of mouse embryo (day 11) cut partly sagittally 4 h after i.v. injection of 15-[¹⁴C]retinoic acid to the mother. The great variations in uptake at different levels of the brain can be seen. Other organs show a comparatively low concentration

13

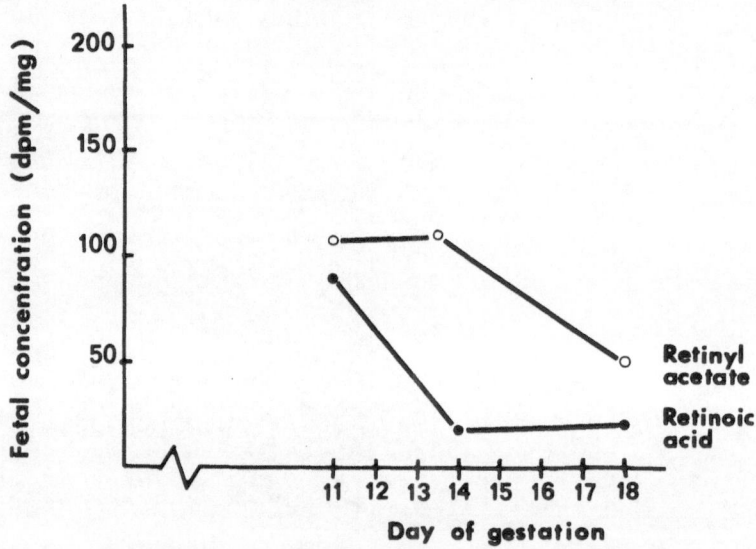

Figure 1.13   Gestational age-related variations in fetal radioactivity concentration 4 h after i.v. injection of 11,12-[³H]retinyl acetate and 11-[³H]retinoic acid, respectively, to the mother. The decrease in retinoic acid concentration from day 11 to 14 corresponds to a decrease in CNS concentration. Retinyl acetate, however, accumulates in for example liver, heart, lung buds and eyes on day 14

other substance hitherto studied, vitamin A included. Different areas of the CNS showed varying degrees of accumulation. While thymidine was incorporated close to the ventricles (internal germinal layer), retinoic acid was most often accumulated in the outer layer(s) of the neuroepithelium. From day 14 and later, the retinoic acid distribution in the CNS was essentially lower, but centres of higher concentration were found e.g. in the myelencephalon and in the spinal cord. The fetus on day 13–14 of gestation did not accumulate retinoic acid in the liver and the lungs like vitamin A.

Liquid scintillation counting experiments (Figure 1.13) showed significant differences between the two vitamin forms. While vitamin A reached the same high fetal level on day 11 and 14 and then decreased to day 18, the retinoic acid concentration decreased significantly from day 11 to day 14. The differences are most probably due to the fact that the retinoic acid brain concentration greatly decreased from day 11 to day 14 (seen by autoradiography), while vitamin A at day 14 was heavily accumulated in the liver, lungs and heart.

COMPARATIVE CONSIDERATIONS

Changes in placental transfer

When the rate of placental passage is discussed it is usually done in the general terms of membrane transport. Calculations of the kinetics of drug molecules across the placenta have most often been done in late gestation. The investigations reported here show that the fetal concentration of a foreign compound injected into the mother generally will be higher in late than in early and midgestation (Figures 1.5 and 1.6). We have however found great variations in this respect between the different molecules. Thus, salicylic acid[8] and hexachlorophene[31] showed high fetal concentrations in both early and late gestation (Figures 1.4–1.7). 2,4,5-T[8] (Figure 1.5) and biphenyl[28], on the other hand, both reached relatively high fetal concentrations in late gestation, but showed a low concentration at day 12 and could not be detected autoradiographically in the embryos during the height of the organogenetic period. It thus seems that the placenta changes its discriminating capacity during gestation both quantitatively – in that the permeability to most foreign molecules successively increases – and more qualitatively by the fact that some molecules which are almost completely inhibited at early gestation will be significantly transported at late gestation. The rationale of these differences is not clear.

One possible explanation for the concentration variations with gestation is a change in the binding capacity of the fetal organs for the drug in question. This apparently explains the variations found for vitamin A and retinoic acid (Figure 1.13). The great variations seen for 2,4,5-T, however, most probably are due to placental changes, as this compound at all fetal ages is found preferably in the serum and body fluids. The placental changes which may contribute to the variations in permeability are for example increased surface area, attenuation of the trophoblasic cell layers and changes in its hemodynamic properties. Some additional factor(s) may exist as well, which may explain the important differences in variations between the drugs discussed above.

Transfer through the vitelline duct

An entrance to the embryo which would have been fairly difficult to calculate is the one via the yolk sac cavity and through the vitelline duct to the embryonic gut endoderm. This passage is open before the chorioallantoic placenta is formed. The accumulation of foreign compounds (cadmium[12], mercury[8], trypan blue[15]) in the visceral endoderm may well be considered a protective mechanism. It is interesting to compare cadmium, which is accumulated here and thus prevented from further transport, with zinc[8] which does not accumulate but passes through and is utilized by the growing embryo (cf. Figures 1.2 and 1.3). From this point of view, it may be considered a 'failure of nature' that part of the endoderm is incorporated into the embryo, forming such essential parts of the organism as the gastrointestinal tract, etc., and that the passage between the gut and the maternal environment via the vitelline duct exists for such a long period (closes at 9.5 and 11 days respectively in the mouse and the rat).

Changes in fetal organ concentration

As has already been pointed out, gestational age-related changes in the total fetal uptake of drugs may be due to concentration variations in the single organs. The advantages of the autoradiographic technique used is perhaps most evident when such variations are to be evaluated.

15

The different drugs reported in this paper which reached the embryo–fetus in considerable amounts throughout gestation (salicylic acid, hexachloro phene, nicotine) all showed a characteristic change in pattern of distribution with fetal maturation. During the organogenetic period a high uptake was seen in the neuroepithelial cells, most often higher than in other tissues. At midgestation (around day 13–15) an even distribution was found with a similar concentration in all tissues. During the days preceeding parturition, a more or less pronounced blood–brain barrier was developed. Physiologically active substances, vitamin A, retinoic acid and zinc similarly showed a high accumulation in the neuroepithelium during the organogenetic period, like thymidine. It may be relevant to consider the accumulation of these nutrients as of importance for cellular multiplication and/or growth.

It would be misleading however to imply that the above mentioned non-physiological drugs (salicylic acid, hexachlorophene, nicotine, and probably many other drugs not yet studied) are taken up into the growing cells of the embryo, especially in the neuroepithelium, because they interact in the processes of cell division or growth. It is however of significance to emphasize that drugs are actually accumulated in these cells. In some cases such an accumulation may be responsible for malformations in the CNS or eyes, if the drugs are given in high doses. This tendency of accumulation may also be related to mental or psychomotor disturbances as measured postnatally. The research in this area is still in its cradle, and it is hoped that the finding of this drug accumulation in neural tissues may stimulate the search for postnatal effects.

Retinoic acid is the substance which has hitherto shown the highest accumulation in the neuroepithelium, generally far higher than vitamin A. Recent research has shown that vitamin A and retinoic acid take part in glycosyl transfer reactions in the formation of the glucoproteins[45]. This may have relevance to the formation of cellular constituents as well as the glucosaminoglucans and collagen of the extracellular matrix. The embryonic spinal cord has been shown to synthesize collagen[46], and collagen is an essential component in tissue interaction[47]. This might be one of the molecular bases for the severe malformations induced by hypervitaminosis as well as deficiency of vitamin A or retinoic acid.

If the molecular mechanism of action for vitamin A and retinoic acid were restricted to the glycosyl transfer reaction, one would expect the two compounds to be more similarly distributed in the embryo. The differences in distribution for the two vitamins may indicate some alternative functional role(s) or at least participation in the formation of different glucoproteins.

One has to be aware of the risks of overinterpretations of results in this field. All accumulations may not necessarily result in tissue damage. Furthermore, due to driving forces and phenomena very specific for the embryonic development, like inductive processes and morphogenetic movements, the accumulation in one tissue may well secondarily disturb the development in another tissue. It is also felt that the technique has to be refined in order to identify structures in more detail. A lot of work remains to map the placental transfer and fetal pattern of accumulation of defined groups of drugs and environmental pollutants as well as nutrients. It is hoped that some clues to the understanding of teratogenic mechanisms may result from future studies along this line.

ACKNOWLEDGEMENTS

Financial assistance of the Swedish Medical Research Council (Grant B78-14X-02876-09C) is gratefully acknowledged.

References

1. Ullberg, A. (1954). Studies on the distribution and fate of S^{35}-labelled benzylpenicillin in the body. *Acta Radiol.* (Stockh), (Suppl.), 118
2. Ullberg, S. (1977). *The technique of whole-body autoradiography.* Cryosectioning of large specimens. *Science Tools.* The LKB Instrument Journal
3. Slanina, P., Ullberg, S. and Hammarström, L. (1973). Distribution and placental transfer of ^{14}C-thiourea and ^{14}C-thiouracil in mice studied by whole-body autoradiography. *Acta Pharmacol. Toxicol.* (Kbh), 32, 358
4. Brandt, I. (1975). The distribution of 2,2',3,4,4',6' and 2,3'4,4',5',6-hexachlorobiphenyl in mice studied by whole-body autoradiography. *Toxicology*, 4, 275
5. Dencker, L., Lindquist, N. G. and Ullberg, S. (1975). Distribution of an ^{125}I-labelled chloroquine analogue in a pregnant Macaca monkey. *Toxicology*, 5, 255
6. Ullberg, S. (1973). Autoradiography in fetal pharmacology. In: L. O. Boréus (ed.). *Fetal Pharmacology*, pp. 55–73. (New York: Raven Press)
7. Waddell, W. J. and Marlowe, G. C. (1976). Disposition of drugs in the fetus. In: B. L. Mirkin (ed.). *Perinatal Pharmacology and Therapeutics*, pp. 119–268. (New York: Academic Press)
8. Dencker, L. (1976). Tissue localization of some teratogens at early and late gestation related to fetal effects. *Acta Pharmacol. Toxicol.* (Kbh), 39 (Suppl.), 1
9. Berlin, M. and Ullberg, S. (1963). Accumulation and retention of mercury in the mouse. *Arch. Envir. Health*, 6, 589
10. Ferm, V. H. (1972). The teratogenic effects of metals on mammalian embryos. In: D. H. M. Woolam (ed.). *Advances in Teratology*, Vol. V, pp. 51–75. (London: Logos Press)
11. Berlin, M. and Ullberg, S. (1963). *The fate of Cd^{109} in the mouse.* An autoradiographic study after a single intravenous injection of $Cd^{109}Cl_2$. *Arch. Environ. Health*, 7, 686
12. Dencker, L. (1975). Possible mechanisms of cadmium fetotoxicity in golden hamsters and mice: uptake by the embryo, placenta and ovary. *J. Reprod. Fertil.*, 44, 461
13. Schlüter, G. (1970). Embryotoxische Wirkungen von Trypanblau bei Mäusen in Abhängigkeit vom Behandlungszeitpunkt. *Naunyn-Schmiedeberg's Arch. Pharmacol.*, 267, 20
14. Wilson, J. G., Beaudoin, A. R. and Free, H. J. (1959). Studies on the mechanism of teratogenic action of trypan blue. *Anat. Rec.*, 133, 115
15. Dencker, L. (1977). Trypan blue accumulation in the embryonic gut of rats and mice during the teratogenic phase. *Teratology*, 15, 179
16. Beck, F., Lloyd, J. B. and Griffiths, A. (1967). Lysosomal enzyme inhibition by trypan blue: a theory of teratogenesis. *Science*, 157, 1180
17. Mulvihill, J. E., Gamm, S. H. and Ferm, V. H. (1970). Facial formation in normal cadmium-treated golden hamsters. *J. Embryol. Exp. Morphol.*, 24, 393
18. Blamberg, D. L., Blackwood, U. B., Supplee, W. C. and Combs, G. F. (1960). Effect of zinc deficiency in hens on hatchability and embryonic development. *Proc. Soc. Exp. Biol. Med.*, 104, 217
19. Hurley, L. S. and Swenerton, H. (1966). Congenital malformations resulting from zinc deficiency in rats. *Proc. Soc. Exp. Biol. Med.*, 123, 692
20. Hurley, L. S. (1974). Zinc deficiency, potatoes, and congenital malformations in man. *Teratology*, 10, 205
21. Burch, R. E. and Sullivan, J. F. (1976). Clinical and nutritional aspects of zinc deficiency and excess. *Med. Clin. North Am.*, 60, 675
22. Ferm, V. H. and Carpenter, S. J. (1967). Teratogenic effect of cadmium and its inhibition by zinc. *Nature (Lond.)*, 216, 1123

23. Lieberman, I. and Ove, P. (1962). Deoxyribonucleic acid synthesis and its inhibition in mammalian cells cultured from the animal. *J. Biol. Chem.*, **237**, 1634

24. Riordan, J. F. and Vallee, B. L. (1976). Structure and function of zinc metalloenzymes. In: A. S. Prasad and D. Oberleas (eds.). *Trace Elements in Human Health and Disease, Volume I. Zinc and Copper*, pp. 227–256. (New York: Academic Press)

25. Swenerton, H., Schrader, R. and Hurley, L. S. (1969). Zinc-deficient embryos: reduced thymidine incorporation. *Science*, **166**, 1014

26. Larsson, K. S. (1971). Action of salicylate on prenatal development. In: H. Tuchmann-Duplessis (ed.). *Malformations Congénitales des Mammifères*. Proceedings of Colloque Pfizer, Amboise, 1970, pp. 171–186. (Paris: Masson)

27. Kimmel, C. A., Wilson, J. G. and Schumacher, H. J. (1971). Studies on metabolism and identification of the causative agent in aspirin teratogenesis in rats. *Teratology*, **4**, 15

28. Brandt, I. (1977). Tissue localization of polychlorinated biphenyls. Chemical structure related to pattern of distribution. *Acta Pharmacol. Toxicol.* (Kbh), **40 (Suppl.)**, 2

29. Neubert, D. and Dillmann, I. (1972). Embryotoxic effects in mice treated with 2,4,5-trichlorophenoxyacetic acid and 2,3,7,8-tetrachlorodibenzo-*p*-dioxin. *Naunyn-Schmiedeberg's Arch. Pharmacol.*, **272**, 243

30. Courtney, K. D. and Moore, J. A. (1971). Teratology studies with 2,4,5-trichlorophenoxyacetic acid and 2,3,7,8-tetrachlorodibenzo-*p*-dioxin. *Toxicol. Appl. Pharmacol.*, **20**, 396

31. Brandt, I., Dencker, L. and Larsson, Y. (1978). Transplacental passage and embryonic-fetal accumulation of hexachlorophene in mice. *Toxicol. Appl. Pharmacol.* (In press)

32. Kimmel, C. A., Moore, W., Jr., Hysell, D. K. and Stara, J. F. (1974). Teratogenicity of hexachlorophene in rats. *Arch. Environ. Health*, **28**, 43

33. Lindquist, N. G. and Ullberg, S. (1972). The melanin affinity of chloroquine and chlorpromazine studied by whole body autoradiography. *Acta Pharmacol. Toxicol.* (Kbh), **31 (Suppl.)**, 2

34. Lindquist, N. G. (1973). Accumulation of drugs on melanin. *Acta Radiol.* (Stockh), **(Suppl.)**, 325

35. Hoffeld, D. R. and Webster, R. L. (1965). Effect of injection of tranquillizing drugs during pregnancy on offspring. *Nature (Lond.)*, **205**, 1070

36. Nair, V. (1974). Prenatal exposure to drugs: effect on the development of brain monoamine systems. In: A. Vernadakis and N. Weiner (eds.). *Drugs and the Developing Brain*, pp. 171–197. (New York: Plenum Press)

37. Lindquist, N. G. and Ullberg, S. (1974). Autoradiography of intravenously injected ^{14}C-nicotine indicates long-term retention in the respiratory tract. *Nature (Lond.)*, **248**, 600

38. Tjälve, H., Hansson, E. and Schmiterlöw, C. G. (1968). Passage of ^{14}C-nicotine and its metabolites into mice foetuses and placentae. *Acta Pharmacol. Toxicol.* (Kbh), **26**, 539

39. Hale, F. (1937). Relation of maternal vitamin A deficiency to microphthalmia in pigs. *Tex. State J. Med.*, **33**, 228

40. Cohlan, S. Q. (1954). Congenital anomalies in the rat produced by the excessive intake of vitamin A during pregnancy. *Pediatrics*, **13**, 556

41. Kalter, H. (1968). *Teratology of the Central Nervous System*, p. 45. (Chicago: University of Chicago Press)

42. Ito, Y. L., Zile, M., Ahrens, H. and DeLuca, H. F. (1974). Liquid-gel partition chromatography of vitamin A compounds; formation of retinoic acid from retinyl acetate in vivo. *J. Lipid Res.*, **15**, 517

43. Kochhar, D. M. (1967). Teratogenic activity of retinoic acid. *Acta Pathol. Microbiol. Scand.*, **70**, 398

44. Howell, J. McC., Thompson, J. N. and Pitt, G. A. J. (1964). Histology of the lesions produced in the reproductive tract of animals fed a diet deficient in vitamin A alcohol but containing vit. A acid. II. The female rat. *J. Reprod. Fertil.*, **7**, 251

45. DeLuca, L. M. (1977). The direct involvment of vitamin A in glycosyl transfer reactions of mammalian membranes. *Vitam. Horm.*, **35**, 1

46. Trelstad, R. L., Kang, A. H., Cohen, A. M. and Hay, E. D. (1973). Collagen synthesis in vitro by embryonic spinal cord epithelium. *Science*, **179**, 295

47. Trelstad, R. L. (1973). The developmental biology of vertebrate collagens. *J. Histochem. Cytochem.*, **21**, 521

2
Antiteratogens as analytical tools

W. LANDAUER

Experimental teratology and its counterpart, the preventive avoidance of misdirected development, grew from quite distinctive roots. One had its origin in frank curiosity, i.e. Virgil's desire of *rerum cognoscere causas*; the other found its starting point in the aspiration of correcting defects that had earlier been traced to deficiencies in the feeding of domestic livestock or of laboratory animals, especially of rats. It was during these latter, chiefly practical, inquiries that attention was first called to the possible importance of using the deliberate correction of aberrant developmental steps, i.e. antiteratogenesis, as an experimental tool for understanding the effects of planned intervention in normal morphogenesis.

It is fair to ask what antiteratogenic research has accomplished to date and what may be its future prospects. In reply to the first question an inventory of results is presented on pages 29–35 (Table 2.2). The relevant information is offered in tabular form and in alphabetic order of the teratogens in combination with which the effects or efficiency of antiteratogenic compounds had been explored. This information is followed by the names of the authors, the dates of their reports, information on the animals or organs that had been used in tests, including dose, route of application and developmental stage of embryos exposed to experimentation; next are listed the principal structural malformations caused by the teratogen in question, and the last column reports the compound or compounds used as antiteratogens. One and the same teratogen may be listed more than once in the tables, either because of different experimental approaches used by the same or another investigator or because different test animals had been employed.

The following abbreviations are used: doses are given in milligrams (mg), micrograms (μg) or in molar terms; times of treatment of chicken or quail embryos in hours of incubation; for mammalian embryos in days of pregnancy. Injections: subcutaneous (s.c.), intravenous (i.v.), intraperitoneal (i.p.), intramuscular (i.m.) or as described.

COMMENT

In gathering data for the Table it was at times difficult to decide on the limits of coverage, the very terms of teratogenicity often lacking clear definition.

For one reason or another, however, it seemed desirable to exclude certain subjects. This, to start with, was felt to be true for hereditary defects even when some or all of their gene-determined symptoms can be removed by chemical means, as is true e.g. for pituitary dwarfism, abnormalities of oto-lith formation or congenital eye defects in mice (Snell, 1929; Smith and MacDowell, 1939; Erway, Hurley and Fraser, 1966; Erway and Mitchell, 1973; Erway, Fraser and Hurley, 1971; Watney and Miller, 1964), for blood abnormalities combined with brachydactylism in rabbits (Petter *et al.*, 1977), and still other conditions. These are important issues, deserving independent discussion. But while the present review is limited to data on *induced* defects, it does not disregard general genetic variance or specific genes that are likely to have a plus or minus effect on incidence or extent of teratogenetic response or on their prevention. The review is also limited to *morphological* aberrations, chiefly because of a scarcity of physiological or biochemical ones – unless these are associated with structural expressions.

A more difficult question is posed by the role of inadequate nutrition in the origin of malformation, whether the deficiency be a general (low-caloric) or a specific one. For, it is clear that most, perhaps all, experimental interventions in embryonic development, resulting in morphological abnormality, may in one way or another have been produced by the failure of essential substances being provided by the organism, i.e. by starvation in the elaboration of quite specific materials needed for parts or organs. Hence, it is not surprising that, for example, restriction of food intake by pregnant mice may be a cause of fetal defects or that it may predispose to the occurrence of faulty develop-mental steps in the presence of agents which otherwise would not have led to the crossing of a teratogenic threshold (Kalter, 1954, 1960; Runner and Miller, 1956; Runner, 1959; Yasuda *et al.*, 1963; Rogoyski, 1967a, b; Rosen-zweig and Blaustein, 1970; Miller, 1973). Reference may also be made to a report by Dyban and Akimova (1966) who found, in progenies of rats, various morphological defects in cases where the maternal diet, prior to treat-ment with thalidomide, had been low in riboflavin and folic acid, but none when the level of B vitamins was raised. Such deliberate, or at any rate preventable, causes of faulty development and their obvious 'cures' need not be discussed in detail in the present context. Reports on the effects of manganese and vitamin A deficiencies and their prevention have, however, been listed in the table because of their intrinsic interest. For reasons similar to the dietary ones our survey does not include work on the experimental interference with glands of internal secretion as far as such steps were re-sponsible among the progeny for abnormalities that could readily be re-paired by providing the requisite hormone (e.g. Rumpler, 1969). Excluded from our summary finally were observations in which supplementation of a particular teratogen affected 'incidence in a nonlinear fashion' without providing significant protection against the defect *per se* (Schardein *et al.*, 1973).

The tabular census of results with antiteratogenic compounds provides abundant evidence to show that inquiries into the nature and effectiveness of such substances have been pursued successfully in several independent directions, the most important ones being competitive replacement of noxious compounds by beneficial ones and the reactivation of enzymes that had become inactivated.

Complete or nearly complete protection against induced developmental defects is clearly one of the principal merits by which the value of work with antiteratogens may be judged, another one presumably being the search for basic information on the developmental steps to which the teratogen-antiteratogen activity relates.

Nicotinamide was the first compound found to have antiteratogenic activity. It was initially and effectively used in 1948 in combination with insulin-treatment of chicken embryos (37).* But the full antiteratogenic value of nicotinamide became evident only when, as supplement to the nicotinamide analogues 3-acetylpyridine and 6-aminonicotinamide, it served to reverse completely the morphologically distinct malformations caused by the two analogues (3, 8, 10, 14).† The extensive and successful use of nicotinamide in combination with a variety of other teratogens is presumably accounted for by its involvement in many metabolic transactions and especially in its serving as versatile support in defects caused by teratogenic disruption of nicotinamide adenine dinucleotide in cellular metabolism (1, 18, 37, 40, 42, 45, 47, 48, 50, 53, 56). In *specific* combinations 1-β-D-2-deoxyribofurano-sylcytosine (19), inosine (32), tryptophan (40) and tetrahydrofolic acid (55) have been equally effective antiteratogens, but have not found the general application of nicotinamide.

In regard to 'complete' protection it will be noted below that morphogenesis of parts may be restored to full normality but often with *reduced* body size, growth requiring a separate stimulus. In many other instances antiteratogenic compounds succeeded only imperfectly in restoring development to normality, the reasons generally remaining to be explored (lack of chemical fit, uncertainty of dose or timing etc.).

Two sources of more generalized antiteratogenic effect have been reported, viz. compounds providing sources of cellular energy and conditions residing in the solvent used for delivering a teratogen to its test object. Experiments on chicken embryos with sodium succinate hexahydrate, DL-α-glycerophosphate hexahydrate disodium, L-ascorbic acid and iso-ascorbic acid sodium when used as supplement to 3-acetylpyridine, 6-aminonicotinamide and sulphanila-mide reduced the teratogenic effects of the latter three compounds to a highly significant extent (5); similarly beneficial effects were obtained when carbachol was supplemented with glucose or sorbose (unpublished). The interpretation of the foregoing data as a result of having provided cellular energy must, however, be accepted with some reservation. For, an experiment in which chicken embryos were treated either with bidrin alone (0.1 mg/egg/96 h) or as supplemented with sodium succinate (10 mg/egg) produced the results (unpublished) shown in Table 2.1. In this instance, therefore, supplementation had no significant effects on the appendicular and visceral skeleton, i.e. the nicotinamide sensitive parts, but greatly reduced ($p < 0.001$) damage to the cervical region and growth of the down. It is true, furthermore, that

* Figures in parentheses refer to the numbering of compounds in the tables.
† Antiteratogenic compounds giving complete or nearly complete protection against the activity of specific teratogens are in the tables marked by an asterisk.

Table 2.1

	Bidrin alone	*Bidrin + Na succinate*
Number treated	130	140
Mortality to 19 days %	16.2	17.9
Survivors	109	115
Normal %	0	2.6
Micromelia %	78.0	81.7
Abnormal beak %	69.7	60.9
Short neck %	96.3	74.8
Short down %	55.0	33.0

the same several supplements were likely to aggravate the effects of teratogens which interfered with carbohydrate utilization of the embryo (e.g.
insulin).

The second problem referred to as of possible interest in analysing the ways
of teratogens to their potential 'hosts' is that of the solvents used. In experiments with chicken embryos identical amounts of nicotine or physostigmine
whether dissolved in water or dimethyl sulphoxide (DMSO) led, for each of
the compounds, to an identical incidence or degree of malformations. In
contrast it was found in tests with 3-acetylpyridine, 6-aminonicotinamide and
sulphanilamide that all teratogenic effects were much reduced in frequency
when the solvent was DMSO rather than water. In a parallel comparison with
the herbicide 3-amino-1,2,4-triazole, it was found that, in addition to more
embryos escaping any teratogenic effects, a highly significant shift occurred
by defective beak formation being less extreme when the solvent was DMSO
rather than water. Finally, reference has already been made to the peculiar
importance the choice of solvent has on discrete parts of a syndrome caused
by treatment with bidrin.

The application of chelating and complexing compounds has as yet found
little use in antiteratogenic research (19, 29), but may offer promising approaches. Interesting observations of 'protection at a price' have, on the other
hand, been encountered when one metal was replaced by another one, e.g.
cadmium by lead (21), the supplementing metal suppressing the effects of the
one used earlier, but causing its own teratogenicity. In other and perhaps still
more impressive combinations of teratogens, it was found that the presence
of one compound completely suppressed the teratogenic effects of the other
one, but led at the same time to an exaggeration of its own interference with
normal development. Such was the case when 6-aminonicotinamide and
3-acetylpyridine or sulphanilamide and 3-acetylpyridine were used in combination (9, 56). Qualitative shifts of a different kind occur when increase in
dose of a compound leads from protection to exaggeration. Thus it was found
that thymidine given in moderate amounts lowers the teratogenicity of
5-fluorodeoxyuridine, but exaggerates it when given in higher doses (31).
Many antiteratogens, as presumably much of any additive, become noxious
above a certain amount; but some antiteratogens also become teratogenic.
This is true, for instance, for nicotinamide which in excess, like its analogue
3-acetylpyridine, is likely to cause in chicken embryos a shortening of the
maxilla.

Problems of peculiar interest are posed by mounting evidence that many syndromes with apparently unitary causation find expression in heterologous systemic consequences (Landauer, 1977b). Thus it has been shown that syndromes produced in chicken embryos by such organophosphorus insecticides as bidrin or parathion or by physostigmine are mediated by two quite independent events and that certain of these events can be prevented by supplementary nicotinamide while others yield to the reactivation of inactivated acetylcholinesterase (45, 47); it is a systemic difference which, in the case of bidrin, finds striking expression even by simply applying the teratogen in two different solvents (Landauer and Salam, 1972). In other heterologous syndromes differential responses remain unresolved. Thus, it is not known why a syndrome which in chicken embryos is caused by insulin treatment and all features of which yield to nicotinamide supplementation should in the presence of glucose-1-phosphate show a significant lowering of beak defects without improvement of malformations of the limbs (Landauer and Rhodes, 1952) or why in the neostigmine-induced syndrome of brevicollis and muscular hypoplasia the presence of toxogonin should only improve the muscular hypoplasia while other reactivators of acetylcholinesterase reduce the expression of both defects (Landauer, 1977a). These are differences crying out for investigation.

An important teratological problem was raised when Zwilling and DeBell (1950) found that micromelia and beak defects of chicken embryos due to treatment with sulphanilamide can be entirely prevented by supplementary nicotinamide, but that body size of the treated embryos did not benefit by the supplementation. The authors felt that their 'study clearly eliminates growth retardation as a causal factor in the production of sulfanilamide induced micromelia'. Greenberg and LaHam (1970) later reported that the effects of treating chicken embryos with malathion were very similar to those produced by sulphanilamide, viz. micromelia, beak defects and greatly reduced body size. As in the case of sulphanilamide the appendicular and facial defects caused by malathion could be completely eliminated by providing supplementary nicotinamide and, as with sulphanilamide, body size was not improved. Greenberg and LaHam found, however, that supplementary tryptophan equally restored morphology and body size to normal. The problem of a relationship between growth and malformation was thereby pushed back to an earlier step of biochemical needs, leaving open the chances of a parallel, if not causal relationship between growth and morphogenesis.

As for the future prospects of antiteratogenic research, they will perhaps to some extent be decided by those who have been reading the preceding review. There have certainly been notable successes in the past. Even more numerous questions openly invite detailed studies and one hopes that answers to them will provide much insight into steps of early normal morphogenesis as well as of those causing deviation and digression from the regular.

ACKNOWLEDGEMENTS

The editor gratefully acknowledges the help of Dr Dinah Sopher and Dr Gillian M. Truslove in the preparation of this manuscript.

References

N.B. *Since the majority of the references quoted in this chapter relate specifically to Table 2.2 the editor has decided that in this special case it would be more appropriate and useful to present them in alphabetic order.*

Beaudoin, A. R. (1961). Teratogenic activity of several closely related diazo dyes on the developing chick embryo. *J. Embryol. Exp. Morphol.*, 9, 14

— (1968). The effect of citric acid on the teratogenic action of trypan blue. *Life Sci.*, 7, I, 635

— (1973). Teratogenic activity of 2-amino-1,3,4-thiadiazole hydrochloride in Wistar rats and the protection afforded by nicotinamide. *Teratology*, 7, 65

— (1976). NAD precursors as antiteratogens against aminothiadiazole. *Teratology*, 13, 95

Chamberlain, J. G. (1966), Development of cleft palate induced by 6-aminonicotinamide late in rat gestation. *Anat. Rec.*, 156, 31

— (1967). Effect of acute vitamin replacement therapy on 6-aminonicotinamide-induced cleft palate late in rat pregnancy. *Proc. Soc. Exp. Biol. Med.*, 124, 888

— (1975). Prevention of 6-aminonicotinamide (6-AN)-induced cleft palate in rats with pyridine nucleotide. *J. Dent. Res.*, 54, 413

Chamberlain, J, G. and Goldyne, M. E, (1970). Intra-amniotic injection of pyridine nucleotides or adenine triphosphate as countertherapy for 6-aminonicotinamide (6-AN) teratogenesis. *Teratology*, 3, 11

Chamberlain, J. G. and Nelson, M. M. (1963a). Congenital abnormalities in the rat resulting from single injections of 6-aminonicotinamide during pregnancy. *J. Exp. Zool.*, 153, 285

— — (1963b) Multiple congenital abnormalities in the rat resulting from acute maternal niacin deficiency during pregnancy. *Proc. Soc. Exp. Biol. Med.*, 112, 836

Chaube, S. (1973). Protective effects of thymidine, 5-aminoimidazole carboxamide and riboflavin against fetal abnormalities produced in rats by 5-(3,3-dimethyl-1-triazeno) imidazole-4-carboxamide. *Cancer Res.*, 33, 2331

Chaube, S., Kreis, W., Uchida, K. and Murphy, M. L. (1968). The teratogenic effect of 1-β-d-arabionfuranosylcytosine in the rat. Protection by deoxycitidine. *Biochem. Pharmacol.*, 17, 1213

Chaube, S. and Murphy, M. L. (1969a). Fetal malformations produced in rats by N-isopropyl-α-(2-methylhydrazino)-p-toluamide hydrochloride (procarbazine). *Teratology*, 2, 23

— — (1969b). Teratogenic effects of 6-hydroxylaminopurine in the rat. Protection by inosine. *Biochem. Pharmacol.*, 18, 1147

— — (1973). Protective effect of deoxycytidylic acid (CdMP) on hydroxyurea-induced malformations in rats. *Teratology*, 7, 79

Dagg, C. P. and Kallio, E. (1962). Teratogenic interaction of fluorodeoxyuridine and thymidine (Abstract). *Anat. Rec.*, 142, 301

Dagg, C, P., Karnofsky, D. A., Lacon, C. and Roddy, J. (1956). Comparative effects of 6-diazo-5-oxo-L-norleucine and o-diazoacetyl-L-serine on the chick embryo. *Proc. Am. Ass. Cancer Res.*, 2, 101

Deol, M. S. (1973). An experimental approach to the understanding and treatment of hereditary syndromes with congenital deafness and hypothyroidism. *J. Med. Genet.*, 10, 235

Dyban, A. P. and Akimova, I. M. (1966). The significance of vitamin B complex and genetic factors in the reaction of rat embryos to thalidomide. *Arch. Anat. Gistol. Embriol.*, 51, 3 (Transl.) (Nat. Lending Library for Science & Technol.)

Ellison, A. C. and Maren, T. H. (1972). The effect of potassium metabolism on acetazolamide-induced teratogenesis. *Johns Hopkins Med. J.*, 130, 105

Erway, L. C., Fraser, A. S. and Hurley, L. S. (1971). Prevention of congenital otolith defect in pallid mutant mice by manganese supplementation. *Genetics*, 67, 97

Erway, L. C., Hurley, L. S. and Fraser, A. (1966). Neurological defect: manganese in phenocopy and prevention of a genetic abnormality of inner ear. *Science*, 152, 1766

— — — (1970). Congenital ataxia and otolith defects due to manganese deficiency in mice. *J. Nutrition*, 100, 643

Erway, L. C. and Mitchell, S. E. (1973). Prevention of otolith defect in pastel mink by manganese supplementation. *J. Hered.*, 64, 111

Ferm, V. H. (1969). The synteratogenic effect of lead and cadmium. *Experientia*, 25, 56

— (1970). Protective effect of ferric dextran on the embryopathic action of indium. *Experientia*, 26, 633

— (1972). The teratogenic effect of metals on mammalian embryos. *Adv. Teratol.*, 5, 51

Ferm, V. H. and Carpenter, S. J. (1968). The relationship of cadmium and zinc in experimental mammalian teratogenesis. *Lab. Invest.*, 18, 429

Forsthoefel, P. F. and Williams, M. L. (1975). The effects of 5-fluorouracil and 5-fluorodeoxyuridine used alone and in combination with normal nucleic acid precursors on development of mice in lines selected for low and high expression of Strong's luxoid gene. *Teratology*, 11, 1

Fujii, Y. (1968). Preventative effects of protein-anabolic steroids upon the congenital defects induced by an anticarcinogen in mice. *Congenital Anomalies*, 8, 219

Fujii, T. (1976). Mitigation of caffeine-induced fetopathy in mice by pretreatment with β-adrenergic blocking agents. *Japan. J. Pharmacol.*, 26, 751

Fujii, T. and Nishimura, H. (1974). Reduction in frequency of fetopathic effects of caffeine in mice by treatment with propranolol. *Teratology*, 10, 149

Greenberg, J. and LaHam, Q. N. (1969). Malathion-induced teratism in the developing chick. *Canad. J. Zool.*, 47, 539

— — (1970). Reversal of malathion-induced teratism and its biochemical implications in the developing chick. *Canad. J. Zool.*, 48, 1047

Greene, R. M. and Kochhar, D. M. (1975). Limb development in mouse embryos: protection against teratogenic effects of 6-diazo-5-oxo-L-norleucine (DON) *in vivo* and *in vitro*. *J. Embryol. Exp. Morphol.*, 33, 355

Grote, W., Claussen, U. and Heinz, D. (1971). Verursachung und Verhütung von Entwicklungsstörungen beim Kaninchen durch Verabreichung von 6-Aminonikotinsäureamid und Nikotinsäureamid. *Arzneimittelforschung*, 21, 825

Hall, B. K. (1972a). Thallium-induced achondroplasia in the embryonic chick. *Dev. Biol.*, 28, 47

— (1972b). Achondroplasia in the embryonic chick: its potentiation by cortisone acetate and alleviation by vitamin C. *Canad. J. Zool.*, 50, 1527

— (1977). Thallium-induced achondroplasia in chicken embryos and the concept of critical periods during development. *Teratology*, 15, 1

Hamly, C.-A., Trasler, D. G. and Fraser, F. C. (1970). Reduction of 6-aminonicotinamide teratogenicity in mice by etherization. *Teratology*, 3, 293

Hodach, R. J., Hodach, A. E., Fallon, J. F., Folds, J. D., Bruyere, H. J. and Gilbert, E. F. (1975). The role of β-adrenergic activity in the production of cardiac and aortic anomalies in chick embryos. *Teratology*, 12, 33

Holmberg, R. E. and Ferm, V. H. (1969). Interrelationships of selenium, cadmium and arsenic in mammalian teratogenesis. *Arch. Envir. Health*, 18, 873

Horii, K-i. (1964). Prevention of congenital malformations in the offspring of alloxan-diabetic mice by the insulin treatment. *Folia Endocrinol. Japan*, 39, 988

Horii, K-i, Watanabe, G-i and Ingalls, T. H. (1966). Experimental diabetes in pregnant mice. Prevention of congenital malformation in offspring by insulin. *Diabetes*, 15, 194

Horvath, C., Szonyi, L. and Mold, K. (1976). Preventive effect of riboflavin and ATP on the teratogenic effects of the phenothiazine derivative T-82. *Teratology*, 14, 167

Hurley, L. S. and Asling, G. W. (1963). Localized epiphyseal dysplasia in offspring of manganese deficient rats. *Anat. Rec.*, 145, 25

Hurley, L. S. and Everson, G. J. (1963). Influence of timing of short-term supplementation during gestation on congenital abnormalities of manganese-deficient rats. *J. Nutrition*, 79, 23

Hurley, L. S., Everson, G. J., Wooten, E. and Asling, C. W. (1961). Disproportionate growth of manganese-deficient rats. I. The long bones. *J. Nutrition*, 74, 274

Hurley, L. S., Wooton, E. and Everson, G. J. (1961). Disproportionate growth in offspring of manganese deficient rats. II. Skull, brain and cerebrospinal fluid pressure. *J. Nutrition*, 74, 282

Kalter, H. (1954). Preliminary studies on the metabolic factors involved in the production of cleft palate in mice (Abstract). *Genetics*, 39, 975

— (1960). Teratogenic action of hypocaloric diet and small doses of cortisone. *Proc. Soc. Exp. Biol. Med.*, 104, 518

Kato, T. (1973). Effect of folate metabolism-related factors on the teratogenic action of sulfonamide in mice. *Congenital Anomalies*, 13, 85

Khan, M. A. (1975). Nicotine sulfate-induced skeletal anomalies in chick embryos (Abstract). *Teratology*, 11, 25A

Kolesari; G. L. and Kaplan, S. (1974). The antiteratogenic effects of hypo- and hyperthermia in trypan blue-treated embryos. *Dev. Biol.*, 38, 383

Landauer, W. (1948). The effect of nicotinamide and α-ketoglutaric acid on the teratogenic action of insulin. *J. Exp. Zool.*, 109, 283

— (1949). Le problème de l'électivité dans les expériences de tératogenèse biochimique. *Arch. Anat. Micros. Morph. Exp.*, 38, 184

— (1952). Malformations of chicken embryos produced by boric acid and the probable role of riboflavin in their origin. *J. Exp. Zool.*, 120, 469

— (1953). Complex formation and chemical specificity of boric acid in production of chicken embryo malformations. *Proc. Soc. Exp. Biol. Med.*, 82, 633

— (1953). On teratogenic effects of pilocarpine in chick development. *J. Exp. Zool.*, 122, 469

— (1957). Niacin antagonists and chick development. *J. Exp. Zool.*, 136, 509

— (1960a). Experiments concerning the teratogenic nature of thallium: polyhydroxy compounds, histidine and imidazole as supplements. *J. Exp. Zool.*, 143, 101

— (1960b). Nicotine-induced malformations of chicken embryos and their bearing on the phenocopy problem. *J. Exp. Zool.*, 143, 107

— (1975a). Cholinomimetic teratogens: studies with chicken embryos. *Teratology*, 12, 125

— (1975b). Cholinomimetic teratogens. II. Interaction with inorganic ions. *Teratology*, 12, 271

— (1976). Cholinomimetic teratogens. III. Interaction with amino acids known as neurotransmitters. *Teratology*, 13, 41

— (1977a). Cholinomimetic teratogens. V. The effect of oximes and related cholinesterase reactivators. *Teratology*, 15, 33

— (1977b). On teratogenic syndromes of unitary causation but heterologous systemic consequences. *Acta Embryol. Exp.*, 3, 335

Landauer, W. and Clark, E. M. (1962). The interaction in teratogenic activity of the two niacin analogs 3-acetylpyridin and 6-aminonicotinamide. *J. Exp. Zool.*, 151, 253

— — (1964). On the role of riboflavin in the teratogenic activity of boric acid. *J. Exp. Zool.*, 156, 307

— — (1964). On the teratogenic interaction of sulfanilamide and 3-acetylpyridine in chick development. *J. Exp. Zool.*, 156, 313

Landauer, W. and Rhodes, M. B. (1952). Further observations on the teratogenic nature of insulin and its modification by supplementary treatment. *J. Exp. Zool.*, 119, 221

Landauer, W. and Salam, N. (1972). Aspects of dimethylsulfoxide as solvent for teratogens. *Dev. Biol.*, 28, 35

— — (1974). The experimental production in chicken embryos of muscular hypoplasia and associated defects of beak and cervical vertebrae. *Acta Embryol. Exp.*, 51

Landauer, W. and Sopher, D. (1970). Succinate, glycerophosphate and ascorbate as sources of cellular energy and as antiteratogens. *J. Embryol. Exp. Morphol.*, 24, 187

Landauer, W. and Wakasugi, N. (1967). Problems of acetazolamide and N-ethylnicotinamide as teratogens. *J. Exp. Zool.*, 164, 499

— — (1968). Teratological studies with sulphonamides and their implications. *J. Embryol. Exp. Morphol.*, 20, 261

Larsson, K. S. (1977). Teratogenic mechanisms of dithiocarbamates (Abstract). *Teratology*, 14, 373

Matthiaschk, G. (1973). Über den Einfluss von L-cystein auf die Teratogenese durch Thiram (TMTD) bei NMRI-Mäusen. *Arch. Toxikol.*, 30, 251

Meiniel, R. (1974). Action protectrice de la pralidoxime vis-à-vis des effets tératogènes du parathion sur le squelette axial de l'embryon de caille. *C.R. Acad. Sci. Paris*, 279 D, 603

— (1975). Prévention par la pralidoxime de certains effets tératogènes induits par le bidrin chez l'embryon de caille. *C.R. Acad. Sci. Paris*, 280 D, 1019

— (1976a). Pluralité dans le déterminisme des effets tératogènes des composés organophosphorés. *Experientia*, 32, 920

— (1976b). Tératogenèse axiale du parathion après action de divers composés connus chez l'adulte ou l'embryon de Vertébrés pour posséder un pouvoir antitoxique ou antitératogène après des expositions aux phosphates organiques. Études chez l'embryon de caille. C.R. Acad. Sci. Paris, 283 D, 1085

— (1976c). Prévention des anomalies induites par deux insecticides organophosphorés (Parathion et Bidrin) chez l'embryon de caille. Arch. Anat. Micros. Morph. Exp., 65, 1

Millen, J. W. and Woollam, D. H. M. (1958). Effect of vitamin B complex on the teratogenic activity of hypervitaminosis A. Nature (Lond.), 182, 940

Miller, T. J. (1973). Cleft palate formation: the effects of fasting and iodoacetic acid on mice. Teratology, 7, 177

Mizutani, M., Ihara, T. and Sugitani, T. (1974). Protective effects of nicotinamide and trypto-phan against the teratogenicity of N,N'-methylene-bis (2-amino-1,3,4-thiadiazole) in the hamster (Abstract). Teratology, 9, A28

Mulherkar, L. S., Joshi, S., Diwan, B. A. and Joshi, P. N. (1967). Reversible effect of chloro-acetophenone by sulhydryl groups on morphogenesis of chick embryos. J. Embryol. Exp. Morphol., 17, 263

Mulherkar, L., Rao, K. V. and Joshi, S. S. (1965). Studies on some aspects of the role of sulfhydryl groups in morphogenesis. J. Embryol. Exp. Morphol., 14, 129

Mulherkar, L., Rao, K. V., Joshi, S. S. and Joshi, P. N. (1966). Reversible effects of chloro-acetophenone on the embryos of Microhyla ornata. J. Anim. Morphol. Physiol., 13, 9

Overman, D. O., Graham, M. N. and Roy, W. A. (1976). Ascorbate inhibition of 6-amino-nicotinamide teratogenesis in chicken embryos. Teratology, 13, 85

Persaud, T. V. N. (1967). Foetal abnormalities caused by the active principle of Blighia sapida (Ackee). W. I. Med. J., 16, 193

— (1968). Teratogenic effects of hypoglycin-A. Nature (Lond.), 217, 471

— (1969a). Attempts in reversing the teratogenic effects of hypoglycin-A with leucine and riboflavin (Abstract). Internat. Conf. Congenit. Malf., Excerpta Medica Intern. Congress, Ser. No. 191, 70

— (1969b). Developmental abnormalities in the rat induced by the amino acid leucine. Naturwissenschaften, 56, 37

— (1970). Congenital abnormalities in relationship to the inhibition of fatty acid oxidation. Wiss. Zschr. Friedrich Schiller Univ. Jena, Math.-Nat. R., 19, 247

— (1972). Teratogenic effect of hypoglycemia. Adv. Teratol., 5, 77

— (1973). Prevention by leucine of hypoglycin B induced teratogenesis in the rat. Exp. Pathol., 8, 283

Petter, C., Bourbon, J., Maltier, J.-P. and Jost, A. (1977). Simultaneous prevention of blood abnormalities and hereditary congenital amputation in a brachydactylous rabbit stock. Teratology, 15, 149

Pierro, L. J. (1961). Teratogenic action of actinomycin D in the embryonic chick. J. Exp. Zool., 147, 203

Proctor, N. H. and Casida, J. E. (1975). Organophosphorus and methyl carbamate insecticide teratogenesis: diminished NAD in chicken embryos. Science, 190, 580

Proctor, N. H., Moscioni, A. D. and Casida, J. E. (1976). Chicken embryo NAD levels lowered by teratogenic organophosphorus and methylcarbamate insecticides. Biochem. Pharmacol., 25, 757

Puricha, N. and Erway, L. C. (1972). Effects of dichlorphenamide, zinc and manganese on otolith development in mice. Dev. Biol., 27, 395

Rao, K. V. (1969). A study of the role of sulfhydryl groups in morphogenesis of the chick embryo. Roux' Arch. Entw. Mech., 163, 161

Roger, J.-C., Upshall, D. G. and Casida, J. E. (1969). Structure–activity and metabolism studies on organophosphate teratogens and their alleviating agents in developing hen eggs with special emphasis on Bidrin. Biochem. Pharmacol., 18, 373

Rogóyski, A. (1967a). The effect of fasting combined with hydrocortisone acetate treatment on the formation of developmental disturbances of the fetus in pregnant mice. Polish Med. J., 6, 1646

— (1976b). Deficient nutrition in pregnant mice as a factor conducive to developmental anomalies in the fetus after simultaneous administration of hydrocortisone acetate. Folia Morphol., 26, 313

Rosenzweig, S. and Blaustein, F. (1970). Cleft palate in A/J mice resulting from restraint and deprivation of food and water. *Teratology*, 3, 47

Ruano Gil, D. (1967). The influence of thalidomide on the development of chick embryos cultivated in vitro. *Acta Anat.*, 66, 226

Ruch, J. V. (1967). Effet antagoniste du D-glucose sur l'action tératogène du phosphate de dexaméthasone chez l'embryon de poulet. *C.R. Soc. Biol.*, 161, 915

Rumpler, Y. (1969). Compensation de l'action tératogène d'un antithyroidien (le 1-méthyle 2 mercapto imidazol ou Basolan) par l'administration simultanée de thyroxine à l'embryon de poulet. *Bull. Assoc. Anat. 53ᵉ Congrès, C.R. Ass. Anat.*, 143, 1450

Runner, M. (1959). Inheritance of susceptibility to congenital deformity. Metabolic clues probided by experiments with teratogenic agents. *Pediatrics*, 23, 245

Runner, M. N. and Miller, J. R. (1956). Congenital deformity in the mouse as a consequence of fasting (Abstract). *Anat. Rec.*, 124, 437

Schardein, J. L., Dresner, A. J., Hentz, D. L., Petrere, J. A., Fitzgerald, J. E. and Kurtz, S. M. (1973). The modifying effect of folinic acid on diphenylhydantoin-induced teratogenicity in mice. *Toxicol. Appl. Pharmacol.*, 24, 150

Scott, W. J., Ritter, E. J. and Wilson, J. G. (1973). DNA synthesis inhibition, cytotoxicity and their relationship to teratogenesis following administration of a nicotinamide antagonist, aminothiadiazole, to pregnant rats. *J. Embryol. Exp. Morphol.*, 30, 257

Semba, R., Yamamura, H. and Murakami, U. (1977). Effect of cadmium pretreatment on teratogenicity and fetolethality of cadmium. *Okajimas Folia Anat. Jap.*, 54, 283

Shih, L. Y., Toliver, C. and Yang, C. S. (1974). Protective effect of ascorbic acid against cleft-palate formation in mice (Abstract). *Teratology*, 9, A 36

Smith, P. E. and MacDowell, E. C. (1939). An hereditary anterior-pituitary deficiency in the mouse. *Anat. Rec.*, 46, 249

Snell, G. D. (1929). *Dwarf*, a new Mendelian recessive character in the house mouse. *Proc. Nat. Acad. Science (USA)*, 15, 733

Sudo, T. (1968). Suppressive effect of vitamin E on experimental micromelia produced by hypervitaminosis A (Japanese, English summary). *Congenital Anomalies*, 8, 163

Sullivan, G. E. (1975). Paralysis and skeletal abnormalities in chick embryo treated with physostigmine. *Austral. J. Zool.*, 23, 1

Sullivan, G. E. and Takacs, E. (1971). Comparative teratogenicity of pyrimethamine in rats and hamsters. *Teratology*, 4, 205

Terada, M. and Nishimura, H. (1975). Mitigation of caffeine-induced teratogenicity in mice by prior chronic caffeine ingestion. *Teratology*, 12, 79

Watney, M. J. and Miller, J. R. (1964). Prevention of a genetically determined congenital eye anomaly in the mouse by the administration of cortisone during pregnancy. *Nature (Lond.)*, 202, 1029

Yamaguchi, T. (1968). Effect of vitamin B_{12} on the incidence of malformations in the rat fetus caused by maternal hypervitaminosis A (Japanese, English summary). *Congenital Anomalies*, 8, 19

Yasuda, M., Nanjo, H. and Suzuki, M. (1963). Teratogenic and lethal effects of fasting during pregnancy upon the offspring of elderly mice (Abstract). *Proc. Cong. Anom. Res. Ass. Japan*, 3, 43

Zwilling, E. (1959). Micromelia as a direct effect of insulin. Evidence from *in vitro* and *in vivo* experiments. *J. Morphol.*, 104, 159

Zwilling, E. and DeBell, J. T. (1950). Micromelia and growth retardation as independent effects of sulfanilamide in chick embryos. *J. Exp. Zool.*, 115, 59

Table 2.2

Number	Teratogen	Authors	Test object	Dose; Route; Stage	Major effects	Antiteratogen
1	Acetazolamide	Landauer and Wagasugi, 1967	Chicken	1–20 mg; yolk sac; 96 h	Short upper beak	ADP 2.5 mg
2	Acetazolamide	Ellison and Maren, 1972	Rat	In maternal diet (0.6%) or 500 mg/kg; s.c.	Defective distal postaxial part right forelimb (both forelimbs after high dose)	KCl or KHCO₃ 2% in drinking water, days 10 and 11
3	3-Acetylpyridine	Landauer, 1957	Chicken	10 µg; yolk sac; 96 h	Abnormal beak, muscular hypoplasia	*Nicotinamide 100 µg, L-tryptophan HCl 5 mg, 3-hydroxyanthranilic acid sodium 2 mg
4	3-Acetylpyridine	Landauer and Salam, 1972	Chicken	0.6 mg; yolk sac; 96 h	as No. 3	Dimethyl sulphoxide as solvent
5	3-Acetylpyridine	Landauer and Sopher, 1970	Chicken	0.6 mg; yolk sac; 96 h	as No. 3	Sodium succinate 7.5 mg; glycerophosphate 10 mg; sodium ascorbate 10 mg
6	Actinomycin D	Pierro, 1961	Chicken	0.0625 µg; yolk sac; 24 h	Axial skeleton defects	DNA 6.25 µg, Deoxyguanosine 25 µg, guanosine 25 µg if pre-mixed with actinomycin
7	Alloxan monohydrate	Horii, 1964; Horii, Watanabe and Ingalls, 1966	Mouse	80 mg/kg; i.v.; day 11	Craniorachischisis, cleft palate, club foot	Lente insulin 0.4 units at 12 h intervals from day 12
8	6-Aminonicotinamide	Landauer, 1957	Chicken	10 µg; yolk sac; 96 h	Abnormal beak, micromelia	*Nicotinamide 100 µg, L-tryptophan HCl 5mg, 3-hydroxyanthranilic acid sodium 2mg
9	6-Aminonicotinamide (6-AN)	Landauer and Clark, 1962	Chicken	2.5 to 10 µg; yolk sac; 96 h	6-AN-determined defects suppressed by supplement	3-Acetylpyridine 0.375 or 0.5 mg
10	6-Aminonicotinamide	Chamberlain, 1966, '67 '75 (Chamberlain and Nelson, 1963a, b)	Rat	8 mg/kg; i.p.; day 15	Abnormal palate	*Nicotinamide 8 mg/kg or fasting (also NAD, NADH)

(continued)

29

Table 2.2 (continued)

Number	Teratogen	Authors	Test object	Dose; Route; Stage	Major effects	Antiteratogen
11	6-Aminonicotinamide	Chamberlain and Goldyne, 1970	Rat	100 μg; intra-amniotic; day 15	Cleft palate, hydrocephalus, lens vacuolation	500 μg NAD-NADP or NADH-NADPH or 100 μg nicotinamide
12	6-Aminonicotinamide	Hamly, Trasler and Fraser, 1970	Mouse	19 mg/kg; i.p.; 9 days 10 h	Cleft lip, etc.	Light ether anaesthesia
13	6-Aminonicotinamide	Landauer and Sopher, 1970	Chicken	5 μg; yolk sac; 96 h	Abnormal beak, micromelia	Sodium succinate 7.5 mg, glycerophosphate 10 mg, sodium ascorbate 10 mg
14	6-Aminonicotinamide	Grote, Clausen and Heinz, 1971	Rabbit	5 mg/kg; i.p.; day 9	Abnormal eyes, ribs, vertebrae, extremities	*Nicotinamide 84 mg/kg daily, days 8 to 12
15	6-Aminonicotinamide	Landauer and Salam, 1972	Chicken	5 μg; yolk sac; 96 h	Reduced teratogenicity	Dimethyl sulphoxide as solvent
16	6-Aminonicotinamide	Shih, Toliver and Yang, 1974 (Abstract)	Mouse	day 13.5	Cleft palate	Ascorbate
17	6-Aminonicotinamide	Overman, Graham and Roy, 1976	Chicken	10 μg, extra-embryonic coelom, days 4–6	Micromelia	Sodium or calcium ascorbate 5–15 mg; ascorbic acid 15 mg
18	2-Amino-1,3,4-thiadiazole	Beaudoin, 1973, '76 Scott, Ritter and Wilson, 1973	Rat	5–100 mg/kg; i.p.; day 11, and 100 mg/kg; i.p.; day 12	Eyes, brain, palate, heart, limbs, tail abnormal	*Nicotinamide 100 mg/kg, NAD, nicotinic acid, quinolinic acid, kynurenine sulphate, L-tryptophan 10–200 mg/kg
19	1-β-D-Arabinofuranosylcytosine	Chaube, Kreis, Uchida and Murphy, 1968	Rat	150–600 mg/kg; i.p.; day 10 to 12	Cleft palate/lip, encephalocoele, abnormal limbs and tail	*1-β-D-2-deoxyribofuranosylcytosine 150–600 mg/kg
20	Boric acid	Landauer, 1952, '53; Landauer and Clark, 1964	Chicken	0.05 or 0.1 ml 5% or 2.5 mg; yolk sac 24 and 92–96 h	Tailless at 24 h, facial defects, short tarsometatarsus, ectrodactyly and syndactyly later	Riboflavin-5-phosphate 100 μg—1 mg at 96 h, sodium pyruvate 19–38 mg at 24 h, D-sorbitol hydrate 10–20 mg at 96 h

No.	Compound	Reference	Species	Dose	Defects	Antiteratogen
21	Cadmium sulphate	Ferm, 1969, '72; Holmberg and Ferm, 1969; Ferm and Carpenter, 1968	Golden hamster	2 mg/kg; i.v.; day 8	Facial, eye and brain defects	Lead 25–50 mg/kg zinc sulphate 2 mg/kg
		Semba, Yamamura and Murakami, 1977	Mouse	5 mg/kg; i.p.; day 7	Exencephaly	Maternal pretreatment with low cadmium doses 0.5 mg/kg
22	Caffeine	Fujii and Nishimura, 1974; Fujii, 1976	Mouse	200 mg/kg; i.p.; day 13	Cleft palate, brachygnathia, digital defects	Propranolol 2.5, 5 or 10 mg/kg; i.v. *prior to* caffeine
		Terada and Nishimura, 1975	Mouse	150 or 250 mg/kg; s.c.; day 13	as above	Chronic pretreatment with 0.5% caffeine in drinking water
23	Chloroacetophenone	Mulherkar, Rao and Joshi, 1965; Mulherkar, Joshi, Diwan and Joshi, 1967; Mulherkar, Rao, Joshi and Joshi, 1966; Rao, 1969	Chicken, *Microhyla ornata* (toad)	2.5 or 5×10^{-4} M; primitive streak, blastoderm, gastrula *in vitro*	Abnormal neural tube; interference with organizer	Cysteine 2.5 to 5 µg
24	Cholinomimetic compounds (Carbachol, decamethonium, neostigmine, succinylcholine etc.	Landauer, 1975a, b; '76; '77a	Chicken	Dose varying with compound; yolk sac; 96 h	Short/twisted neck, muscular hypolasia	Gallamine, benzoquinonium, butyrylcholine, GABA, pyridine-2-aldoxime-methylmethane sulphonate, ambenonium, toxogonin, pyridine-2-aldoxime methiodide
25	Dexamethasone phosphate	Ruch, 1967	Chicken	0.01 mg; egg white; 5th day	Mitotic activity reduced	D-glucose 20–120 mg
26	6-Diazo-5-oxo-6-nor-leucine	Greene and Kochhar, 1975 (Dagg, Karnofsky, Lacon and Roddy, 1956)	Mouse	0.5 mg/kg; i.m.; day 11	Cleft lip, limb defects	5-Aminoimidazole-carboxamide 250 mg/kg (Adenine, hypoxanthine)
27	Dichlorphenamide	Landauer and Wakasugi, 1968	Chicken	2 mg; yolk sac; 96 h	Shortened maxilla, syndactyly	ADP 3 mg
		Puricha and Erway, 1972	Mouse	2.4 or 3.6 mg; s.c.; days 13 to 18	Otolith defects	Additional manganese and zinc in diet

(*continued*)

Table 2.2 (continued)

Number	Teratogen	Authors	Test object	Dose; Route; Stage	Major effects	Antiteratogen
28	5-(3,3-dimethyl-1-triazeno) imidazole-4-carboxamide	Chaube, 1973	Rat	400 mg/kg; i.p.; day 12	Mandible, palate, brain, limb defects	Thymidine 50–1000 mg, 5-aminoimidazole-carboxamide 200–400 mg, riboflavin 10–20 mg, DL-cysteine 200–400 mg
29	DL-Ethionine	Landauer and Salam, 1974	Chicken	4.5 mg; yolk sac; 96 h	Abnormal beak, muscular hypoplasia, short neck	L-methionine 1 mg
30	Ethylenebis (dithio-carbamate) manganese (Maneb)	Larsson 1977 (Abstract)	Rat	oral	Limb and craniofacial defects	Zinc acetate
31	5-Fluorodeoxyuridine	Dagg and Kallio, 1962	Mouse	20 mg/kg; i.p.; day 10	Abnormal hind feet and tail	Thymidine 5–640 mg/kg
32	6-Hydroxylaminopurine	Chaube and Murphy, 1969b	Rat	50–600 mg/kg; i.p.; day 12	Cleft palate, limb and tail defects	*Inosine 200–500 mg/kg, hypoxanthine 50–1000 mg/kg, adenine 50–250 mg/kg
33	Hydroxyurea	Chaube and Murphy, 1973	Rat	500 mg/kg; i.p.; day 11	Defective palate, jaws, limbs and tail	Deoxycytidylic acid 700 mg/kg, cytidine, deoxycytidine
34	Hypoglycin A	Persaud, 1969a, b; '72 ('67, '68, '70)	Rat	30 mg/kg; i.p.; days 1–6	Encephalocoele, syndactyly	Riboflavin phosphate 3 mg/kg
35	Hypoglycin B	Persaud, 1973 ('72)	Rat	100 µg; intra-amniotic; day 14	Visceral defects	Leucine 50 µg
36	Indium	Ferm, 1970	Golden hamster	0.5–2 mg/kg; i.v.; day 8	Deformed limbs	Dextran/iron complex (Imperon) 50 mg/kg
37	Insulin	Landauer, 1948	Chicken	2 IU; yolk sac; 96 h	Abnormal beak and eyes, micromelia	Nicotinamide 18 mg, α-ketoglutaric acid 20 mg

Note: Row between 31 and 32 — Forsthoefel and Williams, 1975; Mouse; 10–20 mg/kg; i.p.; day 10; Polydactyly, ectrodactyly, radial and tibial hemimelia, abnormal caudal vertebrae; Thymidine 20 mg/kg

No.	Compound	Reference	Species	Dose; route; timing	Defect	Antiteratogen
		Landauer and Rhodes, 1952	Chicken	1-2 IU; Yolk sac; 24, 96, 120 h	Early – tailless, late – micromelia and abnormal beak	Early – pyruvic acid 0.5 ml (also DL-lactic and citric acid), late – glucose-1-phosphate 4 mg
		Zwilling, 1959	Chicken, tibial rudiments in vitro			Nicotinamide
38	Iodoacetic acid	Miller, 1973	Mouse	0.5 mg; i.m.; days 11–13 or 12–14	Cleft palate	Succinate
39	N-isopropyl-α-(2-methylhydrazino)-p-toluamide HCl (Procarbazine)	Chaube and Murphy, 1969a	Rat	12–250 mg/kg; i.p.; day 14–17	Limb and tail defects, cleft palate	L-methionine
40	Malathion	Greenberg and LaHam, 1969, '70	Chicken	3.99/6.42 mg; yolk sac; day 4/5	Micromelia, beak defects	*Tryptophan 5 mg, nicotinamide 5 mg
41	Manganese deficiency	Hurley, Everson, Wooten and Asling, 1961; Hurley, Wooten and Everson, 1961; Hurley and Asling, 1963; Hurley and Everson, 1963	Rat		Relatively short long bones, retarded brain growth, change in skull shape	Dietary manganese supplementation
		Erway, Hurley and Fraser, 1966, '70; Erway, Fraser and Hurley, 1971	Mouse		Defective otolith formation	*Mn supplements in diet
42	N,N' methylene bis (2-amino-1,3,4-thiadiazole)	Mizutani, Ihara and Sugitani, 1974 (Abstract)	Hamster	30 mg/kg; oral; day 8	?	Nicotinamide 4 mg/kg at same time or 3 h pre- or post-treatment
43	Mitomycin	Fujii, 1968	Mouse	5 mg/kg; i.v.; day 12	Cleft palate, oligo- and syndactyly	Anabolic steroid 25 mg/kg
44	Nicotine sulphate	Landauer, 1960b, '77a	Chicken	2.5 mg; yolk sac; days 0–12	Short upper beak, short twisted neck, muscular hypoplasia	DL-Lysine mono HCl 10 mg, L-proline 10 mg, leucine HCl 10 mg, glucose 10 mg, sodium pyruvate 10 mg, oximes

(continued)

33

Table 2.2 (continued)

Number	Teratogen	Authors	Test object	Dose; Route; Stage	Major effects	Antiteratogen
45	Organophosphorous and methylcarbamate insecticides (Bidrin and others)	Khan, 1975 (Abstract)	Chicken	5 mg	Fused or partly missing cervical vertebrae	Sodium ascorbate, nicotinamide
		Roger, Upshall and Casida, 1969; Proctor and Casida, 1975; Proctor, Moscioni and Casida, 1976	Chicken	0.1–1.0 mg; yolk sac; day 4	Micromelia, parrot beak, short crooked neck	*Nicotinamide 0.1–1.0 mg (also precursors and derivatives)
		Meiniel, 1974, '75, '76a, b, c	Quail	0.25 mg; prior to incubation by submersion or yolk sac	as previous	*Nicotinamide 1 mg + Pralidoxime 2 mg
46	T-82 phenothiazine derivative	Horvath, Szonyi and Mold, 1976	Rat	2000 mg/kg; i.p.; days 11/12	Cleft palate micrognathia, micromelia, ectopic testis	Riboflavin 100 mg/kg, ATP 12.4 mg
47	Physostigmine (eserine)	Landauer, 1949, '76; Sullivan, 1975	Chicken	0.25 to 1 mg; yolk sac; 96 h	Parrot beak, micromelia, syndactyly, short neck	*Nicotinamide 5 mg, pyruvic acid 0.05 ml, oximes
48	Pilocarpine	Landauer, 1953	Chicken	3 mg; yolk sac; 96 h	Syndactyly	Nicotinamide 5 mg
49	6-n-Propyl-2-thiouracil	Deol, 1973	Mouse	0.1% in mother's drinking water	Abnormal inner ear	Sodium-L-thyroxine 90 mg/100 ml drinking water
50	Pyridine	Landauer and Salam, 1974	Chicken	20 mg; yolk sac; 96 h	Muscular hypoplasia	Nicotinamide 2.5 mg
51	Pyrimethamine	Sullivan and Takacs, 1971	Rat	0.5 mg repeatedly; i.p.; days 5, 6 or 7 to 14	Brachygnathia, cleft palate, limb defects	Folinic acid 6 mg
52	Sodium arsenate	Holmberg and Ferm, 1969	Golden hamster	20 mg/kg; i.v.; day 8	Exencephaly, cleft palate/lip, microphthalmia etc.	Sodium selenite 2 mg/kg
53	Sodium cacodylate	Landauer, 1949	Chicken	2.5 mg; yolk sac; 0 h	Tailless	Nicotinamide 5 mg, pyruvic acid 0.05 ml
54	Sodium methazolamide	Landauer and Wakasugi, 1968	Chicken	2.5 mg; yolk sac; 96 h	Abnormal upper beak, syndactyly	ADP 3 mg
55	Sulphadimethoxine	Kato, 1973 (in Japanese/English, abstract)	Mouse		Cleft palate	*Tetrahydrofolic acid, (nicotinamide, pyridoxine, L-ascorbic acid, to lesser extent)

34

No.	Teratogen	Reference	Species	Dose/treatment	Defect	Antiteratogen
56	Sulphanilamide	Zwilling and DeBell, 1950	Chicken	1.6/1.8 mg; yolk sac; 48 h and day 5	Parrot beak, micromelia	*Nicotinamide 5 mg
		Landauer and Clark, 1964	Chicken	1 mg; yolk sac; 48, 72 or 96 h	Parrot beak, micromelia	Dichlorphenamide 2 mg, 3-acetylpyridine 0.5 mg (at 72, 96 or 120 h), sodium ascorbate 10 mg, ascorbic acid 10 mg, sodium folate 0.75 mg
		Landauer and Sopher, 1970	Chicken	1.3 mg; yolk sac; 96 h		
		Landauer and Wakasugi, 1968	Chicken	1 mg; yolk sac; 96 h	Parrot beak, micromelia	
57	Sympathomimetics: isoproterenol, epinephrine, norepinephrine, phenylephrine	Hodach, Hodach, Fallon, Folds, Bruyere and Gilbert 1975	Chicken	0.4×10^{-9} to 20×10^{-9} mol; on embryonic membrane; day 5	Cardiac and aortic anomalies	Propranolol 10×10^{-9} to 200×10^{-9} pretreatment
58	Thalidomide	Ruano Gil, 1967	Chicken	1 mg; in vitro; stages 4–9	Platyneuria	ATP 0.2 mg
59	Thallium	Landauer, 1960a	Chicken	0.5 mg; yolk sac; 96 h	Parrot beak, micromelia, short tail	Imidazole 0.66–1.76 mg, sorbose 20 mg, arabitol 10 mg
60	Thiram (Tetramethyl-thioperoxy dicarbonic diamide)	Hall 1972b ('72a, '77)	Chicken	tibia in vitro	Micromelia	Vitamin C
		Matthiaschk, 1973	Mouse	10–30 mg; oral; days 5–15	Cleft palate, micrognathia, kyphosis, wavy ribs, abnormal limb bones	L-cysteine 2.5 or 5 mg (oral or i.p.)
61	Trypan blue	Beaudoin 1968 ('61)	Chicken	0.1 ml of 0.1% saline solution; subgerminal or yolk sac; 36–38 h	Tailless	Citric acid 25 mg
		Kolesari and Kaplan, 1974	Chicken	0.2 ml of 1%; into liquid subgerminal yolk; stage 12	Haematomas	Hypo- and hyperthermia
62	Vitamin A	Millen and Woollam, 1958	Rat	40000 IU; oral; days 8–13	Cleft palate	Vitamin B, 0.5 ml B complex (s.c.)
		Yamaguchi, 1968	Rat	30000 IU/100 g; oral; days 10–13	Complex defects	Vitamin B_{12} (s.c.)
		Sudo, 1968	Mouse	10–15000 IU; day 10, 11 or 12	Micromelia	Tocopherol 3-5 mg (days 1–9)

3
Trypan blue induced teratogenesis

F. BECK

INTRODUCTION

Since the discovery of the teratogenic effect of trypan blue by Gillman et al. in 1948[1] experiments attempting to elucidate its mode of action have been regularly reported in the scientific press. Many teratologists are doubtless bemused by this apparently unwarranted concentration on a chemical which is neither a medicine nor a potential environmental pollutant. The dye produces no highly specific malformation syndrome such as that following thalidomide ingestion in man or cortisone treatment in the mouse so it has limited value in studying the pathogenesis of distinctly defined congenital defects. What, then, is the particular value of this agent? Partly, no doubt, its continued use springs from the historical fact that it was one of the earliest ways of inducing malformation in mammals cheaply and the results of its administration were more or less reproducible in a wide variety of rodent species. But it was soon apparent that the trypan blue group of bisazo dyes were somewhat unusual among teratogenic agents. Their peculiarities suggested that a satisfactory explanation of the way in which they produced their effect would throw light on specific physiological processes which, if disturbed, could result in the production of live but congenitally malformed offspring. The special properties of trypan blue which give rise to these speculations are:

1. Teratogenic efficacy, which is maximal at 8.5 days, falls off abruptly at 9.5 days and disappears entirely at 10.5 to 11.5 days of gestation in the rat[2]. Most other teratogenic agents which are not clearly tissue-specific (e.g. ionizing radiations, various antimitotic agents or vitamin A excess) are active well beyond this period of gestation.

2. Although the dye is actively accumulated and clearly demonstrable by light microscopy in a variety of maternal cells particularly in macrophages, it is absent from the cells of malformed embryos. Experiments with radiolabelled trypan blue have confirmed this finding[3] and although some recent papers have demonstrated trace quantities of material in the lumen and cells around

37

the gut in early embryos[4, 5] the balance of evidence strongly suggests that the dye does not exert its effects as a result of penetrating embryonic cells.

3. The dye is accumulated in large quantities in the visceral yolk-sac epithelium, which tissue, it is postulated, forms the major nutritive organ of the rat embryo prior to the development of the chorioallantoic placenta at about 11.5 days of gestation[6]. The switch over from yolk-sac placentation to chorioallantoic placentation therefore roughly coincides with the startling fall-off in teratogenic effect referred to in (1) above, provided one allows a little time for trypan blue to reach and affect the target cells and postulates that the embryo has reserves lasting for a short additional period.

4. A number of teratogenic effects – such as limb malformations – are not seen in rats but may easily be produced by trypan blue in the chick[7]. In birds there is obviously no placental system and the yolk-sac endoderm (which accumulates and digests yolk for embryonic nutrition) remains the major nutritive organ throughout development. Maximal teratogenicity at 36 h and a fall off between 48 and 72 h has been described in the chick[8]. Susceptibility to the dye is thus later than at comparative developmental stages in the rat.

Much of the early experimental work done with trypan blue relied upon impure samples of the dye, and their use produced inconsistent results from various laboratories. The demonstration of numerous contaminants[9] led to some speculation concerning the true nature of the active agent in those dye samples which were highly teratogenic when compared with others which were apparently inert. The matter has now been much clarified by the development of a reliable method of purifying the dye[10]. Pure trypan blue was found to be highly teratogenic in rats[9] and in mice[10, 11]. A red contaminant, characterized chemically by Field et al.[10] failed to produce malformations in rats[9, 12] and in mice[10, 11, 13]. Only Bertini and Sacerdote using mice[14] claim a teratogenic potential for this component. A 'purple' impurity was shown by Field et al.[10] to be a complex mixture which varies in composition from one commercial sample to another; this group of contaminants has been tested in rats[9] and was non-teratogenic. In mice similar results were produced by Field et al.[10] and also by Barber and Geer[11] although the latter authors reported mild teratogenic effects when using one of their purple fractions. The reader is referred to Field et al.[10] for the most comprehensive chemical analysis of the contaminants of commercial samples of trypan blue so far reported and it now seems reasonable to conclude that pure trypan blue is the active teratogenic substance in impure commercially available mixtures. Gillman et al.[15] showed that their trypan blue samples also produced reticuloses and reticulum cell tumours of the liver; in this instance it appears that some as yet unidentified component of the purple fraction is either itself the main tumour-producing agent or potentiates the action of pure trypan blue in this respect[10].

Trypan blue has other effects besides its teratogenic and oncogenic properties and these should be borne in mind when the cause of its teratogenic effect is being sought. *In vivo* the dye is bound to serum proteins, especially to serum albumin[16], and probably very little is present in the free form if given in sublethal doses. Diverse changes in serum protein composition caused by dye injection have been reported[13, 17-23]. Most authors describe a fall in serum albumin and a rise in α- and β-globulins together with the appearance of abnormal electrophoretic bands though the observations vary with the species and the method of protein separation used. The changes are perhaps not surprising in view of the modification of protein catabolism produced by the dye (see below); they may be reflected by alterations in fetal serum proteins[24].

Grabowski and his co-workers[25, 26] have demonstrated an increase in circulating embryonic fluids as well as the presence of haematomata and clear fluid-filled blebs in chick embryos following hypoxia or injection of trypan blue into the yolk-sac. Fluid imbalances in 5-day chicks were associated with serum electrolyte imbalances and a fall in total serum sodium and potassium. Furthermore, an increase in the blood pressure of hypervolemic chick embryos was demonstrated[27]. In another experiment however[28] it was found that although an increase in circulating fluids (which can be caused by trypan blue) raised the blood pressure in chick embryos, treatment with the dye paradoxically *lowered* the blood pressure of rat embryos from treated mothers.

Decreased plasma thyroxine binding capacity following trypan blue treatment has been observed[29] and a decrease in thyroid weight and serum protein bound iodine after repeated dye treatment described[30, 31]. Thereafter[32] it was shown that trypan blue and other disazo dyes displace T4 from plasma protein binding sites. The presence of higher levels of free T4 thus inhibits TSH release presumably by interfering with long and short feedback loops.

It has been observed[33] that the QO_2 (μl oxygen consumed/mg dry weight/h) was raised in preparations of 60 h chick embryos plus area opaca, 12 h after yolk-sac injection of trypan blue. Gunberg (quoted by Kaplan and Johnson[33]) on the other hand found that yolk-sacs from 12.5- and 13.5-day rat fetuses had a decreased oxygen consumption 24 h after maternal treatment with trypan blue (i.e. at 13.5 and 14.5 days).

Other observers[34] have shown an inhibition of incorporation of ^{35}S labelled sodium sulphate into calf costal cartilage *in vitro* following addition of trypan blue to the culture medium. Related azo dyes were also found to inhibit synthesis of radiolabelled acid mucopolysaccharide though the effect was often less than that caused by trypan blue.

Using biochemical assays *in vitro* it has been demonstrated[35] that trypan blue at concentrations of between 10^{-4} and 10^{-3} M inhibited all the lysosomal acid hydrolases (β-glucuronidase, ribonuclease, deoxyribonuclease, acid phosphatase and acid protease) it was exposed to. This observation was confirmed qualitatively *in vivo*[36-38] when it was demonstrated histochemically that a loss of acid phosphatase staining in the visceral yolk-sac of mice treated with trypan blue at 7 days of gestation or later, occurred. Secondary lysosomes obtained from rat liver[39, 40] isolated by cell fractionation digested previously incorporated [125I]albumin but the capacity for protein digestion was inhibited if they also contained trypan blue (i.e. if they had been obtained from a trypan blue treated animal). Explants of 17.5 day yolk-sac from [125I]albumin treated rats are, however, capable of protein digestion at the same rate whether they are explanted from trypan blue treated rats or from control animals not treated with the dye[41]. Williams[42] later demonstrated that teratogenic concentrations of trypan blue added to cultured 17.5 day normal rat yolk-sac inhibited the capacity of that tissue to pinocytize [125I]albumin *in vitro* with the result that its ability to break down radioiodinated protein added to the culture medium was materially decreased. Another effect of trypan blue on the rat yolk-sac relates to the observation[43] that trypan blue uptake is associated with increased ionic absorption. Subsequently[44] it was shown that trypan blue and other colloids increased the amount of valine absorbed by and transported across the adult rat small intestine; the rate of absorption and transport were also increased.

THE POSSIBLE MODES OF ACTION OF TRYPAN BLUE AS A TERATOGEN

From what has been said above it is clear that a substantial body of potentially confusing evidence has accumulated since the writer was co-author of a review on the teratogenic action of bisazo dyes more than a decade ago[45]. In this chapter I shall concentrate on the work carried out by my colleagues and myself but it would be incorrect to imply that we consider our theories of the possible mode of action of bisazo dyes to have been proven. I shall therefore briefly present *a selection* of alternative (or additional) theories in order to emphasize that they require equally serious consideration.

Theories which postulate a primary site of action on the mother

The serum protein changes reported above[13, 17-23] as following trypan blue treatment in a number of mammals make it quite possible that crucial alterations occur in the composition of histiotroph presented to post-implantation embryos (particularly to rodents). The importance of this factor is to some extent supported by the reports of haemorrhages and fluid-filled blebs which occur in treated embryos[46] and have been suggested by some authors as constituting the primary pathogenic effect of trypan blue treatment[25, 47]. Clearly early embryonic changes may also be related (possibly secondarily) to fetal blood pressure alterations[27, 28]. The observation[48] that treatment of pregnant rats with serum α- and β-globulins from trypan blue treated animals fails to produce malformed offspring does not invalidate the possibility that maternal serum changes may cause or contribute to the teratogenic effect of the dye.

Theories which postulate a primary site of action on the embryo

The function and disposition of the extra-embryonic membranes in the rat at 8.5 days of gestation[49], when trypan blue is maximally teratogenic is such that the dye, after penetrating Reichert's membrane, comes into direct contact with the endoderm at the apex of the egg cylinder (Figure 3.1). There is some doubt as to whether this layer of cells does in fact eventually constitute the endoderm of the embryonic gut[50] but if it does contribute to the embryonic tissues or to their morphogenetic mechanisms it is at least a theoretical possibility that a surface action exerted by the dye upon them might be sufficient to produce malformations. There is no doubt that acid bisazo dyes bind to the glycoprotein surface of cells and although there is no direct evidence that a surface effect on embryonic cells is teratogenic such a possi-

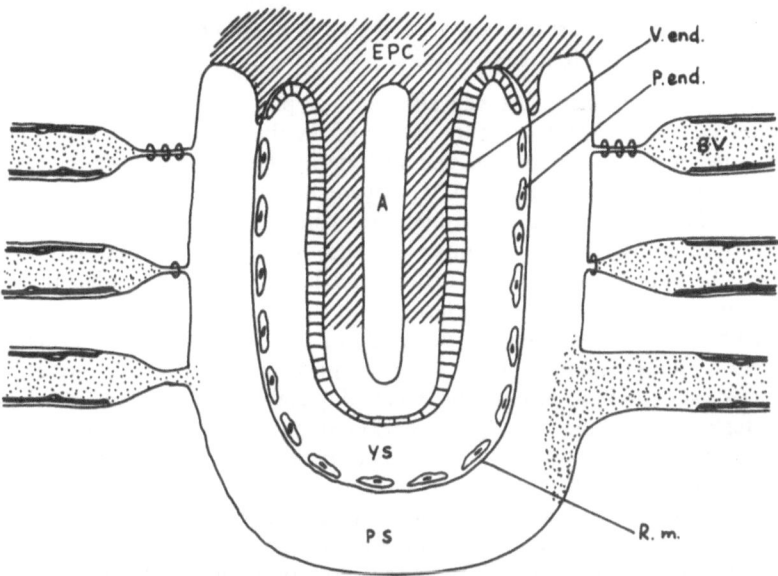

Figure 3.1   Diagram representing early stage in the development of the rat egg cylinder (circa 7 days). The periembryonal sinus (PS) is supplied by decidual vessels (BV); it is separated from the parietal layer of the yolk-sac endoderm (P.end.) by Reichert's membrane (R.m.). Note the disposition of the horseshoe-shaped yolk-sac (YS), the visceral layer of the yolk-sac endoderm (V.end.), the primitive amniotic cavity (A) and the ectoplacental cone (EPC). After Merker and Villegas[49]

bility must be borne in mind[51]. The same type of mechanism could also explain such phenomena as the teratogenic action of the dye in amphibia[52]. Malformations in *Rana pipiens* can be obtained by adding 0.5% trypan blue to spring water in which frog eggs between the stage of appearance of the dorsal lip of the blastopore and the closure of the neural folds were developing. No demonstration of dye incorporation into the cells of the embryo is observed.

Using careful histological techniques the accumulation of trypan blue inside embryonic endoderm cells lining the gut of the rat embryo when the dye was administered before 11 days of gestation (the time of closure of the vitelline duct) was observed[5]. However, it could not be seen in the ectodermal or mesodermal layers. Thus previous suggestions which do not exclude a direct action of the dye on embryonic cells[4, 11] receive some support even though Wilson and his co-workers[3] failed to locate maternally administered ^{14}C labelled dye in the embryo. The possibility of some direct penetration into embryonic cells should also be considered in conjunction with a report[53] of the teratogenic effects following direct injection of trypan blue into the blood stream of 3-day chick embryos.

Theories which postulate a primary site of action on the extra-embryonic membranes

All workers who have observed effects of trypan blue in rodents have been impressed by the massive accumulation of the dye in the vacuolar system of the cells of the inverted yolk-sac endoderm. Furthermore the susceptible period of trypan blue induced teratogenicity roughly corresponds to the period when the yolk-sac constitutes the principal organ of embryonic nutrition[54] (see above) and an all-important site of exchange of material between the mother and fetus. Naturally disturbance in yolk-sac function has therefore been suggested as a cause of the teratogenic activity of the dye.

It has been shown[43] that on days 12, 13 and 14 of gestation trypan blue causes a significant increase in the absorption of $^{45}Ca^{2+}$ $^{35}SO_2^{2-}$ and $^{22}Na^+$ by yolk-sacs and in some cases by developing embryos. Although the drug was not tested during the teratogenic period it obviously altered the transport functions of the yolk-sac at later stages of development. The demonstration[45] of an analogous disturbance caused by trypan blue in the increased absorption and passage of valine across the adult rat gut is also of interest in this connection. Clearly such alterations in ion transport might well be connected with the embryonic ionic changes and alterations in circulating embryonic fluids which Grabowski and his co-workers suggest might be teratogenic by virtue of the production of haematomata and fluid-filled blebs[25-28, 55]. Changes in chick and rat yolk-sac respiratory activity in the presence of trypan blue[33, 56] also imply that the dye alters function at this site and may well be exerting its teratogenic effect here. Grabowski and his group have shown that hypoxia produces hypervolemia and congenital malformations in the chick[25, 26] and Kaplan and Johnson[33] suggest that trypan blue might produce hypoxia by uncoupling oxidative phosphorylation.

Beck and Lloyd and their co-workers have put forward a hypothesis to account for the demonstrable teratogenic effects of trypan blue in mammals; they suggest that the dye causes surface membrane effects upon the cells of the extra-embryonic tissues responsible for histiotrophic nutrition. In the case of rodents the cells affected are the visceral yolk-sac epithelial cells but other tissues are probably involved in other species (see below). For descriptive convenience the hypothesis will first be described as it applies to the rat and will then be extrapolated to other species. It has long been thought that the visceral yolk-sac of the rodent serves an important nutritional role in the postimplantation embryo[6]. Merker and Villegas[49] have shown that the early postimplantation embryo is surrounded to a large part by a periembryonal sinus (Figure 3.1) which contains sluggishly moving maternal blood, there being no readily demonstrable channel of egress for blood delivered to this region by the decidual blood vessels. From the periembryonal sinus blood plasma can readily penetrate Reichert's membrane, this structure frequently being the only barrier between the sinus and the cells of the visceral yolk-sac endoderm[57]. Macromolecules are readily pinocytized by the cells of the visceral yolk-sac epithelium[16] and biopolymers are almost certainly broken down into their monomers by the extraordinarily well-developed lysosomal apparatus contained within these cells. Certainly at later stages of gestation

(17.5 days) *in vitro* preparations of yolk-sac released [125I]tyrosine from previously absorbed [125I]albumin while very little intact labelled protein was released[16]. The development of an alternative organ culture method in which both uptake and digestion of macromolecules takes place *in vitro*[58, 59] has made it possible to investigate the effects of trypan blue upon the catabolic capacity of the yolk-sac in detail. Whole yolk-sacs from 17.5 day pregnant rats were incubated in shaking vessels containing medium 199 and 10% calf serum as nutrient to which was added a radiolabelled substrate. [125I] labelled bovine serum albumin was used as substrate rather than a ^{3}H or ^{14}C labelled protein since the [125I]iodotyrosine liberated is not re-utilized in protein synthesis thus making an analysis of the kinetics of its catabolism a simple procedure. It was demonstrated[42] that the addition of 200 μg per ml of trypan blue to the incubating medium markedly reduced the production of trichloroacetic acid soluble radioactivity by the incubated yolk-sac. At the same time the level of undigested [125I]albumin contained in the cells of the yolk-sac fell dramatically when dye was added. From these data it was possible to calculate

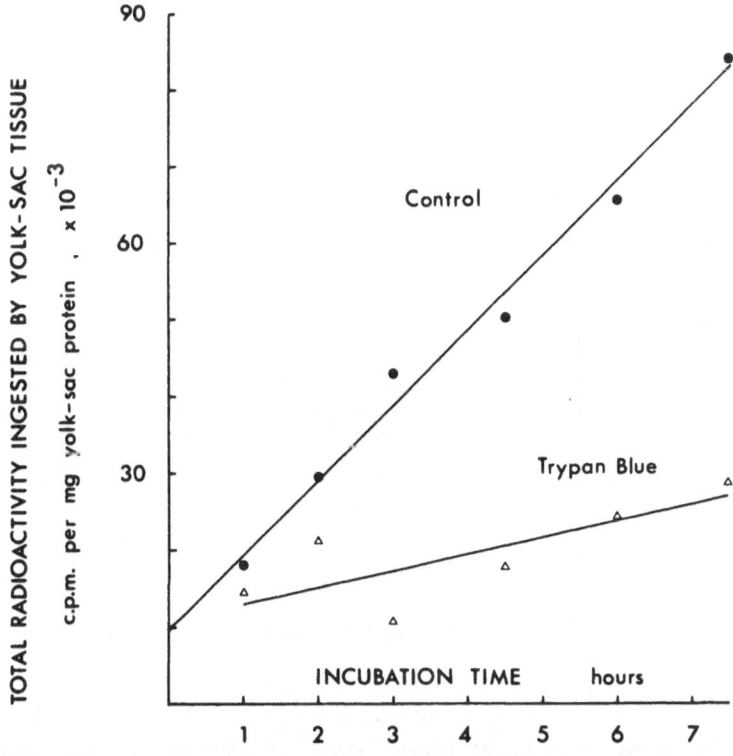

Figure 3.2   Effect of trypan blue, added to the culture medium at a concentration of 200 μg/ml after one hour of incubation, on the quantity of [125I]-labelled albumin ingested by the yolk-sac tissue. ● Data from control tissue. △ Data from an experiment in which trypan blue (200 μg/ml) was added to the medium after one hour of incubation. Results obtained by summation of albumin breakdown products released into the culture medium and undigested residual albumin present in the yolk-sac

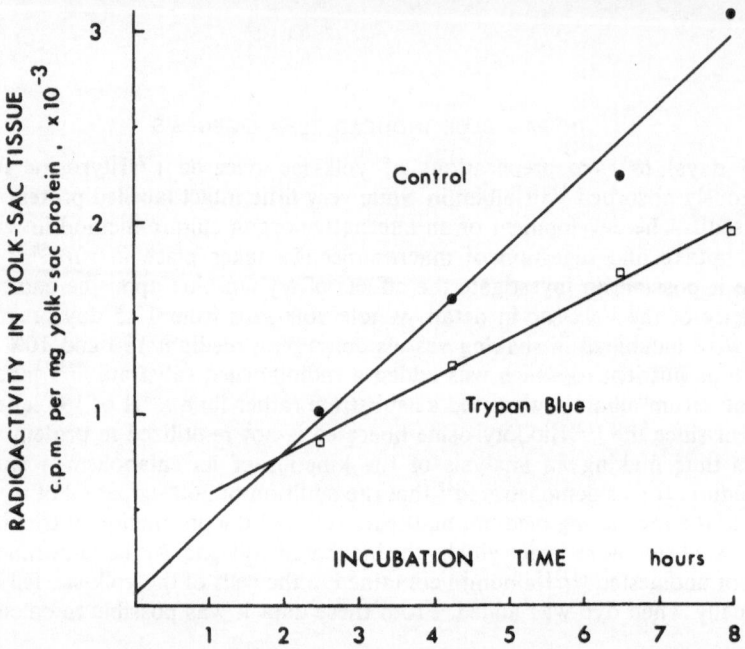

Figure 3.3 Effect of trypan blue, added to the culture medium at a concentration of 200 μg/ml after one hour of incubation, on the amount of [^{125}I]PVP ingested by yolk-sac tissue

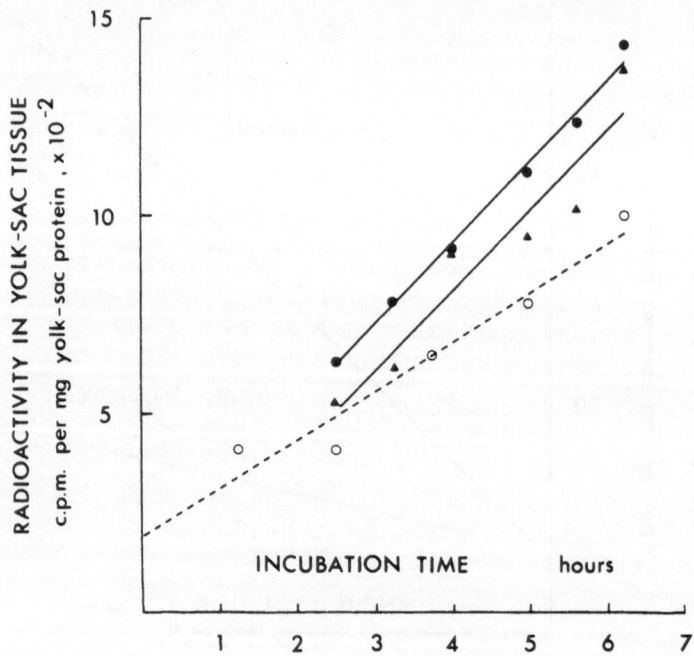

Figure 3.4 Increase in the rate of pinocytic uptake of [^{125}I]PVP when yolk-sacs, exposed to trypan blue *in vitro*, were reincubated in absence of the dye. ▲ uptake of [^{125}I]PVP by yolk-sacs incubated for 2.25 h in medium containing both [^{125}I]PVP and 200 μg/ml of dye, then incubated for 15 min in fresh medium to remove adhering dye, before reincubating in the presence of [^{125}I]PVP alone; ●, control tissues treated as in the above series but in the presence of [^{125}I]PVP alone; O, uptake of [^{125}I]PVP by yolk-sacs continuously exposed to 200 μg/ml of dye

44

that 200 μg of dye per ml of incubation medium decreased the quantity of ^{125}I labelled albumin ingested by the yolk-sac compared with control incubated yolk-sacs (Figure 3.2). It could also be shown that a similar concentration of trypan blue decreased the uptake of [125I]polyvinylpyrrolidone (Figure 3.3). The inhibition of pinocytosis by trypan blue is reversible. If yolk-sacs are first incubated in ^{125}I labelled polyvinylpyrrolidone and trypan blue (200 μg per ml) and then washed for 15 min before re-incubation with [125I]polyvinyl-pyrrolidone alone, full pinocytic capacity is restored (Figure 3.4). Similar recovery is also observed when ^{125}I labelled albumin is the substrate.

Although teratogenic serum levels of trypan blue in rats are in fact in the region of 200 μg per ml[60] it would not be correct to extrapolate the *in vitro* to the *in vivo* situation uncritically. One must bear in mind that for technical reasons 17.5 day old yolk-sac rather than yolk-sac at 8.5 days, when the animal is exclusively dependent upon it for its nutrition, was used. Further-more the incubating medium was not pure rat serum though the latter would more nearly mimic the *in vivo* situation. Trypan blue has been shown to exert a biphasic action on the pinocytic activity of the rat yolk-sac[61]; low levels of the dye (100 μg per ml and less) actually stimulate pinocytosis of [125I]poly-vinylpyrrolidone while levels of 200 μg and more inhibit it.

The situation is even further complicated by the fact that the yolk-sac pinocytizes different substrates at different rates. Thus materials (such as native bovine serum albumin) which are adsorbed onto the plasma membrane prior to endocytosis are interiorized at more than twice the level of others (like polyvinylpyrrolidone) which are minimally bound[62]. Furthermore if albumin is appropriately denatured its endocytic index can be raised to levels which are *twenty* times that of the native protein. Thus although we do not know the exact kinetics of the situation *in vivo* it seems probable that teratogenic levels of trypan blue in the serum inhibit the endocytosis of surface adsorbed macro-molecules in the serum transudate which bathes the visceral yolk-sac epi-thelium at 8.5 days of gestation. Clearly, other disturbances of yolk-sac function occur (e.g. inhibition of lysosomal enzymes, changes in ionic per-meability, possibly changes in respiratory mechanisms, etc.) but it seems entirely possible that the reduction of raw materials available to the embryo consequent upon inhibition of pinocytosis is able to cause developmental defects. Support for this hypothesis comes from the fact that the teratogenic effect ceases at about the time of the establishment of a chorioallantoic placenta when histiotrophic nutrition is largely superseded by haemotrophic exchange of blood solutes across the haemochorial placental barrier.

An *in vitro* method based upon the technique developed by New and Stein[63] has been used for investigating trypan blue teratogenicity[64]. It was concluded that the dye acts directly on embryonic tissues because abnor-malities were obtained when the dye was injected deep into the visceral yolk-sac in a position where it is in direct contact with the embryo. The results are, however, capable of other interpretation. Abnormalities in trypan blue treated explants were also obtained when 10–11 day embryos were exposed to trypan blue placed outside the visceral yolk-sac and at this time the dye has a very low teratogenicity *in vivo*. The explant techniques used at that time were to some extent unphysiological and have now been much improved[65]. It would therefore be most instructive to repeat the experiments with more modern methods along the lines used by New and Brent[66] who, by similar techniques, showed that rat yolk-sac antibodies were probably·teratogenic by virtue of their action on the visceral yolk-sac rather than by direct action on the embryo.

It is instructive to examine the effect of trypan blue on pregnancy in animals that do not possess an inverted yolk-sac placental system. Beck and his co-workers[67, 68] have described the action of the dye in the ferret. This species has a method of histiotrophic nutrition which differs somewhat from that of the rodents. Before the establishment of a chorioallantoic placenta, embryonic nutrition is mediated largely by the breakdown within the cells of the trophoblast of a pabulum composed of degenerating symplasmal cells (formed by proliferation of the endometrial epithelium in response to blastocyst implantation) and the secretion of endometrial glands. The process of endocytosis of ferret histiotroph has not been as well studied as it has in the rat; it may not involve the same cellular mechanisms and the analogy between histiotrophic nutrition in the rat and ferret may eventually prove to be a false one. Without drawing any extravagant conclusions it is, nevertheless, worth

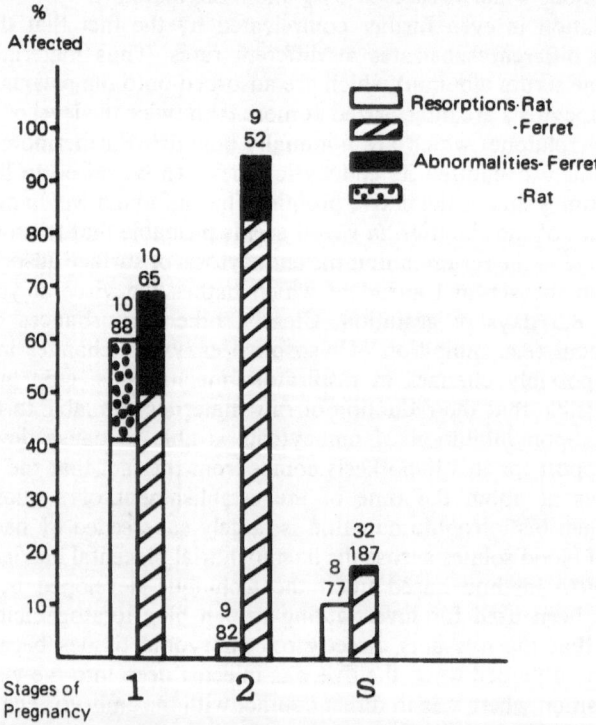

Figure 3.5 Externally visible abnormalities in fetuses following injection of trypan blue (50 mg/kg into ferrets and 100 mg/kg into rats) at similar developmental stages. The number of implantation sites is shown at the top of each block and the number of mothers used above this. The saline treated controls (S) were from mothers given one injection between 8.5 and 13.5 days in rats or between 13 days and 22 days in ferrets

comparing the effects of teratogenic doses of trypan blue in the rat and ferret, first at the primitive streak stage when both forms are dependent upon the intracellular breakdown of histiotroph and secondly at about the 30 somite stage (11.5 days of rat pregnancy and 18 days of ferret pregnancy). At the more advanced stage the rat has developed a chorioallantoic placenta and interference with histiotrophic nutrition no longer produces a teratogenic effect. In the ferret at the same developmental stage nutrition is still largely histiotrophic and trypan blue has a distinct teratogenic effect (Figure 3.5). Judging by morphological appearances the (endotheliochorial) chorioallantoic placenta does not begin to develop in the ferret until about 19+ days of gestation. Teratogenic effects have been produced in the pig using trypan blue at 10 days of gestation[69]; this again is the primitive streak stage at which only histiotrophic nutrition is available to the embryo. Turning to the effect of trypan blue on non-mammalian species it is not inconceivable that similar mechanisms preventing the endocytosis of yolk-sac could be operative in the chick, though defects have also been obtained by direct application of trypan blue impregnated gelatin cubes directly to the branchial arch region[51]. In the latter case, however, some non-impregnated gelatin cubes also caused mal-formations. Maximal teratogenicity following trypan blue injection into the yolk-sac is at 48 h (16 somites) and ceases at 30–36 somites (72 h) when enlarged limb buds are present[8]. It is interesting that limb defects often follow dye treatment in chicks[7] but are very rarely described as primary effects in mammals[22]. Thus, as in ferrets, trypan blue is teratogenic in chicks at a later stage than in the rat, but why it should cease to be teratogenic even at 72 h is not apparent. The same process involving inhibition of yolk-sac utilization could conceivably also be active in trypan blue induced teratogenesis in reptilia[70]. It is perhaps more difficult to explain why inhibition of endocytosis should cause teratogenic effects in amphibia[52] where yolk is already intra-cellularly located when gastrulation begins, though clearly other substances might still require endocytic absorption in order to allow normal development. In passing it is worth mentioning that in amphibia also the teratogenic effects of the dye extend from the time of the appearance of the dorsal lip of the blastopore to stage 14 (the time of closure of the neural folds). Teratogenic susceptibility is therefore again decidedly longer than in the rat and approaches that of the chick.

DISCUSSION

Since it has been shown (see above) that inhibition of pinocytosis is a major toxic effect of trypan blue it is tempting to speculate that this characteristic is of central teratological importance in rodents and possibly in other mam-malian species also. It is conceivable that a similar mechanism may partially or totally explain the effect of the dye in other animals but a greater pro-pensity for the relatively unprotected embryos of lower forms to respond to environmental changes by abnormal development may well mean that other factors (e.g. ionic fluxes, changes in membrane characteristics, etc.) could produce teratogenic effects in a totally different way from that experienced by mammals. It would be most interesting to observe whether closely related azo dyes acted in similar ways; such parameters have been well studied in rodents but much less so in other forms.

In support of the endocytic inhibition theory of teratogenesis in mammals it is relevant to quote the work of Berry[71] who showed that rats treated on day 7 of pregnancy contained less protein than normal embryos up to the 35 somite stage, but that the difference between treated and control groups disappeared soon after the chorioallantoic placenta became functional. Also compatible with these ideas are first the observations of Payne and Deuchar who cultivated 10-day embryos *in vitro* and found that removal of the visceral yolk-sac before culturing led to a failure of growth and development as well as a much lower uptake of [³H]leucine into the embryo, and second the findings of Zawoisky[73] who showed that nutritional supplements of glutamic acid and certain proteins protected pregnant mice against the teratogenic effects of trypan blue.

The observation that the blood flow around the immediately post-implantation rat embryo is sluggish[49] suggests that the embryo is likely to obtain relatively more nourishment from concentrated macromolecular sources than from amino acids and other small molecules. Nevertheless, the extent to which this interpretation is correct is not certain and small molecules crossing the yolk-sac (by 'piggy back' pinocytosis or other means) might to some extent satisfy its nutritional requirements. It has been observed[74] that the yolk-sac may provide an effective barrier to quite small molecules and the full significance of this report awaits extended evaluation.

CONCLUSIONS

Many theories have been put forward to account for the teratogenic action of trypan blue. In the author's opinion the hypothesis that trypan blue inhibits pinocytosis which is necessary for histiotrophic nutrition adequately explains the phenomena observed in mammals. This theory has perhaps been more consistently investigated than any alternative and consequently the body of evidence in its support is considerable. This does not however amount to proof and should not inhibit the systematic testing of other possibilities.

The nature of the trypan blue problem and the promising possibilities of an *in vitro* approach are in the author's opinion worth pursuing for a number of reasons. First, it is important to see whether teratogenesis can result from interference with an early 'placental' system (i.e. with histiotrophic nutrition) and how widely this phenomenon is applicable in mammals. Second, trypan blue appears to be an efficient chemical for interfering with pinocytosis and this in turn is of importance not only in the field of teratogenesis but also in a broader context of cell biology. Third, the ease of tracing moderate quantities of the dye histologically and biochemically gives it certain technical advantages; small quantities will clearly have to be traced by radiolabelling techniques and such methods might well be necessary when the cell surface effects of the dye are studied.

References

1. Gillman, J., Gilbert, C., Gillman, T. and Spence, I. (1948). A preliminary report on hydrocephalus, spina bifida and other congenital anomalies in the rat produced by trypan blue. *S. Afr. J. Med. Sci.*, 13, 47

2. Wilson, J. G., Beaudoin, A. R. and Free, H. J. (1959). Studies on the mechanism of teratogenic action of trypan blue. *Anat. Rec.*, 133, 115

3. Wilson, J. G., Shepard, T. H. and Gennaro, J. F. (1963). Studies on the site of teratogenic action of ^{14}C labelled trypan blue. *Anat. Rec.*, 145, 300

4. Davis, H. W. and Gunberg, D. L. (1968). Trypan blue in the rat embryo. *Teratology*, 1, 125

5. Dencker, L. (1977). Trypan blue accumulation in the embryonic gut of rats and mice during the teratogenic phase. *Teratology*, 15, 179

6. Beck, F., Lloyd, J. B. and Griffiths, A. (1967). A histochemical and biochemical study of some aspects of placental function in the rat using maternal injection of horseradish peroxidase. *J. Anat.*, 101, 461

7. Beaudoin, A. R. and Wilson, J. G. (1958). Teratogenic effects of trypan blue on the developing chick. *Proc. Soc. Exp. Biol. Med.*, 97, 85

8. Beaudoin, A. R. (1961). Teratogenic activity of several closely related disazo dyes on the developing chick embryo. *J. Embryol. Exp. Morphol.*, 9, 14

9. Beck, F. and Lloyd, J. B. (1963). The preparation and teratogenic properties of pure trypan blue and its common contaminants. *J. Embryol. Exp. Morphol.*, 11, 175

10. Field, F. E., Roberts, G., Hallowes, R. C., Palmer, A. K., Williams, K. E. and Lloyd, J. B. (1977). Trypan Blue: identification and teratogenic and oncogenic activities of its coloured constituents. *Chem.-Biol. Interact.*, 16, 69

11. Barber, A. N. and Geer, J. C. (1964). Studies on the teratogenic properties of trypan blue and its components in mice. *J. Embryol. Exp. Morphol.*, 12, 1

12. Kelly, J. W., Feagaus, W. M., Parker, J. C. and Porterfield, J. M. (1964). Studies on the mechanism of trypan blue-induced congenital malformations. *Exp. Mol. Pathol.*, 3, 262

13. Dijkstra, J. and Gillman, J. (1961). Chromatographic separation of biologically active components from commercial trypan blue. *Nature (Lond.)*, 191, 803

14. Bertini, F. and Sacerdote, F. (1970). Malformations caused in mouse embryos by a red dye contained in commercial trypan blue. *Teratology*, 3, 371

15. Gillman, J., Gillman, T. and Gilbert, C. (1949). Reticulosis and reticulum cell tumours of the liver produced in rats by trypan blue with reference to hepatic necrosis and fibrosis. *S. Afr. J. Med. Sci.*, 14, 21

16. Rawson, R. A. (1943). The binding of T-1824 and structurally related disazo dyes by the plasma proteins. *Am. J. Physiol.*, 138, 708

17. Ess, H. and Frederici, L. (1957). Zur Problematik der sog. 'R.E.S.-Blockade' mit Trypanblau. *Z. Ges. Exp. Med.*, 129, 264

18. Langman, J. and Van Drunen, H. (1959). The effect of trypan blue upon maternal protein metabolism and embryonic development. *Anat. Rec.*, 133, 513

19. Yamada, T. (1959). Abnormal serum protein observed in trypan blue treated rats. *Proc. Soc. Exp. Biol. Med.*, 101, 566

20. Beaudoin, A. R. and Ferm, V. H. (1961). The effect of disazo dyes on protein metabolism in the pregnant rabbit. *J. Exp. Zool.*, 147, 219

21. Paoletti, C., Riou, G. and Truhaut, R. (1962). Electrophoretic pattern of plasma proteins in rats treated with trypan blue and ethionine. *Nature (Lond.)*, 193, 784

22. Brown, D. V., Norlind, L. M., Adamovics, A. and Bowen, A. (1963). Studies on serum protein concentrations and organ dye concentrations in trypan blue carcinogenesis. *Proc. Soc. Exp. Biol. Med.*, 114, 290

23. Christie, G. A. (1964). The teratogenic activity of trypan blue and its effect on the thyro-hypophyseal axis in the rat. *J. Anat.*, 98, 377

24. Beaudoin, A. R. and Kahkonen, D. (1963). The effect of trypan blue on the serum proteins of the fetal rat. *Anat. Rec.*, 147, 387

25. Grabowski, C. (1963). Teratogenic significance of ionic and fluid imbalances. *Science*, 142, 1064

26. Kaplan, S. and Grabowski, C. (1967). Analysis of trypan blue induced rumplessness in chick embryos. *J. Exp. Zool.*, 165, 325
27. Grabowski, C. T., Tsai, E. N. C. and Toben, H. R. (1969). The effects of teratogenic doses of hypoxia on the blood pressure of chick embryos. *Teratology*, 2, 67
28. Grabowski, C. T., Tsai, E. N. C. and Chernoff, N. (1971). The effects of trypan blue on the blood pressure of rat embryos. *Teratology*, 4, 69
29. Crispell, K. R., Coleman, J. and Hyer, H. (1957). Factors affecting the binding capacity of human erythrocytes for I^{131} labelled *l*-thyroxine and *l*-triiodothyronine. *J. Chem. Endocrinol. Metab.*, 17, 1305
30. Yamada, T. (1960a). Effect of trypan blue on thyroid function in the rat. *Endocrinology*, 67, 204
31. Yamada, T. (1960b). Mechanism of action of trypan blue in suppressing thyroid function. *Endocrinology*, 67, 212
32. Yamada, T., Whallon, J., Tomizawa, T., Shimoda, S. and Schichijo, K. (1965). Further studies on the mechanism of action of trypan blue in suppressing thyroid activity in the rat. *Metabolism*, 14, 281
33. Kaplan, S. and Johnson, E. M. (1968). Oxygen consumption in normal and trypan blue-treated chick embryos. *Teratology*, 1, 369
34. Kocchar, D. M., Boström, H., Larsson, K. S. and Reio, L. (1967). Influence of trypan blue and related compounds on ^{35}S-sulfate metabolism of cartilage in vitro. *Eur. J. Pharmacol.*, 1, 326
35. Beck, F., Lloyd, J. B. and Griffiths, A. (1967). Lysosomal enzyme inhibition by trypan blue: a theory of teratogenesis. *Science*, 157, 1180
36. Greenhouse, G., Pesetsky, I. and Hamburgh, M. (1969). The effect of teratogenic doses of trypan blue on the yolk sac placenta of the mouse. *J. Exp. Zool.*, 171, 343
37. Nebel, L. and Hamburgh, M. (1966). Observations on the penetration and uptake of trypan blue in embryonic membranes of the mouse. *Z. Zellforsch.*, 75, 129
38. Hamburgh, M., Ehrlich, M., Nathanson, G. and Pesetsky, I. (1975). Some additional observations relating to the mechanism of trypan blue induced teratogenesis. *J. Exp. Zool.*, 192, 1
39. Davies, M., Lloyd, J. B. and Beck, F. (1969). Protein digestion in isolated lysosomes inhibited by intralysosomal trypan blue. *Science*, 163, 1454
40. Davies, M., Lloyd, J. B. and Beck, F. (1971). The effect of trypan blue, suramin and aurothiomalate on the breakdown of ^{125}I-labelled albumin within rat liver lysosomes. *Biochem. J.*, 121, 21
41. Williams, K. E., Lloyd, J. B. and Beck, F. (1971). Digestion of an exogenous protein by rat yolk sac cultured in vitro. *Biochem. J.*, 125, 303
42. Williams, K. E., Roberts, G., Kidston, M. E., Beck, F. and Lloyd, J. B. (1976). Inhibition of pinocytosis in rat yolk-sac by trypan blue. *Teratology*, 14, 343
43. Kernis, M. M. and Johnson, E. M. (1969). Effects of trypan blue and Niagara blue 2B on the in vitro absorption of ions by the rat visceral yolk sac. *J. Embryol. Exp. Morphol.*, 22, 115
44. Kernis, M. M. (1971). The influence of trypan blue, Niagara blue 2B and colloidal carbon on the absorption and transport of valine by rat intestinal segments. *Teratology*, 4, 327
45. Beck, F. and Lloyd, J. B. (1963). The teratogenic effects of azo dyes. *Adv. Teratol.*, 1, 131
46. Waddington, C. H. and Carter, T. C. (1953). A note on abnormalities induced in mouse embryos by trypan blue. *J. Embryol. Exp. Morphol.*, 1, 167
47. Hamburgh, M. (1954). The embryology of trypan blue induced abnormalities in mice. *Anat. Rec.*, 119, 409
48. Beaudoin, A. R. and Roberts, J. (1965). Serum proteins and teratogenesis. *Life Sci.*, 4, 1353
49. Merker, H.-J. and Villegas, H. (1970). Elektronenmikroskopische Untersuchungen zum Problem des Stoffaustausches zwischen Mutter und Kind bei Ratten Embryonen des Tages 7–10. *Z. Anat. Entwickel-Gesch.*, 131, 325
50. Gardner, R. L. and Papaioannon, V. E. (1975). Differentiation in the trophectoderm and inner cell mass. In: M. Balls and A. E. Wild (eds.) *The Early Development of Mammals*, p. 107 (London: Cambridge University Press)

51. Stephan, F. and Sutter, B. (1961). Réaction de l'embryon de poulet au bleu trypan. *J. Embryol. Exp. Morphol.*, 9, 410
52. Greenhouse, G. and Hamburgh, M. (1968). Analysis of trypan blue induced teratogenesis in *Rana pipiens* embryos. *Teratology*, 1, 61
53. Kaplan, S. and Johnson, E. M. (1970). Teratogenic effects of direct injection of aqueous- and protein-bound trypan blue into the bloodstream of 3-day chick embryos. *Teratology*, 3, 269
54. Beck, F. (1976). Comparative placental morphology and function. *Environ. Health Persp.*, 18, 5
55. Grabowski, C. T. (1966). Physiological changes in the bloodstream of chick embryos exposed to teratogenic doses of hypoxia. *Dev. Biol.*, 13, 199
56. Gunberg, D. L. and Wade, F. D. (1961). Effect of trypan blue on oxygen uptake of the rat yolk sac *in vitro. Abst. Teratol. Soc.*, 1, 5
57. Enders, A. C., Given, R. L. and Schlafke, S. (1978). Differentiation and migration of endoderm in the rat and mouse at implantation. *Anat. Rec.*, 190, 65
58. Williams, K. E., Kidston, M. E., Beck, F. and Lloyd, J. B. (1975). Quantitative studies of pinocytosis I. Kinetics of uptake of [^{125}I]polyvinylpyrrolidone by rat yolk sac cultured *in vitro. J. Cell Biol.*, 64, 113
59. Williams, K. E., Kidston, M. E., Beck, F. and Lloyd, J. B. (1975). Quantitative studies of pinocytosis. II. Kinetics of protein uptake and digestion by rat yolk sac cultured *in vitro. J. Cell Biol.*, 64, 123
60. Lloyd, J. B. and Beck, F. (1966). The relationship of chemical structure to teratogenic activity among bisazo dyes: a re-evaluation. *J. Embryol. Exp. Morphol.*, 16, 29
61. Williams, K. E., Lloyd, J. B., Kidston, M. E. and Beck, F. (1973). Biphasic effect of trypan blue on pinocytosis. *Biochem. Soc. Trans.*, 1, 203
62. Lloyd, J. B., Williams, K. E., Moore, A. T. and Beck, F. (1975). Selective uptake and intracellular digestion of protein by rat yolk sac. In: W. A. Hemmings (ed.) *Maternofoetal Transmission of Immunoglobins*, pp. 169–178 (London: Cambridge University Press)
63. New, D. A. T. and Stein, K. F. (1964). Cultivation of post implantation mouse and rat embryos on plasma clots. *J. Embryol. Exp. Morphol.*, 12, 101
64. Turbow, M. M. (1966). Trypan blue induced teratogenesis of rat embryos cultivated *in vitro. J. Embryol. Exp. Morphol.*, 15, 387
65. New, D. A. T., Coppola, P. T. and Cockcroft, D. L. (1976). Improved development of head fold rat embryos in culture resulting from low oxygen and modifications of the culture serum. *J. Reprod. Fertil.*, 48, 219
66. New, D. A. T. and Brent, R. L. (1972). Effect of yolk sac antibody on rat embryos grown in culture. *J. Embryol. Exp. Morphol.*, 27, 543
67. Beck, F. (1975). The ferret as a teratological model. In: D. Neubert and H.-J. Merker (eds.) *New Approaches to the Evaluation of Abnormal Development*, pp. 8–20 (Stuttgart: Georg Thieme)
68. Beck, F., Swidzinska, P. and Gulamhusein, A. P. (1978). The effect of trypan blue on the development of the ferret and rat. *Teratology*, 18, 178
69. Rosenkrantz, J. G., Lynch, P. F. and Frost, W. W. (1970). Congenital anomalies in the pig: Teratogenic effects of trypan blue. *J. Ped. Surg.*, 5, 232
70. Mathur, J. K. and Goel, S. C. (1976). Effects of trypan blue on the development of the garden lizard *Calotes versicolor. Teratology*, 14, 99
71. Berry, C. L. (1970). The effect of trypan blue on the growth of the rat embryo *in vivo. J. Embryol. Exp. Morphol.*, 23, 213
72. Payne, G. S. and Deuchar, E. M. (1972). An in vitro study of functions of embryonic membranes in the rat. *J. Embryol. Exp. Morphol.*, 27, 533
73. Zawoisky, E. J. (1975). Prevention of trypan blue exencephaly and otocephaly in gestating albino mice. *Toxicol. Appl. Pharmacol.*, 31, 191
74. Miller, S. A. and Runner, M. N. (1975). Differential permeability of uterine visceral yolk sac to thymidine and to hydroxyurea. *Dev. Biol.*, 45, 74

4
Congenital skeletal dysplasias – a better understanding via experimental models

A. ORNOY

INTRODUCTION

The two processes involved in bone formation are ossification and calcification. Ossification is the production of bone matrix in which the precipitation of minerals occurs (calcification). Chondrocytes, and more so osteoblasts, are primarily involved in both processes. During life (both during the active period of growth and after cessation of this process) there is a continuous process of bone remodelling – formation and resorption of bone, a dynamic process which causes growth of bone, fracture repair and maintenance of the proper skeletal configuration[1]. Different enzymes are involved in the processes of bone formation, growth and remodelling, and failure of one of these enzymes may cause failure of skeletal growth, thus producing distinct abnormalities – skeletal dysplasias. Although enzymopathies are the regular cause of skeletal dysplasias, and the missing enzymes are usually not well-defined, the biochemical deviation observed may also sometimes not be the primary cause, but rather the results of the skeletal defect[2].

Skeletal dysplasias may be produced in the previously healthy adult animal by different environmental insults such as irradiation[3], lathyrism[4], aminonitriles and other agents. However, these dysplasias are often reversible after the insult has been removed, and are hardly comparable to the congenital skeletal dysplasias.

THE BASIC STRUCTURE OF TUBULAR BONES

Two types of bone formation are known, intramembranous (i.e. of calvarial bones) and endochondral (i.e. tubular bones of extremities). Figure 4.1 shows a longitudinal section of a newborn rat tibia, which is a tubular long bone. The epiphysis, metaphysis and part of the diaphysis are well defined.

In the embryo the mesenchyme of the future long bones condenses, and

53

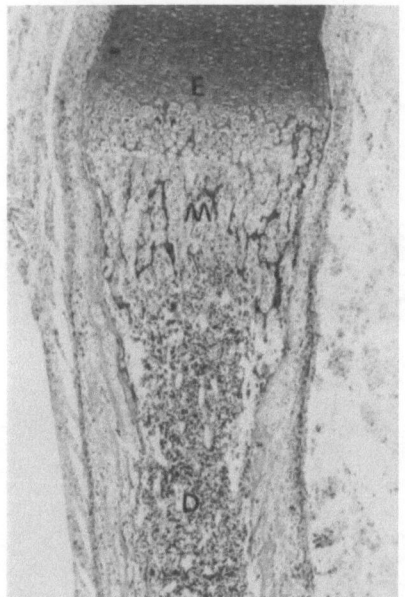

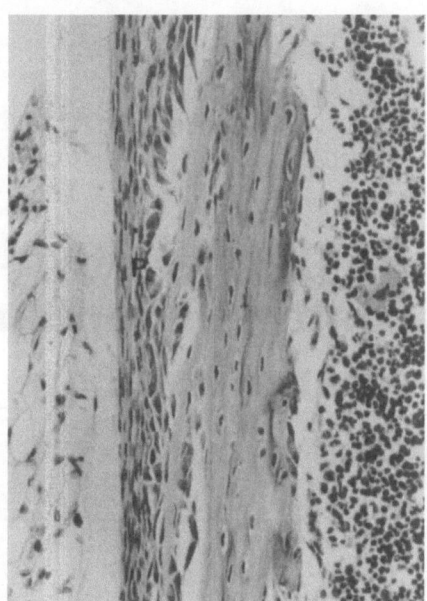

Figure 4.1 Microphotograph of a longitudinal section of a tibia from a newborn rat. Note the epiphysis (E), metaphysis (M) and longitudinally arranged spongy diaphyseal trabeculae (D). Haematoxylin and Eosin. × 38

Figure 4.2 Microphotograph of longitudinal section of tibial diaphysis from a 5-day old young rat. Note the periosteum (P) surrounding the more compact diaphysis, when compared with Figure 4.1. Haematoxylin and Eosin. × 188

the cartilage model is formed, which grows both in length and width. During the time when chondrocytes hypertrophy, the intercellular matrix calcifies, and blood vessels invade the midshaft to form the primary (diaphyseal) ossification centre. The epiphyses ossify later, when secondary ossification centres appear. In between the diaphysis and epiphysis the epiphyseal plate (physis) is found throughout the period of growth[1].

Through the process of bone remodelling which involves formation of new trabeculae and removal of older trabeculae, the cancellous (spongy) bone of the diaphysis is transformed into more compact (cortical) bone (Figures 4.2 and 4.3). Epiphyseal bone, however, remains spongy throughout life.

During growth, longitudinal growth is by continuous proliferation of chondrocytes in the physis (Figure 4.4) while the more mature cartilage is replaced by bone. Growth in width, however, occurs mainly by apposition of new bone from the periosteum. In the bones that ossify directly from mesenchyme (periosteal ossification) bone growth is by apposition of new bone and resorption of the old without involvement of cartilage.

The type of bone involved in skeletal dysplasias i.e. whether it developed through endochondral or periosteal ossification, and in the tubular bones whether primary abnormality is in the physis or diaphysis, is the basis for the various clinically defined entities of bone dysplasias. There is no doubt today, that calcification can also be impaired in some congenital skeletal anomalies,

54

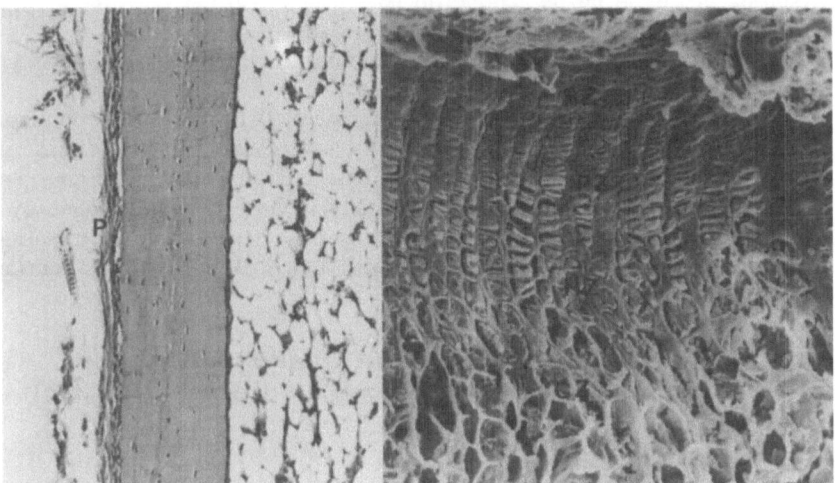

Figure 4.3 Microphotograph of longitudinal section of a metatarsal bone from
a 3-month old rat, to show the compact
diaphysis. P – periosteum. Haematoxylin
and Eosin. × 98

Figure 4.4 Scanning electron microscopic
photograph of epiphyseal plate and metaphysis
from a longitudinally fractured tibia of a 2-month
old rat. Note the different zones of the growing
plate:
RZ – reserve zone; PZ – zone of proliferation;
HZ – zone of maturing and hypertrophic cartilage; CZ – zone of provisional calcification. This
specimen was freeze fractured, critical point
dried and then examined by a Jeol SM35
Scanning Electron Microscope. × 280

either by inappropriate or abnormal bone matrix, or because of impaired
cellular activity[5, 6]. However, it is as yet impossible to differentiate the skeletal
dysplasias on the basis of impairment of calcification.

CLASSIFICATION OF CONGENITAL BONE DYSPLASIAS

The majority of the congenital bone dysplasias are hereditary. Many of them,
however, have a sporadic occurrence, and are probably because of new
mutations. In an attempt to standardize the nomenclature on bone dysplasia
participants of a symposium on bone dysplasias held in Paris during 1969
published a list of 125 distinct skeletal dysplasias (*J. Pediatr.*, **78**, 177, 1971).
Since then, some more previously undescribed entities were added. It is
difficult to assess the incidence of congenital bone dysplasias in human, but it
probably ranges between 1/2000–3000 in liveborn infants. The most common
skeletal dysplasias in man are achondroplasia, osteogenesis imperfecta and
osteopetrosis. Because experimental models of these defects are now available,
these three entities will serve as the basis for our further discussion. Furthermore they represent three dysplasias occurring in different zones of tubular
bones[7]:

(a) Physeal (epiphyseal growth plate) dysplasias: the most common dysplasia in this group is achondroplasia, which also has been reported in some animal species i.e. mice[8, 9], turkeys[10] and rats[11].

(b) Metaphyseal dysplasias: the most common disease being osteopetrosis which was also reported in rabbits[12-14] and rats.

(c) Diaphyseal dysplasias, osteogenesis imperfecta being the most common of these dysplasias, and probably the most severe handicapping disease in humans affected with skeletal dysplasias. This latter anomaly had been found rarely in animals[7, 11], but an environmentally induced congenital bone dysplasia resembling osteogenesis imperfecta has been described in the offspring of rats whose mothers have been treated with high doses of vitamin D_2 during gestation[15].

ACHONDROPLASIA, OSTEOGENESIS IMPERFECTA AND OSTEOPETROSIS, THE MOST COMMON HUMAN SKELETAL DYSPLASIAS

Achondroplasia

This dysplasia results probably from an inborn error in the growth and development of cartilage, specifically a decrease in the rate of epiphyseal plate cartilage maturation[16]. Articular cartilage and the ossification of the epiphyses are usually normal but there is a prominent failure of long bones to grow. The basic defect is, therefore, in endochondral bone formation, and this is manifested in most endochondral bones. Periosteal (intramembranous) bone formation continues normally.

Clinical manifestations

The dwarfism is characterized by shortening chiefly of the extremities but also of truncal length. The mean height attained by males is 129 cm and by females 122 cm. The head is disproportionately large, sometimes due to hydrocephaly. There may be neurological complications to the disproportionate growth of the skull. The incidence of this anomaly is 1/20 000 live births[17]. The inheritance is dominant, but over 80% of cases are sporadic in origin[18]. Homozygotes which constitute a small proportion usually die shortly after birth. The characteristic X-ray findings are shortening of tubular bones; the long bones appear stabby with proximal segments of limbs being more affected than the distal. There is a disproportionate growth of the skull, with a large vault and a base reduced in size.

Histological changes

Sections through the growth plates show the presence of all zones of the epiphyseal plates. The zones of maturing and hypertrophic chondrocytes are thin. Because of reduced rate of cartilage growth endochondral bone formation is reduced, thus longitudinal growth is mainly affected. However, metaphyseal and diaphyseal remodelling are normal[16].

Animal models

One of the commonest congenital skeletal dysplasias in animals is achondroplasia. This congenital dysplasia has been reported in several animal species[7, 9-11]. One of the first descriptions of achondroplasia was by Brown and

Pearce[19], in rabbits. This was a lethal achondroplasia affecting some of the rabbits in their colony. The X-ray findings and clinical picture were similar to human achondroplasia, i.e. marked shortening of the extremities with a generalized involvement of all bones. Histological examination of long bones showed short growth plates, crowded and abnormally oriented chondrocytes. There was also abnormal columnization of the epiphyseal plates[12]. It is interesting that the dachshund and basset hound dogs are achondroplastic, with normal maturation, multiplication and longevity again similar to the human situation[7].

Extensive studies have been performed on achondroplasia found in CN/CN mice. Lane and Dickie[20] described a strain of mice which in the homozygous state have a disorder of endochondral ossification resembling human achondroplasia. This was called the CN trait. Histological and histochemical studies of the bones of these affected mice showed an abnormal maturation of chondrocytes in the physis, but the sequence of endochondral ossification was quite normal. All the zones of the physis were found in the affected mice and the distribution of proteoglycans and collagen fibres was also normal[9]. There was, however, a reduction of height of the maturing and hypertrophic cartilage with some early aging-like changes of chondrocytes and cartilage matrix. This could lead to early reduction of proteoglycan concentration and shortening of cell columns, all of which might in turn lead to some inhibition of the calcification process. Electron microscopic studies performed on the achondroplastic mice[21] did not show any significant differences in the ultrastructure of cartilage and bone matrix between affected and control mice, and this, in spite of the fact that the maturing chondrocytes in the physis of affected animals showed large deposits of glycogen, and numerous vacuoles. The ultrastructure of hypertrophic chondrocytes, and the early phases of calcification in that area were normal. During the stage of active growth of these rats, the columns of hypertrophic cartilage were short, a finding more prominent as the animals grew older (20–30 days). Although the process of calcification continued normally, longitudinal intercartilaginous septa in the zone of maturation were also calcified, in contrast to controls in which calcification occured only in the zones of hypertrophy and of provisional calcification. Many of the hypertrophic chondrocytes did not degenerate normally. The metaphyseal osteoblasts, osteocytes and osteoclasts, as well as bone matrix appeared to be normal. The enzymatic defect in this model of achondroplasia has not been found as yet. Moreover, inheritance of this type is different from that in human (autosomal recessive versus autosomal dominant) which might give rise to a basic difference between these two entities, though the pathology seems similar. In this context it is interesting to note that chondrodystrophy (achondrogenesis) in turkey embryos[10] is transmitted as a dominant sublethal trait, and is probably also different from the classic human achondroplasia.

Osteogenesis imperfecta (osteopsathyrosis)

This hereditary congenital diaphyseal dysplasia is characterized by formation of abnormal bone collagen and matrix, and by abnormal orientation of the collagen fibres. Cartilage proliferation, mineralization of endochondral bone and fracture repair are normal[22, 23].

Clinical features

There are two main variants, the more severe congenital type (osteogenesis imperfecta fetalis) and the tarda type (osteopsathyrosis). In both, the disease is hereditary, transmitted as a dominant trait, although in some cases of osteogenesis imperfecta fetalis there is probably a recessive inheritance.

Congenital type

The fetus may be born dead or death occurs shortly after birth. There are usually multiple fractures, thin skin, thin blue scleras, and poor teeth development. There is a tendency to bleeding in the macula and in other organs due to increased capillary fragility.

In the *tarda form* there are multiple fractures often following only little trauma. There are blue scleras and often teeth abnormalities (dentinogenesis imperfecta). Stature may be stunted.

Radiographic findings are characterized by narrow shafts, thin delicate cortex, and poorly trabeculated spongiosa. The diaphyseal width of long tubular bones is reduced but endochondral bone formation, and longitudinal growth of bones continues normally.

There are radiographic changes in the pelvis as well as the vertebrae, all being osteoporotic with possible accompanying fractures, or even collapse of vertebral bodies. The skull shows severe retardation of calvarial ossification, which in severe cases of osteogenesis imperfecta fetalis may appear as a membranous bag with few isolated ossification centres. Later in childhood ossification of the calvarium may be completed. Many of the affected individuals develop hearing impairment and otosclerosis.

Blood calcium, phosphorus and alkaline phosphatase levels are normal. However, aminoaciduria and low blood uric acid levels are often found[7].

Pathogenesis

A proper hereditary determined experimental animal model is lacking and therefore, this defect had been studied mainly in humans. Generally, in all types of osteogenesis imperfecta periosteal bone is thin, osteoid is deficient, collagen orientation is abnormal and osteocytes are densely packed. Osteocytic lacunae are wide and the calcified osteoid between these densely packed osteocytic lacunae is thin (Figures 4.5 and 4.6)[6, 24, 25].

Although Follis[26] had shown by light microscopy immature collagen in infants affected with osteogenesis imperfecta fetalis, other investigators did not find ultrastructural abnormalities of bone collagen in patients affected by the tarda form[27]. However, Francis et al.[22] had shown defective cross linking in the collagen fibres of skin in such patients, and Smith et al.[28] had shown abnormal bone polymeric collagen in all types of osteogenesis imperfecta patients. Moreover, Dickson et al.[29] and Pentinnen et al.[30] had shown that osteoblasts and fibroblasts of patients with osteogenesis imperfecta produce a large quantity of collagen type III in organ culture, instead of the normal type I collagen of bone. In addition to the abnormal collagen, abnormal

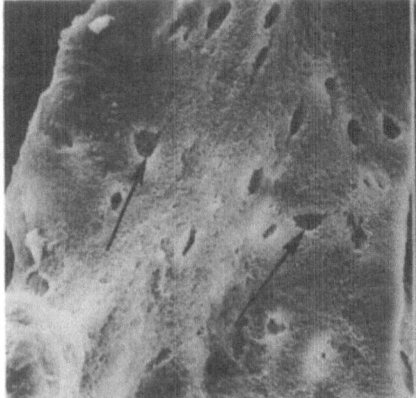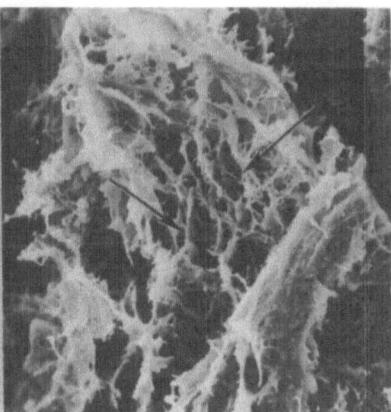

Figure 4.5 SEM photograph of a periosteal surface from a femur of a 7-month old human fetus. The bone was immersed for 24 h in 7% NaOCl, which removed cells and organic material. Note osteocytic lacunae (arrows), and wide calcified osteoid between the cells × 288

Figure 4.6 SEM photograph of a periosteal surface from the femoral diaphysis of a 7-month old stillborn with osteogenesis imperfecta fetalis. The bone was immersed in 10% NaOCl for 30 mins, a procedure that removed cells and organic matrix. Note thin septa between osteocytic lacunae (arrows), the latter being wider than in the controls. × 300 (by kind permission of *Isr. J. Med. Sci.*)

glycosaminoglycans were found in the cartilage and bone of those patients[23], which may be of functional importance to the mineralization process as well as to the abnormal organization of collagen. These defects probably result from defective osteoblasts producing abnormal osteoid, which in turn may result in poor mineralization. The main modelling abnormality in osteogenesis imperfecta is the inability to remodel spongy bone into compact cortical bone. Only seldom are distinct Haversian systems found. This is why long tubular bones are more often fractured. One has to stress, however, that in the milder types of osteogenesis imperfecta the deformities may be minimal, with only few fractures and with roentgenographic signs of osteoporosis. In all cases the early phases of fracture repair appear to be normal[7].

Calcification of epiphyseal cartilage in osteogenesis imperfecta fetalis was studied by Ornoy and Kim[6]. Scanning electron microscopic studies of the epiphyses and metaphyses of long bones showed normal longitudinal inter-cartilaginous septa, covered by homogenous globules of 1 μm in diameter, which are aggregates of hydroxylapatite crystals. On the other hand, dia-physeal trabeculae were thin, osteocytes densely packed and osteocytic lacunae wider than normal, with evidence of increased bone resorption. The fact that bone matrix seems to be abnormal in osteogenesis imperfecta, and that the clinical picture and probably also the mode of inheritance vary, implies that we are dealing here with a heterogeneous group of biochemical disorders of collagen and glycosaminoglycan metabolism.

Experimental models
Only few cases of genetically determined osteogenesis imperfecta in animals have been reported. Among them are offspring of some Siamese and Burmese

cats, which may have osteoporotic bones with a tendency to fracture of extremities and vertebrae, as well as some neurological defects[7].

The possibility of a phenocopy to osteogenesis imperfecta, i.e. induction of an abnormality similar to the genetically determined disease in humans, had been described in a few cases. Warkany et al.[31] described one newborn infant in whom ossification of the skull was lacking. He had slender thin long bones but without fractures. This case, with skeletal changes similar to those found in osteogenesis imperfecta, was accompanied by multiple congenital anomalies attributable to attempted abortion by aminopterin. Similarly, maternal warfarin use during pregnancy was associated with a number of cases in which the newborn infants had chondrodysplasia punctata, which is a type of fetal chondrodystrophy usually of genetic origin[32].

Osteopetrosis: (marble bones, Albers-Schoenberg disease)

Osteopetrosis is characterized by persistence of primary spongy bone. Endochondral bone formation proceeds normally as cartilage proliferation, maturation and calcification appear normal. Metaphyseal trabeculae are formed, but their resorption is impaired. Therefore dense bone ensues. In most cases osteopetrosis is genetically determined by the two common modes of inheritance – autosomal recessive and dominant. There are two types of osteopetrosis in man[7], the congenital and tarda varieties.

Congenital type (malignant osteopetrosis)
This is the more severe and lethal type with symptoms sometimes recognized in utero. In these infants failure to thrive, anaemia, jaundice and hepatosplenomegaly are the prominent symptoms. Death occurs usually shortly after birth. A somewhat milder variant of this type of disease has an infantile onset around the age of one year. Here again, apart from failure of growth and skeletal changes, anaemia and neurological manifestations are prominent. The severity of bone lesions parallels the severity of anaemia and neurological complaints. Inheritance of this type is autosomal recessive[18].

In the adult form (benign type) onset of symptoms is usually in the late childhood or adulthood. Attention is attracted to this variant usually following an incidental fracture or some unexplained anaemia, in which sclerotic bones are found on X-ray. Inheritance may be autosomal dominant or recessive[18].

Radiographic findings
Regularly one finds uniformly sclerotic bones, especially in those ossified in cartilage. There is no distinction between cortical and cancellous bone, so that the fetal spongy bone persists postnatally. Length of tubular long bones is very often normal. Sometimes small areas of rarefaction and fractures are seen. Epiphyseal ossification is of normal contour although the epiphyses are also sclerotic. The calvarium appears normal, but the base of the skull is dense and sclerotic.

Anaemia is a prominent symptom in osteopetrosis due to deficient bone marrow. Haemopoietic foci are usually found in the liver and spleen, and there is a high number of nucleated red blood cells in the circulation. There is a positive calcium balance sometimes accompanied by hypercalcaemia, hypophosphataemia and hypercalciuria, findings which may point to chronic hyperparathyroidism[7].

Pathology

The basic defect is probably reduced osteoclastic metaphyseal bone resorption. Therefore, there is a persistence of the metaphyseal chondro-osseous complex which normally undergoes rapid resorption. Growth usually continues normally, both at the growing plates as well as at the periosteal surface. The molecular structure of bone minerals and organic matrix seems to be normal although there is a high ratio of inorganic/organic material. The arrangement of collagen fibres is abnormal, explaining perhaps why the bones tend to fracture.

Animal models

Osteopetrosis is probably one of the bone dysplasias most studied in animals, because it is found in many animal species.

In rabbits, severe and fatal osteopetrosis was found as a recessive trait by Pearse and Brown[33]. There is persistence of spongy bone and only scanty haemopoietic bone marrow. In addition retardation of growth, tooth abnormalities, progressive anaemia, malnutrition and cachexia develop rapidly and lead to death of the animals. This condition is very similar to the severe congenital variety of human osteopetrosis[34].

In mice, three genetically distinct forms of osteopetrosis have been described all probably occuring due to increased bone formation[13]. In the grey lethal mutant of mice, a skeletal dysplasia similar to the human osteopetrosis tarda exists although growth is not severely disturbed. There is a persistence of diaphyseal bone probably because of a marked decrease in osteoclastic function, which was thought to result from an enzyme deficiency[35]. The longitudinal growth of endochondral bones was, however, normal. The X-rays were of osteosclerotic bone.

Investigations on i.a. osteopetrotic rats had probably given us the best example how a proper experimental model can help in the understanding of the pathogenesis of congenital skeletal dysplasias. The osteopetrotic i.a. (incisor absent) rat was discovered by Greep[36] and further studied by Morse and Greep[37]. This mutation is not lethal and the affected rats have a normal lifespan. The osteopetrosis, manifested shortly after birth, enters a phase of spontaneous remission at the age of 30–50 days. Later on, reduced body weight is probably the single manifestation of this abnormality. The dentition erupts in these rats slowly, and injection of parathyroid hormone (PTH) reverses many of the changes in these animals. However, there is normal endogenous PTH secretion in these animals, as well as a normal response to exogenous parathormone[38]. Marks[13] measured the rate of bone formation and resorption in i.a. rats at various ages. When new bone matrix formation was estimated with the [3H]proline incorporation test, which measures synthesis of new bone collagen, it was found that bone formation was decreased slightly during the first three weeks of life, and slightly increased afterwards.

Bone resorption following administration of PTH was reduced about 20% at the age of three weeks. Further studies demonstrated that osteopetrosis in these rats results from failure of osteoclasts to respond to stimuli promoting resorption. A high concentration of acid phosphatase was found in osteoclasts of i.a. rats, which returned to normal when spontaneous remission occurred[13]. The first signs of the remission are found at 2–3 weeks of age. When calvarial bone from i.a. osteopetrotic rats were studied in organ culture[14] it was found that sera from normal and i.a. rats do not differ as to their ability to support bone resorption in normal animals, but i.a. bone does not respond to para-thyroid hormone in culture. This was interpreted as evidence that the de-creased bone resorption of i.a. rats was because of a cellular defect in osteo-clasts. Ultrastructural studies of osteoclasts from i.a. rats showed that they did not form the ruffled border characteristic for osteoclasts actively involved in bone resorption. It seems therefore that osteoclasts from i.a. rats are unable to increase the synthesis and release of lysosomal enzymes for normal bone resorption.

As a final step to the understanding of the possible mechanism underlying osteopetrosis in i.a. rats Marks has shown recently[39] that after osteopetrotic rats receive sublethal whole body irradiation and are injected with spleen cells from a normal littermate, the dense sclerotic skeleton is completely resorbed by what appeared by electron microscopy to be normal osteoclasts, with a ruffled border and indistinguishable from osteoclasts of normal animals. Concomitantly, osteoclasts originating from the recipient were also found, the latter lacking the characteristic ruffled border. The fact that osteo-clasts from the normal donor completely cured the disease seems to be a final proof that in i.a. rats, osteoclastic function is abnormal. Whether the same mechanism underlies human osteopetrosis remains uncertain.

CONGENITAL SKELETAL DYSPLASIAS PRODUCED BY ENVIRONMENTAL AGENTS

Numerous studies have been published in which skeletal abnormalities have been produced by environmental agents, such as vitamin deficiencies[40], zinc deficiency[41], manganese deficiency[42], several dietary imbalances[43], as well as irradiation[44]. However, the skeleton in these experiments was usually one of the many other fetal systems affected. Deficiency or excess of these agents usually produced multiple malformations, and therefore, the skeletal abnor-malities produced thereby will not be discussed further.

Several experiments were aimed towards induction of skeletal abnor-malities as the main expression of the teratogenic agent used during preg-nancy.

A number of studies have been published in which an attempt was made to interfere with fetal skeletal development by derangements of maternal calcium and phosphorus metabolism. Usually difficulties in breeding were so severe that well-controlled studies were seldom performed[45]. If the imbalance was not severe enough, normal offspring followed[46]. Sontag et al.[47] were able to breed rats fed a low calcium and phosphorus diet, and showed that the offspring have a lower mineral content than controls. When rats were fed a rachitogenic diet (vitamin D deficient, high calcium, low phosphorus) for two weeks, then mated and kept on the same diet throughout pregnancy, the off-spring exhibited general skeletal abnormalities[48]. They had curved bone, tubular bones of extremities and angulation of ribs. Microscopic examination of the long bones showed a wide zone of hypertrophic cartilage cells in the epiphysis with persistence of hypertrophic chondrocytes in the diaphyses. These skeletal changes could be interpreted as fetal rickets.

In a series of experiments the possible effects of high doses of vitamin D_2 during pregnancy on the fetal rat skeleton was studied. A reduction in fetal weight, ash weight, calcium and phosphorus content of the fetal skeleton were found[49]. Microscopical examination of long bones of the extremities showed[50] in 19 and 21 day old fetuses abnormal ossification, i.e. longer epiphyses because of widening of the maturing and hypertrophic zone, and short un-ossified diaphyses (Figures 4.7 and 4.8). When such rats were allowed to deliver, many of the offspring died, but some could be raised and were studied to the age of one month[51]. Growth retardation persisted, and the weight differences even increased (Figure 4.9). Multiple fractures were found and

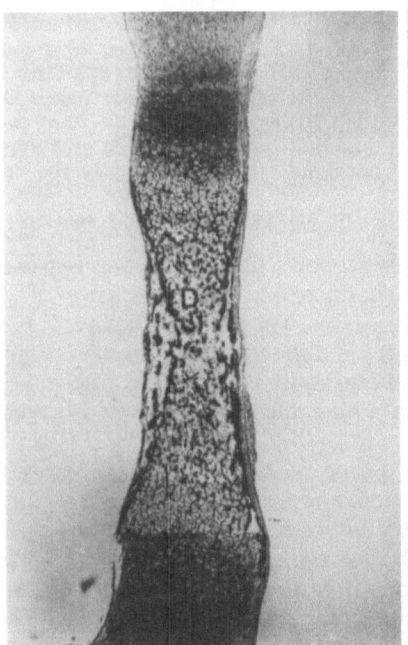 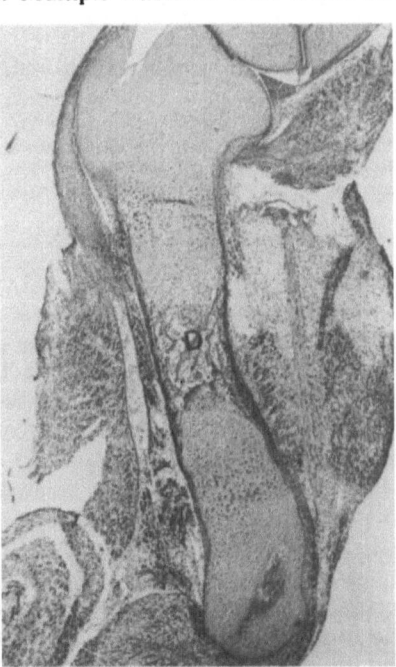

Figure 4.7 Microphotograph of longitudinal section of a tibia from a control 19-day old rat fetus. D – diaphysis. Haematoxylin and Eosin. × 24
(by kind permission of *Isr. J. Med. Sci.*)

Figure 4.8 Microphotograph of longitudinal section of a tibia from a 19-day old rat fetus whose mother was treated daily with 40000 IU vitamin D_2 during days 11–18 of gesta-tion. Note a small diaphysis (D), a wide zone of maturing cartilage and a thin layer of hypertrophic cartilage in the epiphysis, when compared with Figure 4.7. Haematoxylin and Eosin. × 24
(by kind permission of *Isr. J. Med. Sci.*)

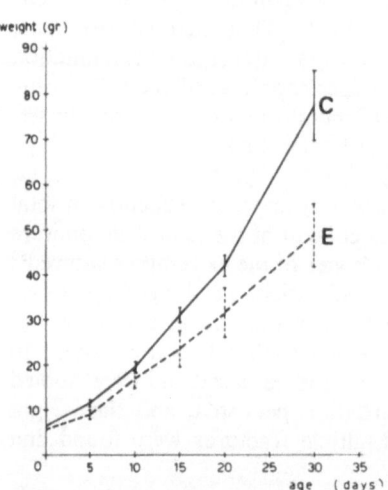

weight (gr)

Figure 4.9 Growth curves of offspring from vitamin D₂ treated rats (E) versus controls (C). Note that the offspring of the animals treated with 40000 IU vitamin D₂ daily during days 10–20 of gestation are smaller, and the weight differences with controls increases with age
(by kind permission of *Isr. J. Med. Sci.*)

Figure 4.10 Microphotograph of a longitudinal section from a femur of a 5-day old young rat whose mother was treated with 40000 IU/day during days 10–20 of gestation. Note thin diaphysis with hypertrophic osteocytes (arrows). Haematoxylin and Eosin. × 182
(by kind permission of *isr. J. Med. Sci.*)

the offspring also developed severe kyphoscoliosis. Ossification was impaired, as evidenced by retarded epiphyseal ossification, and persistence of enchondral bone trabeculae within the diaphyses. The cortical diaphyseal bone was thin with many microscopic fractures in different stages of repair (Figure 4.10). All these *persisting* bone lesions were similar to the human osteogenesis imperfecta, and probably resulted from a fundamental damage to osteogenic tissue by the high doses of vitamin D₂.

When pregnant rats were treated during gestation with high doses of cortisone acetate[52] term fetuses were small, and their skeleton had a high ash content and higher concentration of calcium and phosphorus. Microscopical studies showed persistence of metaphyseal trabeculae and abnormal intercellular substance in cartilage and bone (Figure 4.11). The skeletal lesions resembled human osteopetrosis. When the treated rats were allowed to deliver and the offspring studied at different ages, the skeletal lesions, both morphological and biochemical, slowly disappeared[53]. Concomitant treatment of vitamin D₂ and cortisone acetate exhibited the combined changes produced by each one of the agents individually[15, 52]. When pregnant mice were treated with 1 mg/day of cortisone acetate during days 11–19 of gestation, the 20-day old fetuses were smaller than littermate controls and the long bones were short due to shortening of the diaphysis. The subperiosteal diaphyseal bone was thin, the bone marrow wide and fewer preosteogenic tissue was found. The osteoclast count was higher than in the controls[54]. This congenital

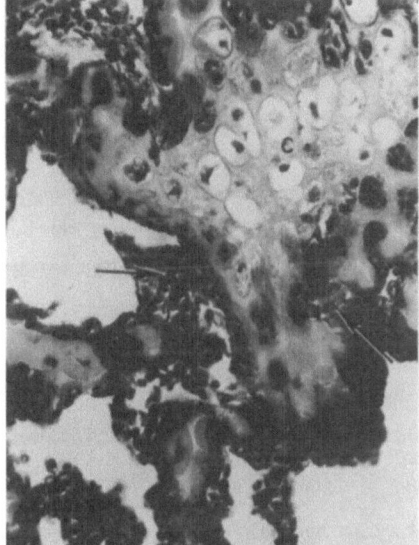 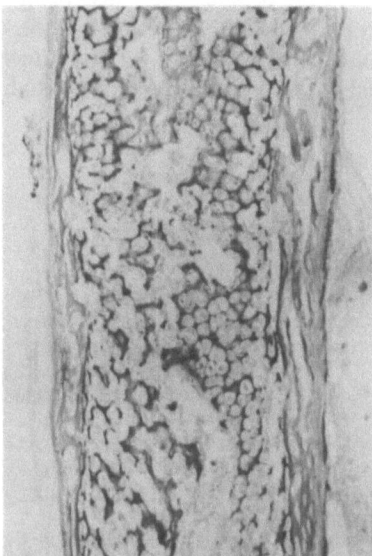

Figure 4.11 Microphotograph of meta-physeal trabeculae from a 21-day old rat fetus whose mother was treated daily with 10 mg cortisone acetate during days 10–20 of gestation. Note persisting hypertrophic chondrocytes (C) surrounded by osteoblasts. Also note abundant preosteogenic tissue between trabeculae (arrows). Haematoxylin and Eosin. × 168

Figure 4.12 Microphotograph of a longi-tudinal section through the tibial diaphysis of a 21-day old rat fetus whose mother was treated with 0.1 mg oestradiol valerianate during days 12, 15 and 19 of gestation. Note persistence of calcified and hypertrophic cartilage within the diaphysis. PAS alcian blue. × 70

bone dysplasia, characteristic for the action of corticosteroids on bone, mimics human congenital osteoporosis or even a mild form of osteogenesis imperfecta. However, one should emphasize again that similar doses of cortisone acetate affected rat fetuses in a different way, by producing osteopetrosis.

High doses of oestrogens were also found to affect transplacentally the fetal skeleton in rats[55]. When 0.2 mg of oestradiol valerianate was injected to pregnant rats on days 12, 15 and 19 of gestation, the 21-day old fetuses and placentae were small, fetal bones were short and contained a high concentration of calcium. Fetal and maternal blood calcium levels were elevated. The prominent microscopic findings was persistence of enchondral trabeculae in the diaphysis, as well as islets of calcified hypertrophic cartilage (Figure 4.12). Thus, the evidence of decreased bone resorption resembled our findings in corticosteroid treated rats resembling human osteopetrosis.

Finally, even high doses of calcitonin were found to affect transplacentally the fetal rat skeleton[56, 57]. The ash content of fetal bones was higher than of controls, but the Ca and P ash percentages were normal. The microscopic findings pointed towards enhanced bone formation and reduced resorption, with a lower number of osteoclasts.

All these experimental models induced by teratogenic agents may have a wider importance than just being phenocopies of human genetically determined skeletal dysplasias. In order to use them as appropriate models in the study of pathogenesis and treatment of human situations, proper biochemical and ultrastructural studies have to be performed.

References

1. Guggenheim, K., Wolinsky, I. and Ornoy, A. (1978). Nutrition and bone formation. (*CRC Press, Inc.*) (In press)
2. Fanconi, G. (1973). Bone diseases of genetic origin. *Triangle*, 12, 129
3. Rubin, P., Andrews, J. R., Swarm, R. and Gump, H. (1959). Radiation induced dysplasias of bone. *Am. J. Roent.*, 82, 206
4. Ponseti, I. V. (1954). Lesions of the skeleton and of other mesodermal tissues in rats fed sweet pea (Lathyrus odoratus) seeds. *J. Bone Jt. Surg.*, 36A, 1031
5. Ornoy, A., Sekeles, E., Smith, P., Simkin, A. and Kohn, G. (1976). Achondrogenesis type I in three sibling fetuses. Scanning and transmission electron microscopy. *Am. J. Pathol.*, 82, 71
6. Ornoy, A. and Kim, J. A. (1977). Scanning electron microscopic findings in osteogenesis imperfecta fetalis. *Israel J. Med. Sci.*, 13, 26
7. Rubin, P. (1969). *Dynamic Classification of Bone Dysplasias*. (Chicago: Yearbook Medical Publishers, Inc.)
8. Johnson, D. R. and Wise, J. M. (1971). Cartilage anomaly (can). A new mutant gene in the mouse. *J. Embryol. Exp. Morphol.*, 25, 21
9. Bonucci, E., Del Marco, A., Nicoletti, B., Petrinelli, P. and Pozzi, L. (1976). Histological and histochemical investigation of achondroplastic mice. A possible model for human achondroplasia. *Growth*, 40, 241
10. Gaffney, L. J. (1975). Chondrodystrophy: an inherited lethal condition in turkey embryos. *J. Hered.*, 66, 339
11. Gruneberg, H. (1963). *The Pathology of Development. A Study of Inherited Skeletal Disorders in Animals.* (Oxford: Blackwell Scientific Publications)
12. Pearse, L. and Brown, W. H. (1945). Hereditary achondroplasia in the rabbit. II. Pathologic aspects. *J. Exp. Med.*, 45, 261
13. Marks, S. C., Jr. (1973). Pathogenesis of osteopetrosis in the rat; reduced bone resorption due to reduced osteoclast function. *Am. J. Anat.*, 138, 165
14. Nyberg, L. M. and Marks, S. C., Jr. (1975). Organ culture of osteopetrotic (i.a.) rat bone: evidence that the defect is cellular. *Am. J. Anat.*, 144, 373
15. Ornoy, A., Horowitz, A., Kaspi, T., Michaeli, Y. and Nebel, L. (1972). Anomalous fetal and neonatal bone development induced by administration of cortisone and vitamin D_2 to pregnant rats. In: *Drugs and Fetal Development*. M. A. Klingberg, A. Abramovici and J. Chemke, (eds.) pp. 219–226 (New York: Plenum Publishing Co.)
16. Rimoin, D. L., Hughes, G. N., Kaufman, R. L., Rosenthal, R. E., McAllister, W. H. and Silberberg, R. (1970). Endochondral ossification in achondroplastic dwarfism. *N. Engl. J. Med.*, 283, 728
17. Harris, R. and Patton, J. T. (1971). Achondroplasia and thanatophoric dwarfism in the newborn. *Clin. Genet.*, 2, 61
18. Warkany, J. (1971). *Congenital Malformations*. (Chicago: Yearbook Medical Publishers, Inc.) Part XIII, pp. 765–886
19. Brown, W. H. and Pearce, L. (1945). Hereditary achondroplasia in the rabbit. I. Physical appearance and general features. *J. Exp. Med.*, 82, 241
20. Lane, P. W. and Dickie, M. M. (1968). Three recessive mutations producing disproportionate dwarfism in mice: achondroplasia, brachymorphic and stubby. *J. Hered.*, 59, 300
21. Bonucci, E., Gherardi, G., Del Marco, A., Nicoletti, B. and Petrinelli, P. (1977). An electron microscope investigation of cartilage and bone in achondroplastic (CN/CN) mice. *J. Submicrosc. Cytol.*, 9, 299
22. Francis, J. O. M., Phil, D. and Smith, R. (1975). Polymeric collagen of skin in osteogenesis imperfecta, homocystinuria and Ehlers-Danlos and Marfan syndromes. *Birth Defects*, XI, 14

23. Engfeldt, T. and Hjerpe, A. (1976). Glycosaminoglycans of cartilage and bone tissue in two cases of osteogenesis imperfecta cóñgenita. *Acta Pathol. Microbiol. Scand.*, 84, 488

24. Engfeldt, B., Engström, A. and Zetterström, R. (1954). Biophysical studies of the bone tissue in osteogenesis imperfecta. *J. Bone Jt. Surg.* (*B*), 36, 654

25. Robichon, J. and Germain, J. P. (1968). Pathogenesis of osteogenesis imperfecta. *Can. Med. Assoc. J.*, 99, 975

26. Follis, R. H. (1951). Osteogenesis imperfecta congenita: a connective tissue diathesis. *J. Pediatr.*, 38, 713

27. Riley, F. C., Jowsey, J. and Brown, D. M. (1973). Osteogenesis imperfecta, morphologic and biochemical studies of connective tissue. *Ped. Res.*, 9, 757

28. Smith, R., Francis, M. J. O. and Bauze, R. J. (1975). Osteogenesis imperfecta, a clinical and biochemical study of a generalized connective tissue disorder. *Qu. J. Med.*, 176, 555

29. Dickson, I. R., Miller, E. A. and Veis, A. (1975). Evidence for abnormality of bone matrix proteins in osteogenesis imperfecta. *Lancet*, 2, 586

30. Pentinnen, R. P., Lichtenstein, J. R., Martin, G. R. and McKusick, V. A. (1975). Abnormal collagen metabolism in cultured cells in osteogenesis imperfecta (collagen types fibroblasts). *Proc. Nat. Acad. Sci. USA*, 72, 586

31. Warkany, J., Beaudry, P. H. and Hornstein, S. (1959). Attempted abortion with 4 aminopteroylglutamic acid (aminopterin); malformations of the child. *Am. J. Dis. Child.*, 97, 274

32. Becker, M. H., Genieser, N. B., Finegold, N., Miranda, D. and Spackman, T. (1975). Chondrodysplasia punctata: Is Maternal Warfarin Therapy a Factor? *Am. J. Dis. Child.*, 129, 356

33. Pearse, L. and Brown, W. H. (1948). Hereditary osteopetrosis of the rabbit. I. General factors and course of disease, general aspects. *J. Exp. Med.*, 88, 579

34. Pearse, L. (1950). Hereditary osteopetrosis of the rabbit. III. Pathologic observations, skeletal abnormalities. *J. Exp. Med.*, 92, 591

35. Barnicot, N. A. (1947). The osteoclast in the grey lethal mouse. *Proc. R. Soc. London, Ser. B*, 134, 476

36. Greep, R. O. (1941). An hereditary absence of the incisor teeth. *J. Heredity*, 32, 397

37. Morse, A. and Greep, R. O. (1960). Histochemical observations on the ribonucleic acid and glycoprotein content of the osteoclasts of the normal and ia rat. *Arch. Oral Biol.*, 2, 38

38. Kenny, A. D., Toepel, W. and Schour, I. (1958). Calcium and phosphorus metabolism in the ia rat. *J. Dent. Res.*, 37, 432

39. Marks, S. C., Jr. (1978). Studies of the mechanism of spleen cell cure for osteopetrosis in ia rats. Appearance of osteoclasts with ruffled borders. *Am. J. Anat.*, 151, 119

40. Nelson, M. M., Ashling, C. W. and Evans, H. M. (1952). Production of multiple congenital abnormalities in young by maternal pteroylgulamic acid deficiency during gestation. *J. Nutr.*, 48, 61

41. Hurley, L. S., Gowan, J. and Swenerton, H. (1971). Teratogenic effects of short-term and transitory zinc deficiency in rats. *Teratology*, 4, 199

42. Hurley, L. S. and Asling, C. W. (1963). Localized epiphyseal dysplasia in offspring of manganese deficient rats. *Anat. Rec.*, 145, 25

43. Hurley, L. S. (1977). Nutritional Deficiencies and Excesses. In: *Handbook of Teratology.* Vol. 1, pp. 261–308 (G. Wilson & H. Fraser, eds.) (New York: Plenum Press)

44. Rugh, R., Duhamel, L., Osborne, A. W. and Vorma, A. (1964). Persistent stunting following X-irradiation of the foetus. *Am. J. Anat.*, 115, 185

45. Toverud, G. (1923). The influence of diet on teeth and bones. *J. Biol. Chem.*, 58, 583

46. Macomber, D. (1927). Effect of a diet low in calcium on fertility, pregnancy and lactation in the rat. *J. Am. Med. Assoc.*, 88, 6

47. Sontag, L. W., Munson, P. and Huff, E. (1936). Effects on the fetus of hypervitaminosis D and calcium and phosphorus deficiency during pregnancy. *Am. J. Dis. Child.*, 51, 302

48. Warkany, J. (1943). Effect of maternal rachitogenic diet on skeletal development of young rats. *Am. J. Dis. Child.*, 66, 511

49. Ornoy, A., Menczel, J. and Nebel, L. (1967). Alteration in the mineral composition and metabolism of rat fetuses and their placentae induced by maternal hypervitaminosis D_2. *Israel J. Med. Sci.*, 4, 827

50. Ornoy, A., Nebel, L. and Menczel, J. (1969). Impaired osteogenesis of fetal long bones induced by maternal hypervitaminosis D_2. *Arch. Pathol.*, 87, 563
51. Ornoy, A., Kaspi, T. and Nebel, L. (1972). Persistent defects of bone formation in young rats following maternal hypervitaminosis D_2. *Israel J. Med. Sci.*, 8, 943
52. Ornoy, A. (1971). The effects of maternal hypercortisonism and hypervitaminosis D of fetal osteogenesis and ossification in rats. *Teratology*, 4, 383
53. Ornoy, A. and Horowitz, A. (1972). Postnatal effects of maternal hypercortisonism on skeletal development in newborn rats. *Teratology*, 5, 153
54. Kaduri, A. J. and Ornoy, A. (1974). Impaired osteogenesis in the fetus induced by administration of cortisone to pregnant mice. *Israel J. Med. Sci.*, 10, 476
55. Ornoy, A. (1973). Transplacental effects of estrogen on osteogenesis in rat foetuses. *Pathology*, 5, 183
56. Ornoy, A., Peshin, J. and Menczel, J. (1973). Transplacental effects of thyrocalcitonin of the osteogenesis and ossification of fetal long bones in rats. *Proc. 9th Europ. Symp. Calcified Tissues*, pp. 149–152 (H. Gzitober and J. Eshberger, eds.)
57. Peshin, J., Ornoy, A. and Menczel, J. (1976). Transplacental effects of thyrocalcitonin on intestinal calcium-binding protein, alkaline phosphatase activity and ossification of long bones in rat fetuses. *Israel J. Med. Sci.*, 12, 248

5
Current concepts on the mechanisms of normal and abnormal secondary palate formation

R. M. SHAH

A childe borne without the uvula or tonsillae but a great passage up the nose from the mouth so that one might almost see out of one into the other: it could not suck but all came out of the nose again, unless if lay backwards.

From Diary of John Ward of
Stratford-on-Avon, 1660

INTRODUCTION

The earliest recorded description of cleft palate seems to be in John Ward's diary. Even though cleft palate has been mentioned only in the recent litera ture, evidence of the deformity far predates the literature, as demonstrated in early Egyptian mummy[1].

In the contemporary world, palatal clefts are among the most common major malformations in humans[2]. The seriousness of the problem is underscored by the statistics that the incidence of cleft palate varies in different countries between 1 in 600 to 1 in 1000 births and is still increasing[3].

It is now generally accepted that direct hereditary tendencies can only be shown in 12–20% of cases[4], and it is therefore possible that exogenous causes may be directly, or jointly, responsible for a large proportion of palatal clefts.

The role of nongenetic factors in the causation of palatal clefts has been elucidated to some degree by numerous experimental studies in animals. However, it was not until Baxter and Fraser[5] observed that cortisone administered early in pregnancy produced cleft palate in embryonic mice, that a reliable experimental system became available for the study of mechanisms that underlie the development of normal and abnormal palate.

Since then, cleft palate has been the target of more research on causes and

69

mechanisms than any other congenital malformations[6]. That was perhaps because palate closure is one of the last important morphogenetic events to occur at a stage when the embryo is relatively large, and correspondingly easier to observe. Furthermore, one can conveniently study various biological phenomena involved during palate morphogenesis, i.e. cellular interaction, movement, differentiation, and adhesion, physiological cell death and extra-cellular matrix formation.

Nevertheless, the evidence of embryological mechanisms by which clefting of human palate occur is scattered, scant and conflicting. The use of tera-togens as experimental tools has begun to contribute to a better understanding of the developmental mechanisms and their failures. Numerous teratogenic agents which produce cleft palate in the laboratory animals have been identi-fied and are aptly summarized in the literature[7-10]. Their use in rodents has greatly increased the understanding of normal and abnormal palatal develop-ment. Discussed in the present chapter are both the problems of normal, and abnormal palatal development.

NORMAL DEVELOPMENT OF THE SECONDARY PALATE

Palatal development was studied as early as 1869 by Dursy[11]. He introduced the term 'secondary palate' to describe that portion of the palate which is formed by the union of two palatal processes. The secondary palate is, there-fore, composed of a portion of the hard palate, and the soft palate. The palatal processes that eventually form the roof of the mouth arise from the inferior surface of the maxilla in the form of a well marked ridge (shelf). At first the shelves hang vertically downwards alongside of the tongue. Subsequently they move from the vertical position into a horizontal position dorsal to the tongue. When both processes are elevated, their medial edges come into contact in the midline of the oral cavity, where they fuse together to form the roof of the mouth and floor of the nasal cavity.

Following publication of Dursy's study many attempts were made to explain the mechanism of palatal closure in humans[12-15]. A review of the older literature on the formation of the palate was given by Peter[16], and later by Lazzaro[17]. On the basis of their reviews, and subsequent studies in laboratory animals and human embryos, one may deduce that the closure of the second-ary palate involves four critical events: (a) formation of the vertical shelf; (b) a change in position of the palatal shelves from hanging vertically at the side of the tongue into one in which they lie horizontally above the tongue; (c) the epithelial fusion of opposing horizontal shelves in the midline and, (d) the removal of the fused epithelium, thereby allowing the mesenchymal union of the palatal shelves.

Formation of the vertical palatal shelves

Very little research is reported on the issue 'why and how do the palatal shelves grow vertically on the side of the tongue, instead of horizontal above the tongue, i.e. their eventual position?'. The question is logical as reorien-tation of the shelves from a vertical to a horizontal plane has been observed in all human and animal embryos studied. Hayward and Avery[18] suggested that, in an embryo, a proportionally large tongue normally occupies the oronasal cavity, causing the palatal shelves to project downward into the floor of the mouth. This suggestion has been supported indirectly by observations on experimental animals[19-24], implying that the palatal shelf grows vertically on the side of the tongue because there is no room for it to grow horizontally over the tongue. Recently, however, several human cases were reviewed[25], in which the palatal shelf has grown vertically when the tongue was either completely missing or very small. One may, therefore, suggest that the vertical growth of the palatal shelf may be controlled by factors other than the tongue.

Most embryology textbooks indicate that palatal epithelium is derived from the ectoderm. However, origin of the palatal mesenchyme is not yet fully understood. Unlike body mesenchyme, the mesenchyme of the upper and midface is derived from the neural crest[26]. This neural crest derived mesenchyme, also known as the ectomesenchyme, forms the skeletal and other connective tissue of the upper and mid-face. Since lateral palatine processes are part of the midfacial complex, it is reasonable to assume that a part of the ectomesenchyme may migrate and proliferate in the maxillary process, to form the primordia of the lateral palatine processes. Thus the factor governing development of palatal shelves may be genetically programmed, and may express itself by controlled cellular migration and proliferation and synthesis of the ground substance.

Reorientation of the palatal shelves

An early enigma of palatogenesis was how the vertical palatine shelves, at first separated by the tongue, are able to move up to a horizontal position. His[12], Inouye[15], Peter[16] and Lazzaro[17] suggested that the palatal shelves 'rotate' from the side of the tongue to a position above the tongue. Pons-Tortella[27] forwarded another explanation for the transposition of palatal shelves. He suggested that the vertical shelf 'regresses', and concomitantly a new horizontal shelf grows from its medial surface at the approximate level of the dorsum of the tongue. With few exceptions[20, 24], observations made on human, mouse, rat and hamster embryos during the past twenty years have, however, rejected both these suggestions. It is now generally accepted by most researchers that a transformation in the 'form' of vertical palatal shelves brings them to the horizontal position[19, 22, 23, 28-30]. The change in the 'form' (remodelling) from a vertical to a horizontal plane is accomplished by 'bulging' of the medial wall inwardly over the tongue, with a simultaneous 'retraction' of the ventral portion of the shelf.

During the past two decades, one of the much discussed questions has been what mechanism(s) is responsible for the remodelling of palatal shelf from vertical to horizontal position. Currently two schools of thoughts are discernible in the literature. One suggests that the shelf remodelling is brought about by 'forces' which reside outside the palate[13, 17, 20, 21, 31-40]; the other implies that the trigger for the reorientation resides within the palatal shelf and involves macromolecular synthesis and cellular reorganization[14, 19, 29, 41-47].

The proposal that external forces such as muscular pressure by the tongue, causing palatal shelf reorientation, has received the attention of many researchers. Lazzaro[17] suggested that withdrawal of the tongue before the shelf movement, and subsequent pressure on the undersurface of the palatal shelves by the tongue, might push the vertical shelf into the horizontal position. Asling et al.[31], and Coleman[20], suggested that the lowering of the tongue is brought about by an increased mandibular growth which precedes shelf transposition. Zeiler et al.[32] and Wragg et al.[36, 37] suggested that the mandible is lowered due to lifting of the head from against the chest. The tongue is simultaneously withdrawn from between the shelves, thus making possible the reorientation of the shelves. Both these explanations were, however, disputed by Humphrey[33, 34], who suggested that the tongue provides an 'active force' by its movement to bring the shelves in the horizontal plane. This view has also been supported by Walker[21, 35] and Walker and Ross[39]. Recently Schweichel and Seinsch[48] and Holt[49] observed morphological and enzymatic evidence suggesting that the myoneural apparatus of the tongue is indeed functional at the time of shelf remodelling. Such an association between the tongue movement and shelf transposition does not, however, establish a cause and effect relationship. Also reorientation of the shelf is observed in numerous cases of aglossia and microglossia[25]. These observations indicate that the tongue may not be involved in palatal shelf remodelling from vertical to horizontal.

From their studies on various inbred strains of mice, Walker and Fraser[19] indicated that the transition of a palatal shelf from vertical to horizontal is too rapid to be due solely to mandibular and tongue growth. They found no sign that the 'tongue dropped' to allow the shelves to become horizontal. Instead, Walker and Fraser hypothesized that shelf remodelling is brought about by an 'intrinsic force' which may reside in the acid mucopolysaccharide component of the ground substance. This proposal has received considerable support. Both histochemical and autoradiographic evidence showed that prior to, and during shelf remodelling, there was an increase in concentration of acidic mucopolysaccharides in the palatal shelf[29, 50–55]. This implies that macromolecule synthesis may have an important role.

Clark[56], on the basis of histochemical staining, suggested that the intrinsic shelf force resided in the elastic fibres, a possibility also suggested by Walker and Fraser[19]. Subsequent histochemical and ultrastructural observations, however, found no evidence of elastic fibres in the shelf tissue[51, 57–60] and the theory is now abandoned.

Recently both ultrastructural and biochemical evidence has accumulated which shows increased collagen synthesis in the palate during shelf reorientation[61–63]. What precise role the collagen fibres may play during shelf remodelling, however, is not known.

Another theory to account for the elevation of the palatine process was put forward by Schorr[14]. He suggested that an increase in cellular proliferation within palatal shelf tissue at the time of elevation was indicative of rapid growth. Mott et al.[64], Jelinek and Dostal[42, 65] and Nanda and Romeo[44] have observed an increased proliferation rate in the mesenchyme prior to reorientation. On the other hand, Walker and Fraser[19], Hughes et al.[66] and Cleaton-Jones[67] did not find any evidence of increased cellular proliferation rate. It is not quite clear why these observations are conflicting. Perhaps these authors examined different areas of palatal shelves at different times. In any case, it has been indicated recently that such a growth does not appear to play a direct role in reorientation of the shelves[68].

The most recent hypothesis to explain remodelling of shelf through the intrinsic mechanism is the molecular interactions of the contractile proteins, actin and myosin. These proteins are synthesized in the mesenchymal cells just prior to reorientation[43]. It has also been observed that prior to, and during reorientation, the mesenchymal cells change their shape from stellate to elongated[47]. These elongated cells contain a system of microfilaments[45, 47], which may represent the acto-myosin complex as described by Lassard et al.[43]. Babiarz et al.[45] have also observed a calcium-dependent ATPase activity in the palate during early development. It is possible that, like smooth muscle, ATPase related contractile protein may take part in the remodelling mechanism. Such a contractile system may be modulated by a neurotransmitter, acetylcholine[46]. Thus, implied in this hypothesis is the morphological movement of the mesenchymal cells by contractile protein. It may be that hyaluronate rich mucopolysaccharide environment[54] may facilitate the cellular movement to bring about reorientation of shelves.

Thus the intrinsic shelf force theory seems more reasonable, and is a more prevalent view now. However, the nature of such a 'force', if any, is still undetermined. No compelling evidence has yet been produced to show how any, or all, of the factors implied as shelf force, bring about palatal shelf reorientation.

Fusion of the opposing palatal shelves

While controversies continued about the mechanisms by which a vertical palatal shelf is reoriented into a horizontal position, very little attention has been given to the process by which the two horizontal palatal shelves unite with one another.

Numerous investigators had noted that in order to form the roof of the mouth, the opposing palatal shelves must fuse with each other[11, 12, 15–17]. It was only during the last fifteen years, however, that both in vivo and in vitro attempts have been made, in different animal species, to determine the sequelae by which the midline epithelia covering the horizontal shelves fused[22, 23, 29, 69–131]. The literature shows considerable difference of opinion regarding the sequence of changes that occur in the midline epithelia of the opposing shelves prior to, and during, fusion.

The mechanism by which opposing epithelial layers are held together (epithelial adherence) during formation of the midline epithelial seam is not well understood. Farbman[85, 86] and Hinrichson and Stevens[109] indicated that initial adhesion between the epithelial cells of the two opposing palatal shelves does not involve desmosome formation. These authors, along with Hayward[91], speculated that a 'sticky substance' on the surface of epithelial cell may be responsible for adhesion. Several researchers, however, observed desmosome formation between the epithelia covering opposing shelves and suggested that desmosomes may be the basis of adhesion[84, 90, 92, 98, 101, 102, 111]. Other workers recently observed the presence of complex carbohydrates on the plasma membrane of the outer epithelial cells, on the vertical and horizontal shelves[104, 116, 118, 128]. They interpreted the presence of carbohydrates on unfused shelves as the 'sticky substance' responsible for epithelial adherence. Recently Shah[130, 131] verified the presence of surface carbohydrates on the surface of the palatal epithelial cells, prior to contact. It was further observed, that except for the presence of occasional granules in the areas which were not contacting, the carbohydrates were virtually absent at the time of actual initial contact. It is highly unlikely that a few small granules of carbohydrates will hold the two shelves together. The contacting areas were forming cell junctions which later appeared to be differentiating into desmosome-like structures. Quantitative analysis indicated a fourfold increase in desmosomal cell junctions from the vertical to epithelial fusion stage. It was suggested that carbohydrates may be involved in the cell surface recognition phenomena, but may not be responsible for actual adhesion. The actual adhesion may be achieved by desmosome-like cell junctions. Earlier Tyler and Koch[119] observed that mouse palatal explants obtained on day 12 of gestation take 72 hours to fuse in culture. The mouse palatal shelves are vertical on day 12, and have a carbohydrate coat on the surface cell plasma membrane[104]. Obviously if only the surface carbohydrates are responsible for adherence, then *in vitro* fusion in Tyler and Koch's experiment would have occurred sooner than 72 hours. It is possible that the 'acquired potentiality' for palatal fusion[71] may be indicative of cell surface recognition by carbohydrates rather than cell adhesion. However, this hypothesis needs further experimental evidence.

Fate of the epithelial seam

Once the seam is formed from the coming together of the two opposing epithelial cell layers, the cells themselves must be removed in order to establish continuity between the mesenchyme of the two palatal shelves. It has been suggested that some healthy epithelial cells in the seam may assume a cannibalistic function and phagocytize their neighbouring cells[85, 86]. However, no evidence of phagocytosis of one epithelial cell by another was observed[74, 101, 102, 110]. It is now generally accepted that a timely intracellular autolytic process is responsible for degeneration of the cells in the epithelial seam[73, 82, 84, 88–92, 94, 101, 102, 110, 111]. As evidence of autophagic cellular degeneration, acid hydrolytic enzymes were localized in the cells of the palatal epithelium, prior to, during, and after the seam formation[88, 89, 91, 110].

What molecular mechanism(s) may regulate the timely degeneration of cells in the epithelial seam is not yet certain. Recently Pratt and Martin[117] observed an elevated level of cyclic AMP during palate closure. Subsequent observations by Waterman and associates[107, 108] indicated that a hormonally regulated adenylate cyclase–cyclic AMP system may play an important role in the timely degeneration of the epithelial cells.

The degenerating cells are exfoliated into the amniotic fluid or extruded into the mesenchyme[74, 101, 102, 110, 111]. In the later location they are then phagocytized by the macrophages[89, 97, 101, 102, 110, 111].

The role played by basal lamina during the fusion between the midline epithelia of the opposing palatal shelves seems more or less settled. Barry[132] speculated that a disruption of the basement membrane on either side of the seam may be the necessary initial event which triggers the subsequent removal of cells from the seam. A few researchers[66, 85, 86] supported Barry's proposal and suggested that an incomplete basal lamina exposed the epithelial seam to some undetermined influence of the mesenchyme which precipitated the seam fragmentation and its subsequent disappearance. However, sufficient evidence has accumulated during the past ten years in different laboratories that an intact basal lamina persists as long as the epithelial cells are present in situ[79, 84, 90, 92, 94–96, 98, 99, 101, 102, 109, 111]. Basal laminae are lost only after disappearance of the epithelial cells. Thus the basal lamina has no role in the fragmentation of the epithelial seam. Perhaps the basal lamina may function as a selective barrier between the epithelium and the mesenchyme and provide mechanical support for the maintenance of epithelial contiguity during palatogenesis.

ABNORMAL DEVELOPMENT OF THE SECONDARY PALATE

Various researchers, on the basis of their observations of the human abortuses have proposed three basic theories of cleft palate formation. The 'classical theory' was originated by Dursy[11] and subsequently supported by His[12]. According to this theory, the primitive face cephalad to the stomodeum was divided into numerous 'peninsular masses' of ectoderm and mesoderm, presumably surrounded by free spaces or clefts. The masses grow, meet, and subsequently fuse in a manner similar to the healing of a wound. Any inhibition in the sequence of events leads to persistence of pre-existing clefts.

The second theory, commonly referred to as the 'mesodermal theory' was proposed by Veau[133]. He suggested that the primary defect results not from lack of fusion between the processes but is due to the failure of mesenchymal tissue to invade and destroy the overlying epithelium. The failure of mesenchymal invasion may be either due to the 'poor quality' of mesodermal tissue or from some unknown inhibitory action on the part of the epithelium. In both instances there is persistence of an epithelial wall between the opposing palatal processes.

Recently Kitamura[134] has suggested that there is an initial epithelial fusion of opposing palatal shelves. However, due to some unknown cause, there is a 'rupture' of the fused epithelia resulting in cleft palate.

Investigations into human cleft palate pathogenesis are scant, perhaps due to the unavailability of the fetal material. Recent studies on the laboratory animals, however, have begun to contribute towards an understanding of the mechanism(s) of cleft palate formation.

In 1950, Baxter and Fraser[5] reported that administration of cortisone to pregnant mice resulted in offspring with cleft palate. Much of the experimental work in the subsequent fifteen years was directed towards 1) methodological aspects of glucocorticoid-induced cleft palate, and 2) identification of other teratogenic agents that would consistently produce cleft palate in large numbers of fetuses. Walker and Fraser[135] were the first to study the mechanism of cortisone-induced cleft palate in mice. Most of the studies since then used glucocorticoid hormones, *in vivo*, to study the pathogenesis of cleft palate in different species[23, 41, 52, 53, 62, 64, 136–155]. Others have used hypervitaminosis A[28, 157–161], irradiation[162], β-aminopropionitrile[163, 164], meclozine[165] and diazo-oxo-norleucine[166–168]. A few *in vitro* studies are also reported in which palatal fusion was prevented following addition of teratogens to the culture[169–174].

In their study on the pathogenesis of cleft palate, Walker and Fraser[135] suggested that cortisone treatment caused a delay in the movement of the shelves from a vertical to a horizontal position. This was attributed to a decrease in the internal shelf force. Since Walker and Fraser's study, most research has been directed towards determining how the shelf force is reduced by a teratogen.

Walker[51], Larsson[41] and Jacobs[52, 53] observed a decreased synthesis of sulphated mucopolysaccharide in the palatal ground substance. This, they interpreted, results in an insufficient development of the internal shelf force. This view, however, has been challenged by Andrew and Zimmerman[141] and Nanda[162] and the issue still remains unsettled.

Shapiro and Shoshan[62] proposed a different mechanism for cleft palate formation subsequent to cortisone administration. Using radioautographic and biochemical techniques, they observed a significant association between the increased specific activity of hydroxyproline in the fetal palatal tissue and cleft palate formation. This observation led the authors to suggest that the cortisone may interfere with biosynthesis of collagen in the fetal palate and thereby inhibit the reorientation of the palatal shelves.

Because glucocorticoids are also known for their growth inhibiting effects, Mott *et al.*[64], and Jelinek and Dostal[151] studied the proliferative pattern in the mesenchymal cells. These authors found a decreased mitosis in the mesen-chyme of the cortisone treated palatal shelves prior to reorientation. They suggested that the palatal clefting was due to a decrease in cell number. Later, however, Zimmerman and associates[140] and Andrew *et al.*[147] noted a positive correlation between the inhibition of RNA synthesis and production of cleft palate. These authors felt that the cleft resulted from a growth inhibitory effect, at the macromolecular level, in the palatal tissue. Unlike previous studies, which suggested that the factor for reduced shelf force resided in the ground substance[41, 51–53], the later group of authors[64, 140, 147, 151] clearly implied that the effect of glucocorticoid is on the cell itself.

The issue of the placental transfer of the teratogen, and/or its metabolite, was quickly resolved. Several authors observed that radioactive glucocorti-coids, when injected into pregnant animals, traverse the placenta and label a variety of tissues in the fetuses[138, 142–144, 150]. These studies have further shown that glucocorticoid induce cleft palate by a direct action on the embryonic tissue.

Recently Bonner and Slavkin[148] suggested that cortisone-induced cleft palate is regulated by genes closely linked to the *H-2* locus. The product of this gene is perhaps a glucocorticoid receptor[153, 154], the level of which has to be raised for the cleft palate to occur following cortisone treatment. In their most recent study, Goldman *et al.*[155] have shown that the mesenchymal cells of palate have specific glucocorticoid receptor protein, with an isoelectric point of approximately 7.0. This protein has high affinity for dexamethasone and triamcinolone. Further mechanism(s) by which steroid-binding protein complexes cause cleft palate are not yet known. But an avenue is now open for further experimentation. For example, it will be interesting to study how the hormone–receptor complex affects the macromolecular synthesis and cell differentiation during cleft palate formation.

In our laboratory, recently we have observed that hydrocortisone when administered to hamsters during pregnancy, prevented epithelial fusion between the opposing palatal processes[23]. Ultrastructural evidence[152] indicated that the initial assault was upon the basal epithelial cells, leading to retardation of protein synthesis, and eventually to an imbalance of intracellular homeostatic mechanism. It was suggested that retardation of protein synthesis by hydrocortisone in turn may inhibit the formation of lysosomal enzymes, and thus alter the controlled process of cellular degeneration during palatogenesis. This latter hypothesis still remains to be verified. The observations, however, support the suggestion of Zimmerman et al.[140] and Andrew et al.[147] that glucocorticoid induces cleft palate by inhibiting RNA synthesis.

The foregoing review of the literature on mechanism(s) of glucocorticoid-induced cleft palate indicates that some evidence is available to support both the 'failure of fusion theory' and the 'mesodermal theory'. Studies with other teratogens also seem to support one or the other of these mechanisms[157–174]. With the exception of one report[156], however, there is no experimental evidence to support Kitamura's[134] 'post fusion rupture' hypothesis.

The precise mechanism(s) for cleft palate formation is not yet known. Although several teratogenic agents have been identified which produce cleft palate in animals, cellular details are recognized only with a few agents. Indeed, it has not yet been determined if the primary teratologic effect is on the palatal epithelium or on the mesenchyme or both.

Autoradiographic, biochemical, ultrastructural and immunogenetic approaches have helped to elucidate some of the issues involved in cleft palate formation. The experimental approach, therefore, should be continued.

ACKNOWLEDGEMENTS

This work was supported by a grant from the Medical Research Council of Canada. The author remains grateful to Mrs Vicki B. Koulours and Miss Vaile Long for their excellent secretarial assistance.

References

1. Smith, G. S. and Dowson, W. R. (1924). *Egyptian Mummies*, p. 106. (London: George, Allen and Unwin Ltd.)
2. Stevenson, A. C. (1969). Findings and lessons for the future from a comparative study of congenital malformation at 24 centers in 16 countries. In: H. Nishimura and J. R. Miller (eds.). *Methods for Teratological Studies in Experimental Animals and Man*, pp. 195–205. (Tokyo: Igaku Shoin)
3. Drillen, C. M., Ingram, T. T. S. and Wilkinson, E. M. (1966). *The Causes and Natural History of Cleft Lip and Palate*, pp. 22–35. (Baltimore: Williams and Wilkins)
4. Poswillo, D. and Roy, L. J. (1965). The pathogenesis of cleft palate. An animal study. *Br. J. Surg.*, 52, 902

5. Baxter, H. and Fraser, F. C. (1950). Production of congenital defects of offsprings of female mice treated with cortisone. *McGill Med. J.*, 19, 245
6. Burdi, A., Feingold, M., Larsson, K. S., Leck, I., Zimmerman, E. F. and Fraser, F. C. (1972). Etiology and pathogenesis of congenital cleft lip and cleft palate. An NIDR state of the art report. *Teratology*, 6, 255
7. Kalter, H. and Warkany, J. (1959). Experimental production of congenital malformations in mammals by metabolic procedures. *Physiol. Rev.*, 39, 69
8. Cohen, R. L. (1964). Evaluation of the teratogenicity of drugs. *Clin. Pharmacol. Ther.*, 5, 480
9. Warkany, J. (1971). *Congenital Malformations. Notes and Comments*, pp. 630–649. (Chicago: Yearbook Publishers)
10. Shah, R. M. (1979). Usefulness of golden syrian hamster in experimental teratology with particular reference to induction of orofacial malformation. In: T. V. N. Persaud (ed.). *Advances in the Study of Birth Defects.* (Lancaster: MTP Press)
11. Dursy, E. (1869). *Zur Entwicklungsgeschichte des Kopfes des Menschen und der höheren Wirbeltiere.* (Tübingen: Verlag Lauppschen)
12. His, W. (1901). Beobachtungen zur Geschichte der Nasen- und Gaumenbildung beim menschlichen Embryo. *Abh. Math. Phys. Kl. Saechs. Ges. Wiss.*, 27, 349
13. Polzl, A. (1904). Zur Entwicklungsgeschichte des menschlichen Gaumens. *Anat. Hefte.*, 27, 245
14. Schorr, G. (1908). Entwicklungsgeschichte des sekundären Gaumens bei einigen Saugetieren und beim Menschen. *Anat. Hefte*, 36, 69
15. Inouye, M. (1912). Die Entwicklung des sekundären Gaumens einiger Säugetiere mit besonderer Berücksichtigung der Bildungsvorgänge am Gesichte und des Unlagerungsprozesses der Gaumenplatten. *Anat. Hefte*, 46, 1
16. Peter, K. (1924). Die Entwicklung des Säugetiergaumens. *Ergeb. Anat. Entwicklungsgesch.*, 25, 448
17. Lazzaro, C. (1940). Sul meccanismo di chiussura del palato secondario. *Monit. Zool. Ital.*, 51, 249
18. Hayward, J. R. and Avery, J. (1957). A variation in cleft palate. *J. Oral Surg.*, 15, 320
19. Walker, B. E. and Fraser, F. C. (1956). Closure of the secondary palate in three strains of mice. *J. Embryol. Exp. Morphol.*, 4, 1976
20. Coleman, R. D. (1965). Development of the rat palate. *Anat. Rec.*, 151, 107
21. Walker, B. E. (1971). Palate morphogenesis in the rabbit. *Arch. Oral Biol.*, 16, 275
22. Greene, R. M. and Kochhar, D. M. (1973). Palatal closure in the mouse as demonstrated in frozen sections. *Am. J. Anat.*, 137, 477
23. Shah, R. M. and Travill, A. A. (1976). Morphogenesis of the secondary palate in normal and hydrocortisone treated hamsters. *Teratology*, 13, 71
24. Holmstedt, J. O. V. and Bagwell, J. N. (1977). Morphogenesis of the secondary palate in the Mongolian gerbil (*Meriones unguiculatus*). *Acta Anat.*, 97, 443
25. Shah, R. M. (1977). Palatomandibular and maxillo-mandibular fusion, partial aglossia and cleft palate in a human embryo. Report of a case. *Teratology*, 15, 261
26. Ross, R. B. and Johnston, M. C. (1972). *Cleft Lip and Palate*, pp. 3–16. (Baltimore: Williams and Wilkins)
27. Pons-tortella, E. (1937). Über die Bildungsweise des sekundären Gaumens. *Anat. Anz.*, 84, 13
28. Kochhar, D. M. and Johnson, E. M. (1965). Morphological and autoradiographic studies of cleft palate induced in rat embryos by maternal hypervitaminosis A. *J. Embryol. Exp. Morphol.*, 14, 233
29. Anderson, H. and Matthiessen, M. (1967). Histochemistry of the early development of the human face and nasal cavity with special reference to the movement and fusion of the palatine processes. *Acta Anat.*, 68, 473
30. Iizuka, T. (1973). Stage of the closure of the human palate. *Okaj. Fol. Anat. Jap.*, 50, 249
31. Asling, C. W., Nelson, M. M., Dougherty, M. D., Wright, H. V. and Evans, H. M. (1960). The development of the cleft palate resulting from maternal pteroylgutamic (folic) acid deficiency during the later half of gestation in rats. *Surg. Gynecol. Obstet.*, 111, 19
32. Zeiler, K., Weinsten, S. and Gibson, R. D. (1964). A study of the morphology and the time of closure of the palate in the albino rat. *Arch. Oral Biol.*, 9, 545

33. Humphrey, T. (1969). The relation between human fetal mouth opening reflexes and closure of the palate. *Am. J. Anat.*, 125, 317
34. Humphrey, T. (1971). Development of the oral and facial motor mechanisms in human fetuses and their relation to craniofacial growth. *J. Dent. Res.*, 50, 1428
35. Walker, B. E. (1969). Correlation of embryonic movement with palatal closure in mice. *Teratology*, 2, 121
36. Wragg, L., Klein, M., Steinvorth, G. and Warpeha, R. (1970). Facial growth accomodating secondary palate closure in rat and man. *Arch. Oral Biol.*, 15, 705
37. Wragg, L., Diewert, V. and Klein, M. (1972). Myoneural maturation and function of the fetal rat tongue at the time of secondary palate closure. *Arch. Oral Biol.*, 17, 673
38. Wragg, L., Diewert, V. and Klein, M. (1972). Spatial relations in the oral cavity and the mechanisms of secondary palate closure in the rat. *Arch. Oral Biol.*, 17, 683
39. Walker, B. E. and Ross, L. M. (1972). Observations of palatine shelves in living rabbit embryos. *Teratology*, 5, 97
40. Diewert, V. (1974). A cephalometric study of orofacial structures during secondary palate closure in the rat. *Arch. Oral Biol.*, 19, 303
41. Larsson, K. S. (1962). Closure of the secondary palate and its relation to sulfo-mucopolysaccharides. *Acta Odontol. Scand.*, 20, 4
42. Jelinek, R. and Dostal, M. (1973). The role of mitotic activity in the formation of the secondary palate. *Acta Chir. Plast.*, 15, 216
43. Lessard, J., Wee, E. and Zimmerman, E. F. (1974). Presence of contractile proteins in mouse fetal palate prior to shelf elevation. *Teratology*, 9, 113
44. Nanda, R. and Romeo, D. (1975). Differential cell proliferation of embryonic rat palatal processes as determined by incorporation of tritiated thymidine. *Cleft Palate J.*, 12, 436
45. Babiarz, B., Allenspach, A. and Zimmerman, E. F. (1975). Ultrastructural evidence of contractile systems in mouse palates prior to rotation. *Dev. Biol.*, 47, 32
46. Wee, E., Wolfson, L. and Zimmerman, E. F. (1976). Palate shelf movement in mouse embryo culture: Evidence for skeletal and smooth muscle contractility. *Dev. Biol.*, 48, 91
47. Shah, R. M. (1978). A cellular mechanism for the palatal shelf reorientation from a vertical to a horizontal plane. Light and electron microscopic study. (In preparation)
48. Schweichel, J. U. and Seinsch, W. (1973). Die Entwicklung des Bewegungsapparates der Rattenzunge im Licht- und Elektronenmikroskop. *Z. Anat. Entwicklungsgesch.*, 140, 153
49. Holt, T. J. (1975). A morphologic and histochemical study of the developing tongue musculature in the mouse: Its relationship to palatal closure. *Am. J. Anat.*, 144, 169
50. Larsson, K. S., Bostrom, H. and Carlsoo, S. (1959). Studies on the closure of the secondary palate. I. Autoradiographic study in the normal mouse embryo. *Exp. Cell. Res.*, 16, 379
51. Walker, B. E. (1961). The association of mucopolysaccharides with morphogenesis of the palate and other structures in mouse embryos. *J. Embryol. Exp. Morphol.*, 9, 22
52. Jacobs, R. M. (1964). Histochemical study of morphogenesis and teratogenesis of the palate in mouse embryos. *Anat. Rec.*, 149, 691
53. Jacobs, R. M. (1964). S^{35}-liquid-scintillation count analysis of morphogenesis and teratogenesis of the palate in mouse embryos. *Anat. Rec.*, 150, 271
54. Pratt, R. M., Goggins, J., Wilk, A. and King, C. T. (1973). Acid mucopolysaccharide synthesis in the secondary palate of the developing rat at the time of rotation and fusion. *Dev. Biol.*, 32, 230
55. Ferguson, M. W. J. (1977). The mechanism of palatal shelf elevation and the pathogenesis of cleft palate. *Virchows Arch. A*, 375, 97
56. Clark, K. H. (1956). Histological investigation of cortisone induced cleft palate in mice. *Genetics*, 41, 637
57. Stark, R. and Ehrman, N. (1958). The development of the center of the face with particular reference to surgical correction of bilateral cleft lip. *Plast. Reconstr. Surg.*, 21, 177
58. Isaacson, R. and Chaudhry, A. P. (1962). Cleft palate induction in strain A mice with cortisone. *Anat. Rec.*, 142, 479
59. Loevy, H. (1963). Developmental changes in the palate of normal and cortisone treated strong A mice. *Anat. Rec.*, 142, 375
60. Frommer, J. and Monroe, C. (1969). Further evidence for the absence of elastic fibers during movement of the palatal shelves in mice. *J. Dent. Res.*, 48, 155

61. Pratt, R. M. and King, C. T. (1971). Collagen synthesis in the secondary palate of the developing rat. *Arch. Oral Biol.*, 16, 1181
62. Shapiro, Y. and Shoshan, S. (1972). The effect of cortisone on collagen synthesis in the secondary palate of mice. *Arch. Oral Biol.*, 17, 1699
63. Hassell, J. R. and Orkin, R. W. (1976). Synthesis and distribution of collagen in the rat palate during shelf elevation. *Dev. Biol.*, 49, 80
64. Mott, W. J., Toto, P. and Hilgers, D. (1969). Labelling index and cellular density in palatine shelves of cleft palate in mice. *J. Dent. Res.*, 48, 263
65. Jelinek, R. and Dostal, M. (1974). Morphogenesis of cleft palate induced by exogenous factors, VII. Mitotic activity during formation of the mouse secondary palate. *Folia Morph.*, 22, 94
66. Hughes, L., Furstman, L. and Bernick, S. (1967). Prenatal development of the rat palate. *J. Dent. Res.*, 46, 373
67. Cleaton-Jones, P. (1976). Radioautographic study of mesenchymal cell activity in the secondary palate of the rat. *J. Dent. Res.*, 55, 437
68. Greene, R. M. and Pratt, R. M. (1976). Developmental aspects of secondary palate formation. *J. Embryol. Exp. Morphol.*, 36, 225
69. Moriarty, T. M., Weinstein, S. and Gibson, R. D. (1963). The development *in vitro* and *in vivo* of fusion of the palatal processes of rat embryos. *J. Embryol. Exp. Morphol.*, 11, 605
70. Konegni, J. S., Chan, B., Moriarty, T. M., Weinstein, S. and Gibson, R. D. (1965). A comparison of standard organ culture and standard transplant techniques in the fusion of the palatal processes of rat embryos. *Cleft Palate J.*, 2, 219
71. Pourtois, M. (1966). Onset of the acquired potentiality for fusion in the palatal shelves of rats. *J. Embryol. Exp. Morphol.*, 16, 171
72. Reeve, W. L., Porter, K. and Lefkowitz, W. (1966). *In vitro* closure of the rat palate. *J. Dent. Res.*, 45, 1375
73. Mato, M., Aikawa, E. and Katahira, M. (1966). Appearance of various types of lysosomes in the epithelium covering lateral palatine shelves during secondary palate formation. *Gunma J. Med. Sci.*, 15, 46
74. Mato, M., Aikawa, E. and Katahira, M. (1967). Alteration of fine structure of the epithelium on the lateral palatine shelf during the secondary palate formation. *Gunma J. Med. Sci.*, 16, 79
75. Mato, M., Aikawa, E. and Katahira, M. (1967). The characteristic cell reaction in the epithelial layer at the lower surfaces of the nasal septum during a secondary palate formation. *Gunma J. Med. Sci.*, 16, 244
76. Mato, M., Aikawa, E. and Katahira, M. (1967). Studies on cell reaction of the nasal epithelium during the fusion of palatine shelves. *Anat. Anz.*, 121, 504
77. Mato, M., Aikawa, E. and Katahira, M. (1968). Further studies on 'cell reaction' at the lower surface of the nasal septum of human embryos during the fusion of the palate. *Acta Anat.*, 71, 154
78. Mato, M., Aikawa, E. and Smiley, G. R. (1972). Invagination of human palatal epithelium prior to contact. *Cleft Palate J.*, 9, 335
79. Mato, M., Smiley, G. R. and Dixon, A. D. (1972). Epithelial changes in the presumptive regions of fusion during secondary palate formation. *J. Dent. Res.*, 51, 1451
80. Vargas, V. (1967). Palatal fusion *in vitro* in the mouse. *Arch. Oral Biol.*, 12, 1283
81. Vargas, V. (1968). Fusion of the palatine shelves with heterotypic explants in the mouse. *Arch. Oral Biol.*, 13, 845
82. Vargas, V., Nasjleti, C. and Azcurra, J. (1972). Cytodifferentiation of the mouse secondary palate *in vitro*: morphological, biochemical, and histochemical aspects. *J. Embryol. Exp. Morphol.*, 27, 413
83. Myers, G. S., Patrakis, N. L. and Lee, M. (1968). Factors influencing fusion of rat palates grown *in vitro*. *Anat. Rec.*, 162, 71
84. De Angelis, V. and Nalbandian, J. (1968). Ultrastructure of mouse and rat palatal processes prior to and during secondary palate formation. *Arch. Oral Biol.*, 13, 601
85. Farbman, A. I. (1968). Electron microscope study of palate fusion in mouse embryos. *Dev. Biol.*, 18, 93
86. Farbman, A. I. (1969). The epithelium—connective tissue interface during closure of the secondary palate in rodent embryos. *J. Dent. Res.*, 48, 617

87. Smiley, G. R. and Dixon, A. D. (1968). Fine structure of midline epithelium in the developing palate of the mouse. *Anat. Rec.*, 161, 298
88. Angelici, D. and Pourtois, M. (1968). The role of acid phosphatase in the fusion of the secondary palate. *J. Embryol. Exp. Morphol.*, 20, 15
89. Koziol, C. A. and Steffek, A. (1969). Acid phosphatase activity in palates of developing normal and chlorcyclizine treated rodents. *Arch. Oral Biol.*, 14, 317
90. Brusati, R. (1969). Ultrastructural study of the processes of formation and involution of the epithelial sheet of the secondary palate in the rat. *J. Submicrosc. Cytol.*, 1, 215
91. Hayward, F. (1969). Ultrastructural changes in the epithelium during fusion of the palatal processes in rats. *Arch. Oral Biol.*, 14, 661
92. Morgan, P. (1969). Recent studies on the fusion of the secondary palate. *London Hosp. Gaz.*, 72, 6
93. De Angelis, V. (1969). The distribution of glycogen in mouse and rat palatal processes during secondary palate formation: An ultrastructural study. *Arch. Oral Biol.*, 14, 385
94. Shapiro, B. L. and Sweney, L. (1969). Electron microscopic and histochemical examination of oral epithelial mesenchymal interaction (programmed cell death). *J. Dent. Res.*, 48, 652
95. Smiley, G. R. (1970). Fine structure of mouse embryonic palatal epithelium prior to and after midline fusion. *Arch. Oral Biol.*, 15, 287
96. Sweney, L. and Shapiro, B. L. (1970). Histogenesis of Swiss white mouse secondary palate from nine and one-half days to fifteen and one-half days in utero. *J. Morphol.*, 130, 435
97. Matthiesen, M. and Anderson H. (1972). Disintegration of the junctional epithelium of human hard palate. *Z. Anat. Entwicklungsgesch.*, 137, 153
98. Smiley, G. R. and Koch, W. E. (1971). Fine structure of mouse secondary palate development *in vitro*. *J. Dent. Res.*, 59, 1671
99. Smiley, G. R. and Koch, W. E. (1972). An *in vitro* and *in vivo* study of single palatal processes. *Anat. Rec.*, 173, 405
100. Smiley, G. R. and Koch, W. E. (1975). A comparison of secondary palate development with different *in vitro* techniques. *Anat. Rec.*, 181, 711
101. Chaudhry, A. P. and Shah, R. M. (1973). Palatogenesis in hamster. II. Ultrastructural observations on the closure of palate. *J. Morphol.*, 139, 329
102. Chaudhry, A. P. and Shah, R. M. (1978). Light and electron microscopic observations on closure of the secondary palate with the primary palate and nasal septum. *Acta. Anat.* (In press)
103. Hudson, C. and Shapiro, B. L. (1973). A radioautographic study of deoxyribonucleic acid synthesis in embryonic rat palatal shelf epithelium with reference to the concept of programmed cell death. *Arch. Oral Biol.*, 18, 77
104. Greene, R. M. and Kochhar, D. M. (1974). Surface coat on the epithelium of developing palatine shelves in the mouse as revealed by electron microscopy. *J. Embryol. Exp. Morphol.*, 31, 683
105. Waterman, R., Ross, L. and Meller, S. M. (1973). Alterations in the epithelial surface of A/Jax mouse palatal shelves prior to and during fusion: A scanning electron microscopic study. *Anat. Rec.*, 176, 361
106. Waterman, R. and Meller, S. M. (1974). Alterations in the epithelial surface of human palatal shelves prior to and during fusion: A scanning electron microscopic study. *Anat. Rec.*, 180, 111
107. Waterman, R., Palmer, G., Palmer, S. J. and Palmer, S. M. (1976). Catacholamine-sensitive adenylate cyclase in the developing golden hamster palate. *Anat. Rec.*, 185, 125
108. Waterman, R., Palmer, G., Palmer, S. J. and Palmer, S. M. (1977). *In vitro* activation of adenylate cyclase by parathyroid hormone and calcitonin during normal and hydrocortisone induced cleft palate development in the golden hamster. *Anat. Rec.*, 188, 431
109. Hinrichsen, C. and Stevens, G. (1974). Epithelial morphology during closure of the secondary palate in the rat. *Arch. Oral Biol.*, 19, 969
110. Shah, R. M. and Chaudhry, A. P. (1974). Light microscopic and histochemical observations on the development of palate in golden Syrian hamster. *J. Anat.*, 117, 1
111. Shah, R. M. and Chaudhry, A. P. (1974). Ultrastructural observations on the closure of soft palate in hamster. *Teratology*, 10, 17

112. Goss, A. and Avery, J. K. (1975). Epithelial mesenchymal interaction in palatal shelf fusion. An *in vitro* study. *Austr. Dent. J.*, **20**, 152
113. Goss, A. (1975). Human palatal development *in vitro*. *Cleft Palate J.*, **12**, 210
114. De Paola, D., Drummond, J. F., Lorente, C., Zarbo, R. and Miller, S. A. (1975). Glycoprotein biosynthesis at the time of palatal fusion by rabbit palate and maxilla cultured *in vitro*. *J. Dent. Res.*, **54**, 1049
115. Brinkley, L., Basenhour, G., Branch, A. and Avery, J. (1975). A new *in vitro* system for studying secondary palate development. *J. Embryol. Exp. Morphol.*, **34**, 485
116. Pratt, R. M. and Hassell, J. R. (1975). Appearance and distribution of carbohydrate rich macromolecules on the epithelial surface of the developing rat palatal shelf. *Dev. Biol.*, **45**, 192
117. Pratt, R. M. and Martin, G. R. (1975). Epithelial cell death and cyclic AMP increase during palatal development. *Proc. Nat. Acad. Sci. USA*, **72**, 874
118. Souchon, R. (1975). Surface coat of the palatal shelf epithelium during palatogenesis in mouse embryo. *Anat. Embryol.*, **147**, 133
119. Tyler, M. S. and Koch, W. E. (1975). *In vitro* development of palatal tissue from embryonic mice. I. Differentiation of the secondary palate from 12-day mouse embryos. *Anat. Rec.*, **182**, 297
120. Tyler, M. S. and Koch, W. E. (1977). *In vitro* development of palatal tissues from embryonic mice. II. Tissue isolation and recombination. *J. Embryol. Exp. Morphol.*, **38**, 19
121. Tyler, M. S. and Koch, W. E. (1977). *In vitro* development of palatal tissues from embryonic mice. III. Interactions between palatal epithelium and heterotypic oral mesenchyme. *J. Embryol. Exp. Morphol.*, **38**, 37
122. Newall, D. and Edwards, J. (1976). *In vitro* fusion of the human secondary palate. *Cleft Palate J.*, **13**, 54
123. Cleaton-Jones, P. (1976). A scanning electron microscope study of the developing rat secondary palate. *S. Afr. J. Med. Sci.*, **41**, 1
124. Tassin, M. and Weil, R. (1974). Contribution à l'étude de la fusion de crètes palatines de souris: cultures pendant des temps courts. *J. Biol. Bucc.*, **2**, 237
125. Tassin, M. and Weil, R. (1977). Mesenchymal fusion of the palatal processes and of heterotypic explants. *Roux's Arch. Dev. Biol.*, **181**, 95
126. Tassin, M. and Weil, R. (1977). Changements de l'épithelium médian des bourgeons palatins de souris au stade préfusion. *Roux's Arch. Dev. Biol.*, **181**, 357
127. Hassell, J. R. and Pratt, R. M. (1977). Elevated levels of cAMP alters the effect of epidermal growth factor *in vitro* on programmed cell death in the secondary palatal epithelium. *Exp. Cell. Res.*, **106**, 55
128. Meller, S. M. and Barton, L. H. (1978). Extracellular coat in developing human palatal processes: Electron microscopy and ruthenium red binding. *Anat. Rec.*, **190**, 223
129. Gartner, L. P., Hiatt, J. L. and Provenze, D. V. (1978). Succinic dehydrogenase activity during palate formation in the Mongolian gerbil. *J. Anat.*, **125**, 133
130. Shah, R. M. (1978). Role of desmosomes and ruthenium red bound carbohydrates during hamster palatogenesis. *J. Dent. Res.*, **57**, 317
131. Shah, R. M. (1978). The role of ruthenium red bound surface carbohydrates and desmosomes during palate closure in hamster. (In preparation)
132. Barry, A. (1961). Development of the branchial region of human embryos with special reference to the fate of epithelia. In: S. Pruzansky (ed.). *'Congenital Anomalies of the Face and Associated Structures'*, pp. 46–62, (Springfield: C. C. Thomas)
133. Veau, V. (1938). Hasenscharten menschlicher Keimlinge auf der Stufe 21-23 mm s. st. L. *Z. Anat. Entwicklungsgesch.*, **108**, 459
134. Kitamura, H. (1966). Epithelial remnants and pearls in the secondary palate in the human abortus: A contribution to the study of the mechanisms of cleft palate formation. *Cleft Palate J.*, **3**, 240
135. Walker, B. E. and Fraser, F. C. (1957). The embryology of cortisone induced cleft palate. *J. Embryol. Exp. Morphol.*, **5**, 201
136. Kohno, J. (1960). Embryological studies on experimental cleft palate due to hydrocortisone. *J. Osaka City Med. Center*, **9**, 253
137. Ross, L. and Walker, B. E. (1967). Movement of palatine shelves in untreated and teratogen-treated mouse embryos. *Am. J. Anat.*, **121**, 509

138. Nasjleti, C., Avery, J., Spencer, H. and Walden, J. (1967). Tritiated cortisone distribution and induced cleft palate in mice. *J. Oral Ther. Pharmacol.*, 4, 71
139. Shapiro, Y. (1969). An autoradiographic study of 3H-proline uptake in the palate of normal mice and in the palate of mice treated with hydrocortisone. *J. Dent. Res.*, 48, 1039
140. Zimmerman, E. F., Andrew, F. and Kalter, H. (1970). Glucocorticoid inhibition of RNA synthesis responsible for cleft palate in mice: A model. *Proc. Nat. Acad. Sci. USA*, 67, 779
141. Andrew, F. and Zimmerman, E. (1971). Glucocorticoid induction of cleft palate in mice; no correlation with inhibition of mucopolysaccharide synthesis. *Teratology*, 4, 31
142. Waddel, W. J. (1971). The distribution of cortisone-C^{14} in pregnant mice. *Teratology*, 4, 35
143. Wood, N., Marks, A., Schmitz, D., Bowman, D. and Toto, P. (1972). Radioautographic labelling in placentas and fetal palatal shelves after maternal injections of tritiated cortisone in A/Jax mice. *J. Dent. Res.*, 51, 67
144. Reminga, T. and Avery, J. (1972). Differential binding of ^{14}C-cortisone in fetal placental and maternal liver tissue in A/Jax and C57BL mice. *J. Dent. Res.*, 51, 1426
145. Zimmerman, E. and Bowen, D. (1972). Distribution and metabolism of triamcinolone acetonide in inbred mice with different cleft palate sensitivities. *Teratology*, 5, 335
146. Greene, R. M. and Kochhar, D. (1973). Spatial relations in the oral cavity of cortisone-treated mouse fetuses during the time of secondary palate closure. *Teratology*, 8, 153
147. Andrew, F., Bowen, D. and Zimmerman, E. (1973). Glucocorticoid inhibition of RNA synthesis and the critical period for cleft palate induction in inbred mice. *Teratology*, 7, 167
148. Bonner, J. J. and Slavkin, H. (1975). Cleft palate susceptibility linked to histocompatibility − 2(H-2) in the mouse. *Immunogenetics*, 2, 213
149. Holst, P. and Mills, B. (1975). Tissue phosphatase changes following triamcinolone associated with cleft palate in rats. *Teratology*, 11, 57
150. Spain, K., Kisieleski, W. and Wood, N. (1975). Cleft palate induction: Quantitative studies of ^{3}H corticoids in A/Jax mouse tissues after maternal injections of ^{3}H cortisol. *J. Dent. Res.*, 54, 1069
151. Jelinek, R. and Dostal, M. (1975). Inhibitory effect of corticoids on the proliferative pattern in mouse palatal processes. *Teratology*, 11, 193
152. Shah, R. M. and Travill, A. (1976). Light and electron microscopic observations on hydrocortisone induced cleft palate in hamsters. *Am. J. Anat.*, 145, 149
153. Solomon, D. and Pratt, R. M. (1976). Glucocorticoid receptors in murine embryonic facial mesenchyme cells. *Nature (Lond.)*, 264, 174
154. Goldman, A., Katsumata, M., Yaffe, S. and Gasser, D. (1977). Palatal cytosol cortisol-binding protein associated with cleft palate susceptibility and H-2 genotype. *Nature (Lond.)*, 265, 643
155. Goldman, A., Shapiro, B. and Katsumata, M. (1978). Human fetal palatal corticoid receptors and teratogens for cleft palate. *Nature (Lond.)*, 272, 464
156. Angelici, D. R. (1968). Re-opening of fused palatal shelves. *Cleft Palate J.*, 5, 205
157. Walker, B. E. and Crain, B. (1960). Effects of hypervitaminosis A on palate development in two strains of mice. *Am. J. Anat.*, 107, 49
158. Steffek, A., King, C. and Derr, J. (1967). The comparative pathogenesis of experimentally induced cleft palate. *J. Oral Ther. Pharmacol.*, 3, 9
159. Kocchar, D. (1968). Studies of vitamin A induced teratogenesis: effects on embryonic mesenchyme and epithelium and on incorporation of ^{3}H-thymidine. *Teratology*, 1, 299
160. Nanda, R. (1970). The role of sulfated mucopolysaccharides in cleft palate production. *Teratology*, 3, 237
161. Nanda, R. (1971). Tritiated thymidine labelling of the palatal processes of rat embryos with cleft palate induced by hypervitaminosis A. *Arch. Oral Biol.*, 16, 435
162. Callas, G. and Walker, B. (1963). Palate morphogenesis in mouse embryos after X-irradiation. *Anat. Rec.*, 145, 61
163. Pratt, R. M. and King, C. (1972). Inhibition of collagen cross-linking associated with β-aminopropionitrile induced cleft palate. *Dev. Biol.*, 27, 322
164. Mato, M., Uchiyama, E. and Smiley, G. (1975). Ultrastructural changes in rat palatal epithelium after β-aminopropionitrile. *Teratology*, 11, 153

165. Morgan, P. (1976). The fate of the expected fusion zone in rat fetuses with experimentally induced cleft palate – an ultrastructural study. *Dev. Biol.*, 51, 225
166. Pratt, R. M. and Greene, R. M. (1976). Inhibition of palatal epithelial cell death by altered protein synthesis. *Dev. Biol.*, 54, 135
167. Morgan, P. and Pratt, R. M. (1977). Ultrastructure of the expected fusion zone in rat fetuses with diazo-oxo-norleucine (DON) induced cleft palate. *Teratology*, 15, 281
168. Greene, R. M. and Pratt, R. M. (1977). Inhibition by diazo-oxo-norleucine (DON) of rat palatal glycoprotein synthesis and epithelial cell death *in vitro*. *Exp. Cell. Res.*, 105, 27
169. Myers, G. S., Petrakis, N. and Lee, M. (1967). Effects of 6-aminonicotinamide and of added vitamin A on fusion of embryonic rat palates *in vitro*. *J. Nutr.*, 93, 25
170. Pourtois, M. (1971). The fate of rat palatal shelves cultivated *in vitro* in the presence of periodic acid. *Arch. Oral Biol.*, 16, 503
171. Lahti, A., Antila, E. and Saxen, L. (1972). The effect of hydrocortisone on the closure of the palatal shelves in two inbred strains of mice *in vivo* and *in vitro*. *Teratology*, 6, 37
172. Saxen, I. (1973). Effects of hydrocortisone on the development *in vivo* of the secondary palates in two inbred strains of mice. *Arch. Oral Biol.*, 18, 1469
173. Fairbanks, M. and Kollar, E. (1974). Inhibition of palatal fusion *in vitro* by hadacidin. *Teratology*, 9, 169
174. Hassell, J. R. (1975). The development of rat palatal shelves *in vitro*. An ultrastructural analysis of the inhibition of epithelial cell death and palate fusion by the epidermal growth factor. *Dev. Biol.*, 45, 90

6
Genetic studies of teratogen-induced cleft palate in the mouse

F. G. BIDDLE

INTRODUCTION

This review of genetic studies of teratogen-induced cleft palate in the mouse will be concerned with two questions. What is the nature of strain differences in response to the teratogens? Why should strain differences be studied?

A premise of modern biology is that a biological process is most easily studied with 'mutations' that alter the process. The mutations can be characterized and inferences can be drawn about the processes that they affect. (Some would say there is certainty that a process is important to the organism as well as being interesting to the observer only when it is altered by mutation[1].) This approach is standard procedure with microbial and cell culture systems and, to a large degree, at the organismal level with experimental species that lend themselves to being mutagenized and selected for specific traits (e.g. *Drosophila*). In the mouse, with recent developments in the technology of fast analysis, specific mutations are being screened at the molecular level[2, 3].

The selected mutation approach is not yet possible for experimental embryology or teratology in the mouse. Nevertheless, it may be useful during this review to consider the different stocks of the mouse as a resource in which to search for potential 'mutations' in the cleft palate response to teratogens. If the philosophy of the selected mutation approach is kept in mind, how can different 'mutations' in the cleft palate response be identified and used as tools to study developmental mechanisms?

Many different strains and stocks of the mouse exist[4]. Most have never been used to examine teratogenic responses and the few strains in which the cleft palate response has been examined represent, most likely, a very limited sampling of the genome. Therefore, the interpretation of genetic studies in this review will be cautious.

It has been stated several times that strain differences in the cleft palate response to a teratogen cannot be used as tools to investigate the mechanisms of the malformation; instead, they can be used to study the mechanism of the strain difference[5, 6]. This review will try to give perspective to this statement and to suggest how genetic tools can be sought to investigate mechanisms of cleft palate induction.

No attempt will be made to discuss studies of molecular and morphogenetic mechanisms of cleft palate induction. Several reviews have been published in this area[7, 8].

HISTORICAL BACKGROUND

Maternal treatment with cortisone acetate induced cleft palate in mouse embryos[9]. This provided the embryologist with a useful model in which to study palatogenesis before the malformation occurred. As different strains were examined, strain differences in the frequency of the cleft palate response were found[10].

To the geneticist; presented with stable strain differences, and, therefore, by definition, genetic differences, the question is: what is the cause of the difference? Is the trait controlled by a single gene difference; are a small number of gene loci involved; is the trait more complex, that is polygenic?

The first attempt to analyse strain differences in the frequency of cortisone-induced cleft palate used the A/J and C57BL/6J strains[11]. A/J is more sensitive than C57BL/6J but both respond. Reciprocal crosses between the two strains demonstrated a maternal effect in addition to the embryonic effect; F_1 embryos in A/J dams expressed a higher frequency of cortisone-induced cleft palate than F_1 embryos in C57BL/6J dams. Further matings of reciprocal F_1 females to A/J males (to recover genes for sensitivity) ruled out the involvement of cytoplasmic factors in the maternal effect trait. The concepts of maternal and cytoplasmic effect traits in teratology and in the cleft palate response in particular have been reviewed recently[12].

No simple mode of inheritance could explain the breeding data for the difference between A/J and C57BL/6J in cortisone-induced cleft palate. It was suggested that the trait of the strain difference could be described in general terms as a 'polyfactorial, quantitative character' that involved genetic variation in both the mother and the embryo[11].

In retrospect, the first attempt to search for the genetic cause of the strain difference may have been complicated by the types of matings that were made. Segregation of the trait was followed through females that were crossed and backcrossed to A/J males. The maternal effect and embryonic effect would be segregating, but the maternal effect would be one generation behind the embryonic effect.

The frequency of cortisone-induced cleft palate was examined in a few other strains and strains crosses[13-15] and more variation in the frequency of response was uncovered. Cortisone acetate was administered in a standard dosage of either 1.25 or 2.5 mg/mouse on each of days 11 through 14 of gestation. The strains tested could be ranked in decreasing frequency of response: A/J, A/Strong > DBA > C3H > C57BL/6J > CBA.

A minor controversy arose when A/Strong (similar in response to A/J) was mated reciprocally with C3H and tested with cortisone[14, 15]. A maternal effect was identified between A/Strong and C3H, but the direction was opposite to expected. F_1 embryos were more sensitive in C3H dams than in A/Strong dams. In retrospect, it appears that the maternal effect trait was expected to follow the strain order of sensitivity. One hypothesis to explain the difference in maternal effect is sex-linked (X-chromosome linked) embryonic sensitivity factors[16] and these were demonstrated to be involved. A second hypothesis is that the embryonic effect and maternal effect traits could be separate, genetically independent systems[12]. This hypothesis remains to be tested. The difficulty with the study of most maternal effect traits is that they are defined in terms of the frequency of the embryonic response for which genetic variation is also observed.

A different teratogen, 5-fluorouracil (5-FU), was found to induce cleft palate, in addition to other skeletal defects, and the frequency varied widely between inbred strains[17]. With two of the strains, 129/Dg and BALB/cDg, a backcross mating scheme was used to search for the number of genes involved in the response difference[18]. A backcross test mating scheme with one strain of female maintained a constant maternal effect and permitted segregation of the embryonic response to be followed through the sires. The principles of this breeding scheme will be discussed later. For the embryonic difference in frequency of 5-FU-induced cleft palate, a single-gene difference was rejected and more complex models, given certain assumptions, were evaluated. A minimal estimate of three loci was suggested to control the strain difference in embryonic cleft palate response to 5-FU and some of the data suggested the cleft palate response and other skeletal defects were genetically separable.

Many agents have been reported to induce cleft palate in mice and the list continues to grow[5, 19–21]. Unfortunately, there has been no systematic testing of different strains for cleft palate susceptibility with this list of teratogens. A difference in frequency of response between the A/J and C57BL/6J strains for a few of these compounds has been enigmatic; A/J appears to be more sensitive than C57BL/6J. Galactoflavin-induced (riboflavin deficiency) cleft palate appears to be the sole exception[13].

A threshold model was developed to explain the strain difference in cleft palate response between A/J and C57BL/6J[22, 23]. In the absence of teratogens, A/J embryos appeared to close their palates at a later gestational age than C57BL/6J[24]. Cortisone treatment appeared to delay palate closure and a conceptual model was put forth to suggest that closure of the palatal shelves must occur before a critical stage in embryogenesis or cleft palate would result[25]. If a teratogen induced cleft palate by causing a delay in palate closure, it would induce a greater frequency in A/J than C57BL/6J because this genotype exhibited, normally, a later palate closing.

RATIONALE FOR THE PRESENT STUDY

With this background, the present study set out to test whether genes that controlled the difference between A/J and C57BL/6J in the frequency of cleft palate response to cortisone and to 6-aminonicotinamide (6-AN) were the same or different[26]. Was A/J more susceptible than C57BL/6J to both teratogens for the same or different reasons? 6-AN was selected in addition to cortisone because it has been extensively studied with respect to the strain difference between A/J and C57BL/6J[27–31], but no genetic analysis had been initiated. A systematic group of experiments was initiated to search for a conceptual framework that would permit both a genetic and physiological interpretation of the strain differences in cleft palate response to the two teratogens.

CONCEPTUAL FRAMEWORK

Norms of reaction

Most reports of strain distributions of teratogen-induced cleft palate have compared only the frequencies of cleft palate produced by a specific dose of the teratogen. The use of absolute frequencies is valid only if the teratogen produces cleft palate by the same mechanism in all genotypes and there is an additive, linear relationship between genotypes on the frequency scale.

It is not adequate to demonstrate that increasing dosage causes an increase in the frequency of cleft palate without some method of relating dose and response. What happens in the strain that apparently does not express cleft palate or that expresses cleft palate in all treated embryos when other strains express intermediate frequencies of response?

To be meaningful, comparisons between strains or genotypes should be based on prior knowledge of what is termed the 'norm of reaction'. It is an expression attributed by Mayer[32] to Woltereck in the word 'Reaktionsnorm', the standard mode of reaction that an individual inherits. The concept has been discussed most succinctly by Lewontin[33]. In genetics, the object for study is the relationship between genotype, environment and phenotype. The norm of reaction can be described as a 'table of correspondence between phenotype, on the one hand, and genotype-environment combinations on the other'[33].

To consider the concept of the norms of reaction with teratogen-induced malformations and strain differences in response, attention is focused on dose–response curves. Their importance is illustrated by comparing two sets of hypothetical response curves (Figure 6.1a, b) for two different genotypes, G1 and G2. The regression lines are drawn linearly. For the cleft palate response, the phenotype is not cleft palate *per se* but rather the frequency of the cleft palate response. One type of environment, or range of environments, may be the dose of teratogen.

In Figure 6.1a, one strain (G1) has a flat dose–response curve, another strain (G2) has a steep dose–response curve and the curves intersect. Therefore, depending on dosage, the strains will either have similar responses, measured by the frequency of cleft palate, or very different ones. This is an example of interaction between genotype and dosage of teratogen. If a single dose is used, the genetic interpretation of the difference in frequency of response will depend on the dosage chosen.

In Figure 6.1b, the dose–response curves are parallel. The difference between genotypes, measured on the phenotypic scale, is the same for all doses. Also, the difference between dosages is the same for all frequencies of cleft palate. Thus, there is additivity between genotype and dosage, measured

TERATOGEN-INDUCED CLEFT PALATE

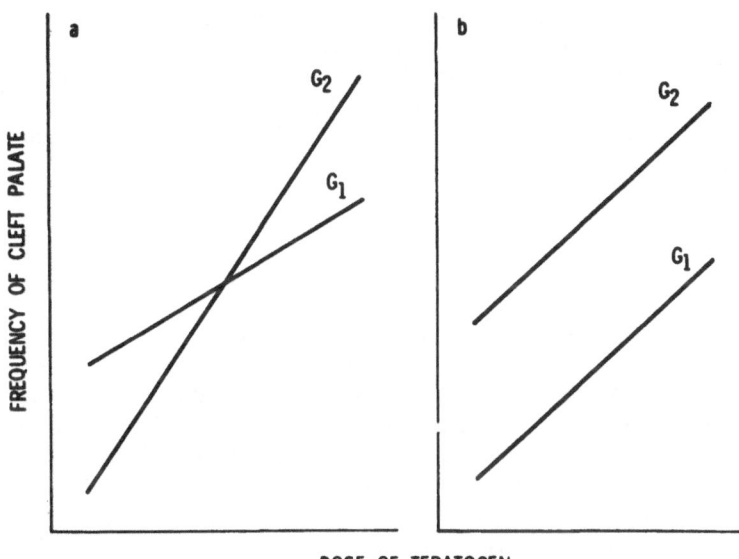

Figure 6.1   Two sets of hypothetical cleft palate dose–response curves for different genotypes (G1 and G2). a. The dose–response curves intersect and demonstrate interaction between genotype and teratogen. b. The dose–response curves are parallel and demonstrate an additive relationship between genotype and teratogen

by the phenotype. The importance of the information in Figures 6.1a and b and the distinction between the two will become clear after a discussion of appropriate dose–response models.

In both figures (6.1a and b), the cause of the difference between genotypes is the subject for genetic study. Is the difference due to one gene with major effect or to many genes with small effects? The approach to the genetic study and our interpretation of it will depend entirely on the observed dose–response behaviour.

Figures 6.1a and b illustrate another point. If the dose–response curves for different genotypes are as overtly different as in Figure 6.1a, the interaction between genotype and teratogen is sufficient grounds to say the genotypes arrive at the cleft palate phenotype by different mechanisms. Distinction is made here, as elsewhere in this review, concerning the concept of gene–teratogen interaction. The appropriate usage seems to be the one familiar in statistics 'where the effects of two treatments applied together cannot be predicted from the average responses of the separate factors'[34]. This is quite different from several proposed examples of gene–teratogen interaction that are conceptually 'additive' (e.g. review and comments by Fraser[6]).

Dose–response models and transformations

In keeping with the discussion of 'norms of reaction' and the question 'what is a teratogenic response?', it is necessary to consider *appropriate* dose–response models and *appropriate* transformations of dose–response data. The word appropriate is used intentionally; the experimenter must judge (and justify) the models and transformations. At least two general criteria seem important: (1) the model and associated transformation for statistical evaluation should be compatible not only with the experimental data but also with the biology, and (2) the model should permit further hypotheses (and their testing) concerning the nature of the teratogenic response.

Probit dose–response model

The probit dose–response model was used to re-investigate mouse strain differences in frequency of cleft palate response to cortisone and 6-aminonicotinamide. The assumption of the probit model is that the underlying trait is dosage tolerance rather than the binomial one of cleft palate (presence and absence). Tolerance is assumed to be continuously and normally distributed on a log-dose scale[35, 36]. A further assumption is that the embryo is the responding unit (each embryo has an independent chance of responding with cleft palate). The variance of the tolerance distribution is due to random errors: for example, not every embryo is precisely at the same stage of morphological and physiological development; the amount of teratogen is subject to errors in formulation and titration.

Probit analysis was developed by C. I. Bliss to summarize dosage–mortality data in toxicology[37, 38] and its use in other fields has been traced historically by Finney[36]. It has been used in teratological studies to describe strain differences in frequency of malformation and mortality responses[39], but little attention has been paid to it in experimental teratology until recently[40–42].

A somewhat fanciful description will serve to illustrate the principles and assumptions of the probit model and, perhaps, remove some of the mystery for those not familiar with it. Consider the ideal mouse embryo that can be titrated with a teratogen to the point just sufficient for it to express cleft palate. Not every embryo will express cleft palate at exactly the same dose and, if cleft palate is a typical biological response, the frequency distribution of the response when plotted on a linear dosage scale will be skewed to the right (Figure 6.2a). If dosage is transformed to a log scale, the frequency

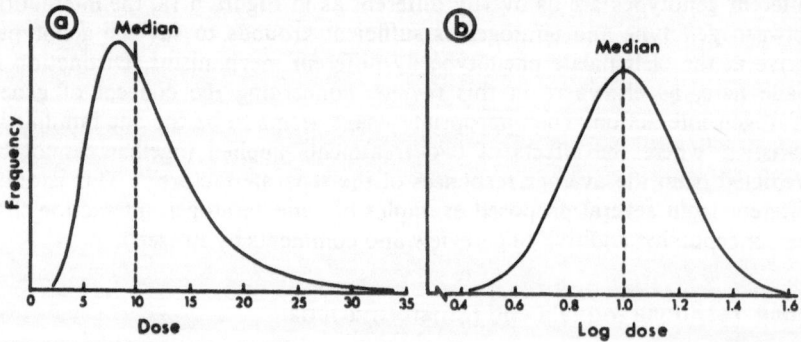

Figure 6.2 Frequency distribution of the cleft palate tolerance of the hypothetical population of mouse embryos that is titrated with a teratogen. a. Frequency is plotted against linear dose. b. Frequency is plotted against the logarithm of dose

distribution of the embryos that have responded will be normally distributed about the median dose for the response (Figure 6.2b). At the mean dose, if the frequency distribution is symmetrical, 50% of the embryos will have responded and 50% will not have responded.

If the frequencies of embryos that have responded are cumulated across the dosage scale and plotted against dosage, the normal sigmoid curve results (Figure 6.3a). The inflexion point will coincide with the dosage by which 50% of the embryos have responded. Instead of using a scale that is linear in percent, cumulative frequency can be plotted on a scale that is linear in probits (Figure 6.3b). The cumulative frequency of response is now linear on log dosage.

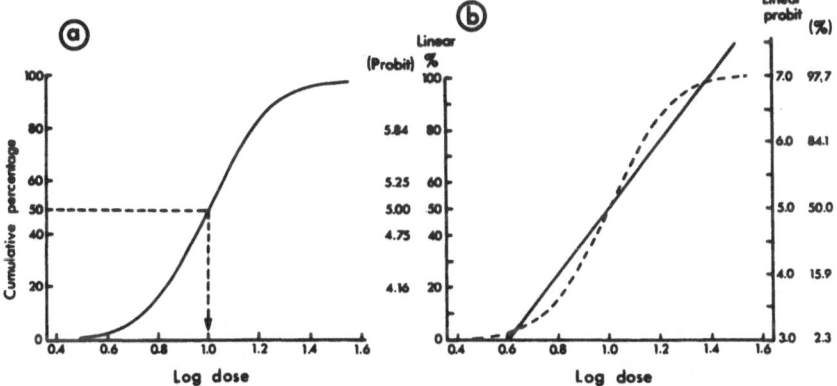

Figure 6.3   a. Cumulative frequency of embryos that have responded with cleft palate is plotted against log dose. The arrow indicates the dose below which half the population has responded (median effective dose). b. The cumulative frequency is transformed to a scale that is linear in probits

The probit scale is derived from the standard normal distribution with a mean of zero and unit standard deviation (unit variance). Cumulative frequency (the proportion of the area under the curve) is expressed in terms of units of standard deviation and an arbitrary value of 5 is added to the normal equivalent deviate scale to convert all practical values to positive numbers. For example: at probit value of 4.0 (−1.0 standard deviation), 15.87% of the population has responded; at 5.0 (0 standard deviation), 50% of the population has responded. The values equivalent to percents have been tabulated[36] or can be derived quickly from tables of areas under the standard normal curve[34, 43].

Once the cumulative frequency of response versus dosage has been linearized, the mean dosage for the cleft palate response can be estimated by regression analysis with maximum likelihood estimation[36] or graphical approximation[44]. A number of doses (more than two because two degrees of freedom are used in regression estimates), below and above the expected mean dose, can be picked, groups of embryos can be treated with different single doses, and the frequency of responders in each dosage group can be recorded. If the frequency of responders (on a probit scale) can be fitted to linear regression on dosage (log scale), the mean dose for the response (median effective dose) can be estimated. Also, if this can be done, there is no evidence with which to reject the dosage tolerance model and the validity of its assumptions. The slope of the log-probit dose–response curve is a measure of the variance (spread) of the median effective dose; however, more will be said about this when different teratogens that induce cleft palate are considered.

A systematic departure from linear regression will indicate that the log transformation of dosage may not be appropriate. Another transformation of dosage can be tried. Non-random departures from linearity (heterogeneity) will suggest that the responding units (the embryos) may not be independent. Several alternatives seem possible: the experiment can be repeated; the embryos can be examined in the natural grouping of litters with parametric[45] and non-parametric[46, 47] tests if a monotic increasing frequency of response to dosage is being tested; different dose–response models can be explored if a tolerance model is being sought.

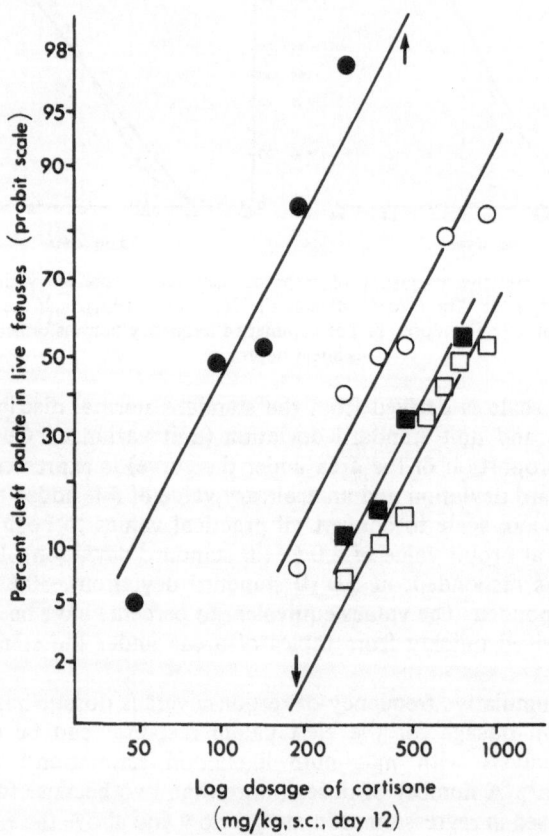

Figure 6.4 Probit regression of the cleft palate response to cortisone in the A (●), B6 (■), A.B6 F_1 (○) and B6.A F_1 (□) embryos. The dose–response curves were fitted to a common slope. By convention, the arrows indicate doses resulting in either 0 or 100% response (infinite empirical probits). (Reprinted with permission from *Genetics*[40])

CLEFT PALATE INDUCED BY CORTISONE AND 6-AN

Dose–response comparisons with cortisone

The cleft palate response to dosage of cortisone acetate was determined for the A/J (A) and C57BL/6J (B6) strains, the F_1 embryos (A.B6 F_1 and B6.A F_1) from reciprocal crosses, and the SWV strain[40]. (By convention, the maternal genotype is written first in hybrid abbreviations.) Cortisone was administered subcutaneously on day 12 of gestation that was determined to be the day of maximum response of A/J embryos to a single dose of cortisone. The day of finding the copulation plug is defined as day 0.

Figures 6.4 and 6.8 summarize the results of the probit analyses. The A, B6, A.B6 F_1 and B6.A F_1 embryos differed significantly in median effective dose (log ED_{50}); SWV and A were similar. The slopes of the regressions for the five genotypes did not differ significantly from parallelism.

The conceptual triad of 'stimulus, response and reactivity', discussed extensively by Mayer[32] and used widely in the biological and physical sciences can be employed. The genotypes that were examined do not differ in response to the cortisone stimulus because the slopes of the response curves were parallel. They differ in reactivity (also referred to as tolerance or dosage tolerance) or the amount of cortisone that is required to induce cleft palate. This suggests that the mechanism of cleft palate response to cortisone may be the same in all genotypes that were examined.

Cleft palate reactivities to other glucocorticoids

The teratology literature was searched for cleft palate dose–response data for other glucocorticoids and other genotypes of the mouse. The apparent fit (by eye) of the limited data to the slope of the cortisone cleft palate response curve was very surprising.

Figures 6.5a–c compare two different glucocorticoids that were reported to have been administered in the same way in a single test strain. Note that different dosage schedules were used in each figure. The differences between glucocorticoids appear to be in relative effective doses and not in the cleft palate response. Cleft palate induction may be another useful biological assay for relative potency of glucocorticoids.

Figure 6.5d presents the cleft palate responses to cortisone of C3H, A.C3H F_1 and C3H.A F_1 embryos that were reported by two different laboratories[13, 14]. The A strain (A/Strong[14] and A/J[13]) exhibited 100% cleft palate at both doses. As discussed previously, the maternal effect of A relative to C3H is in the opposite direction to the maternal effect of A relative to C57BL6. Twice as much cortisone is required to induce cleft palate in an F_1 embryo in B6 dams relative to A dams (Figure 6.4), but only half as much cortisone is required in C3H dams relative to A (estimated from ED_{50}s in Figure 6.5d). A search for major gene effects on maternal metabolism of cortisone[48] which may underlie the maternal effect must account for this limit of effects.

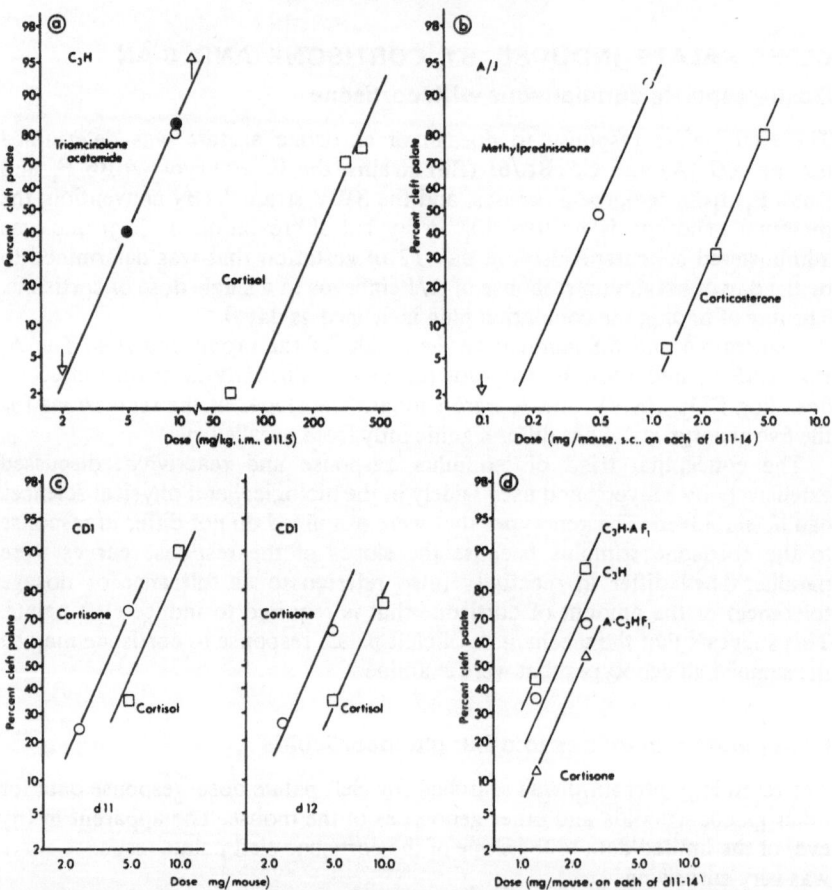

Figure 6.5 Comparison of cleft palate reactivities of different strains of mice to different glucocorticoids. a. C3H was treated with triamcinolone acetomide (O, ●) and cortisol (□); open symbols[92]; closed symbols[48]. b. A/J was treated with methylprednisolone[93] (O) and corticosterone[94] (□). c. CD1 was treated with cortisone[95] (O) and cortisol[96] (□) on either day 11 or 12. d. C3H (O), C3H.A F_1 (□) and A.C3H F_1 (△) were treated with cortisone (1.25 mg dosage[14] and 2.5 mg dosage[13]). (Reprinted with permission from *Genetics*[40])

Genetic model for reactivity differences to cortisone

The genetic model for the difference in embryonic tolerance of the A and B6 strains to cortisone—induced cleft palate is illustrated in Figure 6.6. 'Cortisone tolerance' (median effective dose for cleft palate) is plotted against embryonic genotype as a proportion of A-strain genes. Since the dose—response curves did not differ from parallelism, there was no evidence for interaction and additivity between genotype and teratogen can be assumed. The maternal effect, determined by the relative difference between the ED_{50}'s of the reciprocal F_1 embryos can be subtracted from the B6 strain to give an adjusted B6 mean tolerance. The difference between the adjusted B6 mean and the A-strain mean represents the genetic range between the two strains without the

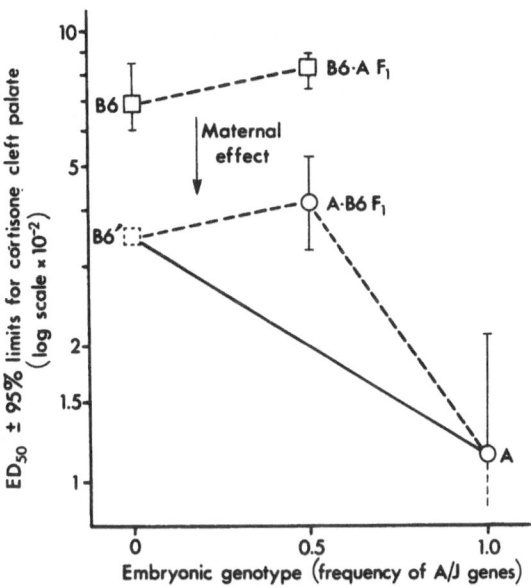

Figure 6.6 Genetic model to interpret the difference between the A and B6 embryos to cortisone-induced cleft palate. Embryonic tolerance to cortisone (ED_{50}) is plotted against embryonic genotype. The B6 tolerance is corrected for maternal effect (B6′) and exhibits dominance to that of A. (Reprinted with permission from *Genetics*[40])

maternal effect. The solid line is used to define the expected values for hybrid genotypes based on an additive genetic model; that is, the alleles at each locus, regardless of number of loci, have an equal and additive effect to shift the mean tolerance between the B6 and A-strain limits. For example, the F_1 embryo has half the A-strain genes and, with an additive genetic model, would be expected to have an intermediate tolerance to cortisone. However, mean tolerance of the F_1 embryo is significantly greater than expected with dominance in the direction of B6. There may be overdominance since the B6.A F_1 mean was significantly greater than B6 (unadjusted).

Dose–response comparisons with 6-AN

A complete dose–response study of 6-AN-induced cleft palate was not done for the A and B6 strains and the reciprocal F_1 embryos[49]. Analysis of data for A and B6, kindly supplied by Dr A. C. Verrusio[30], suggested that the two strains differed only in reactivity (Figure 6.7a).

There was a departure from a normal tolerance distribution in the lower dose range (departure from linear log-probit regression in Figure 6.7a). It is unexplainable in view of its absence when the A strain was re-examined (Figure 6.7b) and other genotypes were tested (AB6.A BC_1 and B6A.A BC_1, Figure 6.7b; SWV strain, Figure 6.8).

The bulk of the dose–response data for 6-AN-induced cleft palate suggested that, for the genotypes of embryos that were examined, genetic differences most likely changed only the dosage tolerance. Therefore, a genetic

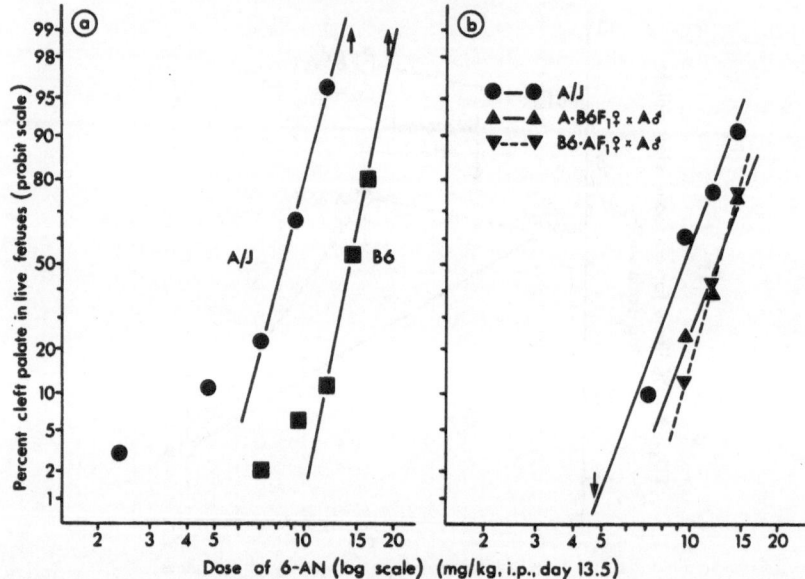

Figure 6.7 Probit regression of the cleft palate response to 6-AN. a. Data from the A and B6 strain responses are from Verrusio[30] and are fitted to separate linear probit regressions for doses above the apparent departures from linear regression. b. Data from the A strain and the AB6.A BC$_1$ and B6A.A BC$_1$ embryos are fitted to separate linear probit regressions. (Reprinted with permission from *Teratology*[49])

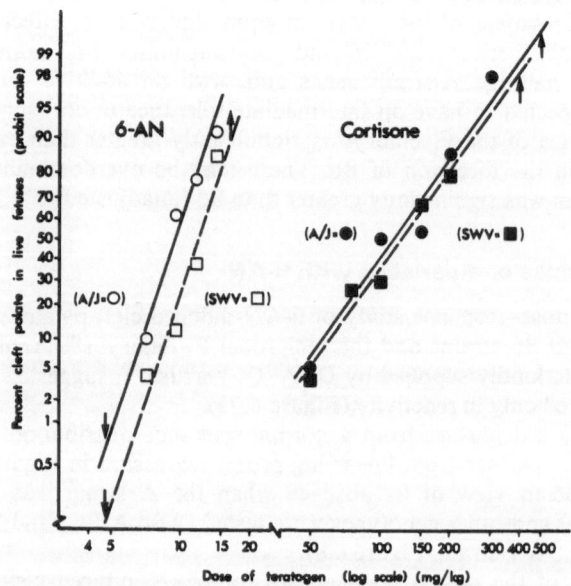

Figure 6.8 Comparison of the cleft palate dose–response curves for cortisone and 6-AN with the A and SWV strains. (Reprinted with permission from *Teratology*[51])

study of embryonic reactivity to 6-AN-induced cleft palate could be undertaken with a single dose of 6-AN. The cleft palate frequency difference could be used to follow segregation of genes for reactivity and to estimate the number of loci involved.

Comparison between cortisone and 6-AN traits

It was immediately obvious that the slopes for the families of cleft palate dose–response curves for cortisone and for 6-AN were significantly different. Figure 6.8 compares the two compounds with the A and SWV strains[50, 51]. Returning again to the conceptual triad of 'stimulus, response and reactivity', the difference in slopes indicates that the response to the two compounds is different. This appears to be the logical way to demonstrate that the mechanisms of the cleft palate response to the two teratogens are different without resorting to secondary consequences of teratogen treatment.

GENETIC STUDIES

Breeding scheme

The breeding scheme that was used to search for the genetic control of the embryonic reactivity difference to cortisone and 6-AN is outlined in Figure 6.9. It has been used to investigate genetic differences in response to other

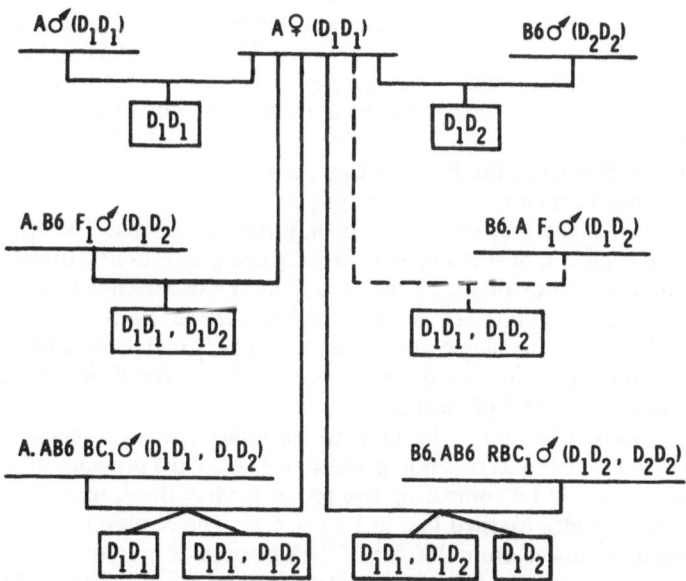

Figure 6.9 Test mating scheme used to search for the genetic control of the difference in embryonic reactivity to cortisone- and 6-AN-induced cleft palate. The segregation of hypothetical alleles D_1 (A strain) and D_2 (B6 strain) at a single autosomal locus is superimposed on the mating scheme and demonstrates that segregation of embryonic reactivities occurs with test-mating progeny of F_1 sires but is not detected until the test-mating of backcross sires. (Modified with permission from *Genetics*[55])

teratogens[18, 52]. For the cortisone and 6-AN traits, it permitted not only a test of whether single gene differences were involved in each trait but also a test for genetic independence of the two traits.

To interpret the scheme, suppose that the cortisone trait is controlled by a single, autosomal gene difference. At the hypothetical locus, D, the A strain has allele D_1 and B6 has allele D_2. The segregation of the alleles can be followed in Figure 6.9. Since the cortisone trait of B6 was dominant to that of the A strain (Figure 6.6), test matings were made with the A-strain female to recover gene(s) for the recessive A-strain reactivity.

In Figure 6.9, A.B6 F_1, A.AB6 BC_1, and B6.AB6 RBC_1 males were collected without teratogen treatment and, together with A and B6 males, they were each test mated with several A-strain females. The test-mated females were treated with cortisone and the cleft palate frequency in live embryos (without cleft lip) was tabulated for each male. (The A strain has a spontaneous cleft lip trait, approximately 8% of term embryos, that may have an associated cleft palate. By convention, embryos with cleft lip, regardless of palate condition, were excluded from the data.) The test mating scores (sire scores) defined the phenotype of each sire for the embryonic trait of cortisone reactivity. The single maternal environment of the A-strain female allowed the maternal genetic variation in this trait to be ignored.

Segregation of the genes for embryonic reactivity would be expected in the test-mating progeny for the F_1 males. However, these embryos are effectively in a single litter and all that would be observed is an increase in cleft palate frequency above that of the B6 males and tending towards that of A males. The BC_1 males should be of two genotypes (D_1D_1 or D_1D_2) and their test-mating scores will provide evidence for segregation of the difference in reactivity. If the trait is controlled by a single gene difference, half the BC_1 sires should have test-mating scores like F_1 males and half should be like A strain males. Similarly, the RBC_1 males should be of two types: half like B6 males, half like F_1 males.

Bimodality of the BC_1 sire scores is only part of the test. The critical test for the single-gene model is beyond the BC_1 sire generation[53]. Bimodality of the test-mating scores of BC_2 sires as well as the occurrence of segregating and non-segregating families of BC_2 males, derived from individual BC_1 males, will confirm the single gene model. The BC_2 males must be obtained from a random sampling of the BC_1 sires in order for them to provide a second, independent test of the model[54].

The breeding scheme can be used to test other specified genetic models, such as a two-gene model or a polygenic model. Also, the presence of X-linked genes can be sought by comparing sire scores from A.B6 F_1 males with those from B6.A F_1 males (dashed line in Figure 6.9). The X chromosomes of the two types of F_1 males are different: in the one case, it is from the A female; in the other, it is from the B6 female. In contrast, the autosomes of the two types of F_1 males are identical (Figure 6.9). The test for X-linked genes is made by finding a difference between the reactivity of female test-mating progeny from the F_1 sires and no difference between the male progeny.

Cortisone trait

The test mating scores for the cortisone trait[55] are displayed in Figure 6.10. The mean score for each generation is plotted against genotype as a proportion of A-strain genes. Since the embryonic tolerance to cortisone of B6 embryos was dominant to that of A, the fit of the means of the RBC_1, F_1 and BC_1 sires to the regression line suggests there is no interaction (epistasis) between loci if more than one locus is involved.

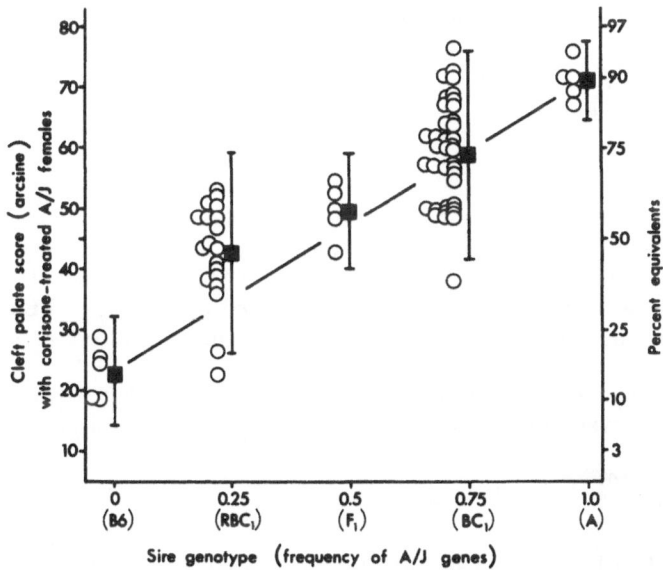

Figure 6.10   Regression of mean sire scores (solid squares, arcsin, ± 2 s.d.) for cortisone cleft palate on sire genotype as frequency of A-strain genes. The individual sire scores in each generation are plotted as open circles. The percent scale, equivalent to the arcsin transformation, is shown. (Reprinted with permission from *Genetics*[55])

The distributions of the BC_1 and RBC_1 sire scores are not bimodal. The single gene model can be rejected. As expected, the single-gene model was rejected also by the test matings of BC_2 sires[55]. The distributions of the BC_1 and RBC_1 sires are not single normal distributions and, therefore, a polygenic model of many genes with equal and additive effects can be rejected also.

A search of the BC_1 and RBC_1 sire distributions suggested that two or three loci with independent effects may explain the breeding data. However, at this point, a new breeding study must be undertaken with a design appropriate for testing these models.

No evidence for X-linked genes with major effect on embryonic reactivity to cortisone was found with the test matings of the reciprocal F_1 sires. Contrary to a previous suggestion of X linkage[16], no support for the hypothesis was found in this study.

Other genetic differences between A and B6 were also segregating in the BC_1 and RBC_1 sires. Of the three traits that were measured and tested (albino, *c*; brown, *b*; histocompatibility-2, *H*-2), association between cortisone re-activity and *H*-2 was suggested. Further study is required to distinguish between linkage of the cortisone trait to *H*-2 and a pleiotropic effect of *H*-2 on cortisone reactivity.

The association between *H*-2 and a component of the embryonic cortisone trait is intriguing because the maternal effect trait on cortisone reactivity is also associated with *H*-2[56]. A gene acting in the embryo on cortisone re-activity may function later in the adult female. Strain differences in the kinetics of glucocorticoid-binding proteins may play a role in the cleft palate reactivity trait[57] and the variants of the binding proteins may be associated with *H*-2[58]. However, since a single-gene model could not explain the embry-onic reactivity trait, more than glucocorticoid-binding variants are involved in the strain difference.

6-AN trait

The genetic study of the reactivity difference between A and B6 embryos to 6-AN-induced cleft palate was done parallel to the cortisone study[49]. The response of reciprocal F_1 embryos did not suggest a significant maternal effect and, when the F_1 embryos were compared with the A and B6 strains, there was only a slight indication of dominance deviation from additivity in the direction of the B6 genotype.

The absence of a significant maternal effect on the cleft palate response when 6-AN was used alone is in direct contrast to the observations when

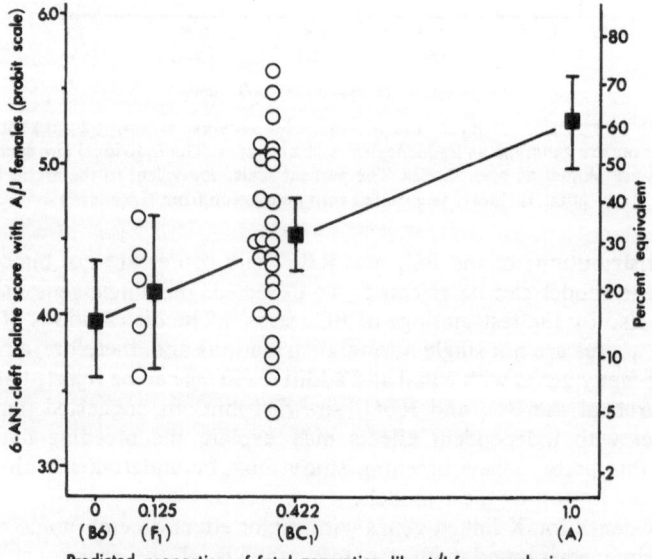

Figure 6.11 Regression of mean sire scores (solid squares, ±95% limits) for 6-AN cleft palate on the proportion of test-mating progeny that are genetically like the A strain with the three-locus epistatic model. The individual A.B6 F_1 and BC_1 sire scores are plotted as open circles. (Reprinted with permission from *Teratology*[51])

6-AN treatment is followed by a 'protective' dose of nicotinamide[27]. Is the maternal effect present with nicotinamide and absent when 6-AN is used alone? It may be that the maternal effect is a variant in response to nicotinamide rather than to 6-AN. A dose–response study of this trait in reciprocal F_1 embryos (between A and B6) may be useful, but, in contrast to other dose–response studies of 6-AN and nicotinamide[30], the effect of a series of specified doses of nicotinamide on the ED_{50} for 6-AN-induced cleft palate should be determined. The functional relationship between nicotinamide, 6-AN and genotype needs to be established in a manner similar to that for toxicity[59].

An extensive test-mating study of several A and B6 males with A females was not done. Instead, the test-mating scores of A.B6 F_1 and A.AB6 BC_1 males were compared with the mean values for A and B6 males mated with A females. The F_1 and BC_1 sire means differed from the expected values with either a single locus model or more than one locus but with no epistasis (interaction between loci). A search of some simple epistatic models was made[49]. The F_1 and BC_1 sire means could be fitted to a three-locus genetic model and the sire scores are plotted against embryonic genotype from this model in Figure 6.11.

Tests for association between the 6-AN cleft-palate scores of the BC_1 sires and the albino (c) brown (b) and H-2 marker genes were made. In contrast to the findings with the cortisone trait, there was no association between 6-AN reactivity and H-2. There was an indication of association with brown (b) in the direction suggesting either linkage or pleiotropy.

Test for association between the cortisone and 6-AN traits

The association of H-2 with the embryonic cortisone trait but not with the 6-AN trait was a prelude to the direct test between the cortisone and 6-AN traits. For the A.AB6 BC_1 sires that had test-mating scores for both cortisone and 6-AN, there was no evidence for association between the two teratogenic reactivities[60]. Two types of statistical evaluation were made with the cortisone and 6-AN sire scores; a rank-ordered correlation (Spearman rank correlation[61]) that is appropriate if the same single gene controlled both reactivities; a variation of the contingency test[62] that is appropriate for other genetic models in which clustering but no genetic association between the two traits is expected. It appears that the reactivity differences between the A and B6 strains to the two teratogens, both measured by cleft palate, are due to two, separate genetic systems.

MULTIFACTORIAL/THRESHOLD MODEL

The demonstration that the embryonic responses of cortisone- and 6-AN-induced cleft palate occur by different mechanisms and that the embryonic cleft palate reactivities (differences in tolerance) to the two teratogens are genetically independent may appear, at first glance, to reject the multifactorial/threshold model[22] that has been proposed to explain the strain difference between A and B6 mice in cleft palate induction by many teratogens. This is not the case.

Palate closure has been expressed in terms of a normalized frequency distribution over gestational age[63]. The rate of palate closure and the mean times for different stages of palate closure[24] to be reached can be quantified with a simple biometrical model. The rate of palate closure in the A and B6 strains appeared to be the same, but the mean times of closure were significantly different. Cortisone appeared to delay this mean time of closure by a constant amount in several different genotypes[64].

Is there an association between genetic differences in reactivity (tolerance) to teratogens, the effect of the teratogens on the delay of palate closure, and the genetic differences in mean closure time during normal palate development (in the absence of teratogens)? The test of the threshold model for the cleft palate trait seems now to be a genetic one.

Since palate development, like any other morphogenetic process, is envisaged as the interplay between interrelated and interdependent events, the following scenario can be constructed. When the traits of teratogenic reactivity and mean palate closure times are allowed to segregate, low and high reactivity and early and late times of palate closure may be genetically independent. The metabolic differences that may determine a low tolerance and possibly a large delay in palate development for a specific teratogen may be found in an embryo that genetically closed its palate either early or late. Therefore, the effects of the teratogen would be superimposed on this normal developmental pattern.

To permit this to be seen, comparisons must move beyond the two strains of A and B6. An examination of segregating generations is possible but this has a biological limitation. Sample sizes needed to make all the desired tests for association may exceed the reproductive capacity of the mouse[60]. Alternatively, fuller use of the genetic resources of the mouse could be made, beyond two strain comparisons. The more variables that are being correlated in two fixed genotypes, the greater is the probability that some of them will be associated by chance rather than by cause. (Similar discussions have been presented concerning physiological genetics under the heading 'uses and abuses of genetic variation'[65].)

GENETIC RESOURCES IN THE MOUSE

A greater sampling of genotypic combinations, at least the viable combinations, will be made when more inbred strains are examined. Three genetic questions could be asked of a survey for cortisone-induced cleft palate:

(1) Do all strains respond with cleft palate and with the same dose—response characteristics? A difference in the expected slope would suggest a genetic interaction (a 'mutation' in the teratogenic response?).

(2) If all strains respond with the same slope and differ only in ED_{50}, how many different ED_{50}'s are there? If the ED_{50}'s are continuously distributed over a large dose range, reactivity differences may be controlled by many gene loci or possibly a multiple allelic series at relative few loci. However, if the range of ED_{50}'s is described by only a few discrete values, a small number of loci, possibly with few alleles may determine the reactivity differences.

(3) Once the rank ordering of strains for cortisone reactivity has been established, tests can be made for association of this trait with other cleft palate teratogens and normal physiological or morphogenetic parameters. The hypotheses concerning the mechanism of the strain differences could be tested.

Genetic relationships between inbred strains

The degree of genetic relationship was estimated among 27 common inbred strains for which pedigree relationships were largely unknown[66]. The strains were clustered into essentially three groups on the basis of genetic similarity at 16 polymorphic loci. For teratological purposes, the 16 loci can be assumed to be a relatively unbiased sampling of the genome and the three clusters of strains can be put to practical use.

If genetic differences in a teratogenic response are being sought, they will be found more likely between representatives of the three clusters of strains. If genetic variation is found only between the clusters, its cause will be due more likely to many gene differences. On the other hand, large differences in response between strains within a genetically similar cluster would suggest that relatively few genes control the difference. This approach was used to attempt a search for genetic differences in reactivity to 6-AN-induced cleft lip[67] and is being used to identify differences in reactivity to cadmium-induced ectrodactyly[68].

Recombinant inbred strains

The recombinant inbred (RI) strains were developed as a tool for immuno-genetics[69]. The RI strains are produced by crossing two highly inbred strains, raising an F_2 generation, and then deriving a series of inbred lines from F_2 pairs by sib matings for at least ten generations. Viable chance recombinations of genes that are present in the two parental strains will be fixed in the RI strains.

RI strains are useful for traits where the genetic analysis depends on repeated observations with many individuals. Conventional breeding tests with F_2 and BC progeny are not possible if each individual is tested destructively, and therefore only once. RI strains represent replicate populations of recombinant genotypes present in the two founding strains. If only two distributions of phenotypes for the trait in question are found with the RI strains, and correspond to those of the two parental strains, the trait is likely to be determined by a single gene (or closely linked genes). New phenotypic distributions in the RI strains will permit an estimate of the minimum number of genes involved in the trait.

RI strains can also be used to search for marker loci for linkage tests of new variants. The similarity of the distribution pattern of the new variant in the RI strains with the distribution of alleles at previously identified loci will suggest potential linkage.

Use of RI strains in experimental teratology has been advocated[12].

Congenic sublines

Congenic sublines were developed initially for immunogenetics[70–72]; many others are in existence. The term congenic indicates stocks that are produced by the introduction of a gene into an inbred strain by a series of backcrosses. The congenic stocks differ from the inbred background strains at different loci and, depending on the number of generations of backcrossing, they will also contain an indeterminant amount of genetic material that is linked to the specific loci. In contrast to this, coisogenic stocks arise by chance mutations within inbred strains. Use of the congenic sublines is expanding in many fields of biology[73].

Two congenic sublines of the A and C57BL/10 strains, containing H-2 antigenic haplotypes of the opposite strains, were used to examine the cortisone cleft palate response[56]. The maternal effect difference between A and C57BL/10 on the embryonic cleft palate response appeared to be associated with H-2 haplotype. A broader survey of the congenic sublines of the A and C57BL/10 strains, as well as those for other strains, seems warranted.

Known physiological variants

Modification of responses to specific teratogens can be predicted for some of the known genetic variations or mutations in metabolism in the mouse. For example, in the establishment of the mouse model for the fetal alcohol syndrome, strains of mice were picked that differed in alcohol dehydrogenase activity[74]. Association of the metabolic variation with any observed differences in teratogenic response would serve as immediate genetic tools to dissect the mechanism of the strain difference. Similarly, genetic variation in inducible enzyme activity that is controlled by the Ah locus was used to predict strain differences in the developmental toxicity of various polycyclic hydrocarbons[75]. The extensive endocrine variation in the mouse[65] has yet to be employed effectively in the study of glucocorticoid-induced cleft palate.

MODIFICATION OF THE CLEFT PALATE RESPONSE

Alteration of teratogenic responses by other chemical agents, diet, stress, and so forth has received considerable attention and has been reviewed recently[6,12]. For the cleft palate response to teratogens, the interactions between different genotypes and the modifying agents may be an important method to search out genetic variation in the cleft palate response. It would seem prudent, however, to have biometrical procedures with which to evaluate changes in response and biological models with which to interpret the changes.

Presently, the general approach in teratology seems akin to a 'black box'. A teratogen is administered and the 'response' is recorded as a frequency (or severity) of malformation, a biochemical variable or a change in an embryonic structure. The effects of other treatments are summarized by deviations in the 'response'.

There is a need to expand the 'black box' approach, but at the same time, provide a conceptual integration of the teratogenic endpoint with physiology and embryology. The approaches used in pharmacology to investigate the joint effects of drug mixtures[76] or the 'isobole' technique to distinguish between the additive and synergistic effects of joint treatments[77, 78] seem directly applicable to experimental teratology. The polychotomous response models used in ecology and epidemiology[79-81] or the method of ridit analysis[82, 83] can be used to summarize qualitative descriptions of embryological changes. In examining these approaches, a conclusion of a recent review of the biometrics of response surface methodology should be kept in mind that a '... field in which we believe valuable work needs to be done is the joint development by the biologist and the statistician of particular biologically reasonable models for particular practical research problems'[84].

Glucocorticoid-induced cleft palate

Diet is known to modify the cleft palate response to cortisone. The diet effects varied between different doses and different maternal weights in A/J and C57BL/6[85]. Two diets caused a consistent difference in the frequency of

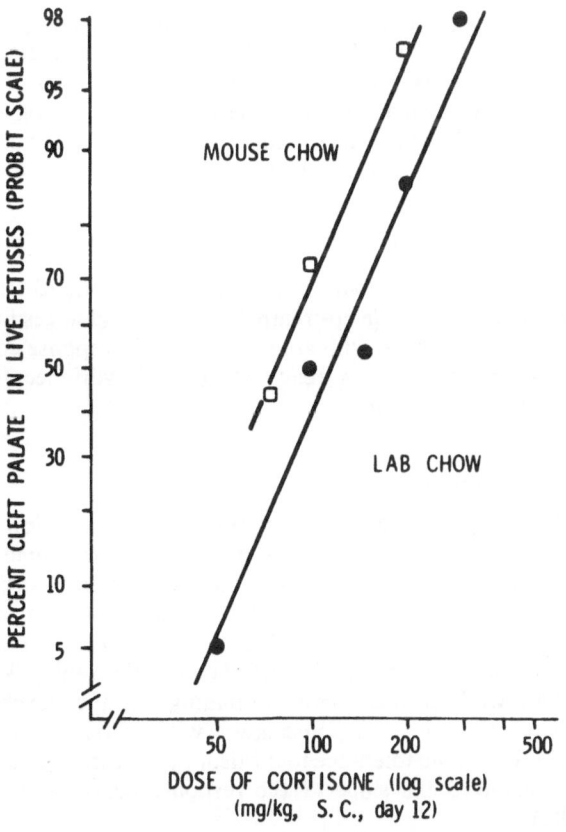

Figure 6.12 Dose–response curves for cortisone-induced cleft palate in the A strain on two diets. The regressions do not differ from parallelism and are fitted to a common slope

cortisone-induced cleft palate in C57BL/6J that were in either sham operated or control groups[86]. Pyridoxine-deficiency was suggested to act synergistically at low doses of cortisone but additively at high doses on the cleft palate response[87], but this conclusion may be an artefact created by the binomial variance in untransformed frequency data. After transforming the frequency data to a probit scale, the dose–response data[87] suggest that pyridoxine deficiency acts additively to reduce the tolerance to cortisone-induced cleft palate.

A change in the maintenance diet from Lab Chow to Mouse Chow was required to overcome breeding difficulties during part of the cortisone cleft-palate study[55]. A limited dose–response study for the A/J strain was made with Lab Chow and Mouse Chow (Figure 6.12). The ED_{50} for cortisone-induced cleft palate was reduced significantly from 115 mg/kg with Lab Chow to 79 mg/kg with Mouse Chow. It was assumed that the diet effect on other genotypes would be similarly additive; in a single-dose comparison with C57BL/6J, the reactivity increased by an approximately equal amount.

A recent study with two other strains, SW/Fr and CL/Fr, has revealed an intriguing interaction of genotype with diet for the cortisone cleft palate trait[88]. The slopes of the dose–response curves did not change with the two diets; instead, the reactivities (ED_{50}'s) of the two strains changed in opposite directions. This emphasizes that 'comparisons of norms of reaction in different strains should be made only if the tested animals are on the same diet' and that 'deductions about which components of the diet are important in altering a norm of reaction can be misleading if only one genotype has been tested'[88].

6-AN-induced cleft palate

The cleft palate response to 6-AN in the A/J and C57BL/6J strains is modified by diet or by nicotinamide co-treatment[28–31]. The changes in frequency of cleft palate need to be defined with appropriate dose–response comparisons to distinguish between changes in reactivity (ED_{50}), variance (slope) of the tolerance distributions or both.

Embryology of palate closure

Epidermal growth factor (EGF) stimulates the differentiation of epidermal structures and, because it increased the frequency of cortisone-induced cleft palate in A/J embryos, EGF was suggested to potentiate cortisone-induced cleft palate by acting 'synergistically' with it[89]. The histological effects of EGF and cortisone on the one hand and the changes in frequency of cleft palate on the other are difficult to conceptualize. EGF causes thickening (keratinization) of the epithelium while cortisone causes thinning. For the morphogenetic endpoint of cleft palate, a dose–response study with cortisone would determine whether EGF lowered the tolerance to cortisone (and therefore acts additively) or whether it changed the slope of the response curve (and therefore acts synergistically).

The dramatic changes in the histological structure of the palatal epithelium (thickness, cell size, nuclear density), induced by EGF and/or cortisone and presented in a series of light micrographs[89], suggest these variables may be quantified to establish a relationship with dosage. It might be possible, within a dose range that does not result in 100% cleft palate, to evaluate the relationships between treatments, histological changes and the morphogenetic endpoint of cleft palate. One approach may be similar to that used to quantitate the events of palate closure in terms of a simple biometrical model[63]. It has improved the understanding of genetic differences in the timing of embryonic palate closure and the effects of cleft palate-inducing teratogens on it[64, 90].

SUMMARY

With the genetic information that is available for either cortisone (and possibly all glucocorticoids) or 6-aminonicotinamide, cleft palate appears to be a normal reaction of mouse embryos. The genotypes that have been tested to date differ only in reactivity (the dosage that is required for the response). A broader survey of the genetic resources of the mouse and more attempts to modify the norm of reaction are required to obtain genetic tools to dissect the biology of the cleft palate response.

The demonstration that there may be simple genetic causes for the embryonic differences in cleft palate reactivities to cortisone and 6-aminonicotinamide must be interpreted cautiously. The differences are the net differences between only two genotypes. There may be different ones for other genotypic pairs.

If the net differences between genotypes in cleft palate reactivities is in fact small and therefore due to genes of major effect, there will be a chance to identify the functions of these genes at a level that is closer to the primary gene action. It is with this goal that genetic studies, similar to those described in this review, should be repeated and expanded.

In addition to the three principles of teratology of the timing of teratogenic treatment, the dosage and the genotype[91], a fourth concern about an appropriate dose–response model may be added. Dose–response models appear to be essential for the interpretation, comparison and further study, especially genetic study, of teratogenic responses. Little importance should be placed on the specific dose–response model, discussed in this chapter, for cortisone- and 6-aminonicotinamide-induced cleft palate unless it leads to a conceptual framework with which to better understand the biology of teratogen-induced cleft palate and to detect genetic variation in both reactivity and response.

ACKNOWLEDGEMENTS

During the writing of this chapter, the financial assistance of a postdoctoral fellowship from the Dean of Medicine, Memorial University, and Dr P. W. Allderdice is gratefully acknowledged. My association with Dr F. C. Fraser and other colleagues at McGill University, Dr D. M. Juriloff and Dr W. M. Layton aided in the formulation of the experimental designs and interpretations of data. The secretarial assistance of Judy Power and the Medical Audio-Visual Department, Memorial University, helped to complete the chapter.

References

1. Postlethwait, J. H. and Schneiderman, H. A. (1973). Developmental genetics of *Drosophila* imaginal discs. *Annu. Rev. Genet.*, 7, 381
2. Cumming, R. B. (1978). The genetics of formamidase 5 (brain formamidase) in the mouse. *Genetics*, 88, S17 (abstract)
3. Russell, L. B., Russell, W. L., Popp, R. A., Vaughan, C. and Jacobson, K. B. (1976). Radiation-induced mutations at mouse hemoglobin loci. *Proc. Nat. Acad. Sci. USA*, 73, 2843
4. Staats, J. (1976). Standardized nomenclature for inbred strains of mice: sixth listing. *Cancer Res.*, 36, 4333
5. Fraser, F. C. (1969). Gene–environment interactions in the production of cleft palate. In: H. Nishimura and J. R. Miller (eds.). *Methods for Teratological Studies in Experimental Animals and Man*, pp. 34–49. (Tokyo: Igaku Shoin)
6. Fraser, F. C. (1977b). Interactions and multiple causes. In: J. G. Wilson and F. C. Fraser (eds.). *Handbook of Teratology*, Vol. 1, pp. 445–463 (New York: Plenum Press)
7. Burdi, A., Feingold, M., Larsson, K. S., Leck, I., Zimmerman, E. F. and Fraser, F. C. (1972). Etiology and pathogenesis of congenital cleft lip and cleft palate, an NIDR state of the art report. *Teratology*, 6, 255
8. Greene, R. M. and Kochhar, D. M. (1975). Some aspects of corticosteroid-induced cleft palate: a review. *Teratology*, 11, 47
9. Baxter, H. and Fraser, F. C. (1950). The production of congenital defects in the offspring of female mice treated with cortisone. A preliminary report. *McGill Med. J.*, 19, 245
10. Fraser, F. C. and Fainstat, T. D. (1951). Production of congenital defects in the offspring of pregnant mice treated with cortisone. *Pediatrics*, 8, 527
11. Kalter, H. (1954). The inheritance of susceptibility to the teratogenic action of cortisone in mice. *Genetics*, 39, 185
12. Biddle, F. G. and Fraser, F. C. (1977b). Maternal and cytoplasmic effects in experimental teratology. In: J. G. Wilson and F. C. Fraser (eds.). *Handbook of Teratology*, Vol. 3, pp. 3–33. (New York: Plenum Press)
13. Kalter, H. (1965). Interplay of intrinsic and extrinsic factors. In: J. G. Wilson and J. Warkany (eds.). *Teratology. Principles and Techniques*, pp. 57–80. (Chicago: University of Chicago Press)
14. Loevy, H. (1963). Genetic influences on induced cleft palate in different strains of mice. *Anat. Rec.*, 145, 117
15. Loevy, H. (1968). Cortisone-induced teratogenic effects in mice. *Proc. Soc. Exp. Biol. Med.*, 128, 841
16. Francis, B. M. (1973). Influence of sex-linked genes on embryonic sensitivity to cortisone in three strains of mice. *Teratology*, 7, 119
17. Dagg, C. P. (1963). The interaction of environmental stimuli and inherited susceptibility to congenital deformity. *Am. Zool.*, 3, 223
18. Dagg, C. P., Schlager, G. and Doerr, A. (1966). Polygenic control of the teratogenicity of 5-fluorouracil in mice. *Genetics*, 53, 1101
19. Dagg, C. P. (1966). Teratogenesis. In: E. L. Green (ed.). *Biology of the Laboratory Mouse*, pp. 309–328. (New York: McGraw-Hill)
20. Deuschle, F. M. and Kalter, H. (1962). Observations on the mandible in association with defects of the lip and palate. *J. Dent. Res.*, 41, 1085
21. Shepard, T. H. (1976). *Catalog of Teratogenic Agents*. 2nd Ed. (Baltimore: The Johns Hopkins University Press)
22. Fraser, F. C. (1976). The multifactorial/threshold concept – uses and misuses. *Teratology*, 14, 267
23. Fraser, F. C. (1977a). Relation of animal studies to the problem in man. In: J. G. Wilson and F. C. Fraser (eds.). *Handbook of Teratology*, Vol. 1, pp. 75–96. (New York: Plenum Press)
24. Walker, B. E. and Fraser, F. C. (1956). Closure of the secondary palate in three strains of mice. *J. Embryol. Exp. Morphol.*, 4, 176
25. Walker, B. E. and Fraser, F. C. (1957). The embryology of cortisone-induced cleft palate. *J. Embryol. Exp. Morphol.*, 5, 201

26. Biddle, F. G. and Fraser, F. C. (1974). Are the genes determining susceptibility tc cortisone and 6-aminonicotinamide induced cleft palate the same or different? *Teratology*, 9, A14 (abstract)
27. Goldstein, M., Pinsky, M. F. and Fraser, F. C. (1963). Genetically determined organ specific responses to the teratogenic action of 6-aminonicotinamide in the mouse. *Genet. Res.*, 4, 258
28. Pollard, D. R. and Fraser, F. C. (1968). Further studies on a cytoplasmically transmitted difference in response to the teratogen 6-aminonicotinamide. *Teratology*, 1, 335
29. Pollard, D. R. and Fraser, F. C. (1973). Induction of a cytoplasmic factor increasing resistance to the teratogenic effect of 6-aminonicotinamide in mice. *Teratology*, 7, 267
30. Verrusio, A. C. (1966). Biochemical basis for a genetically determined difference in response to the teratogenic effects of 6-aminonicotinamide. Ph.D. Thesis. (Montreal: McGill University Library)
31. Verrusio, A. C., Pollard, D. R. and Fraser, F. C. (1968). A cytoplasmically transmitted, diet-dependent difference in response to the teratogenic effects of 6-aminonicotinamide. *Science*, 160, 206
32. Mayer, E. (1963). *Introduction to Dynamic Morphology*, p. 33. (New York: Academic Press)
33. Lewontin, R. C. (1974). Annotation: the analysis of variance and the analysis of causes. *Am. J. Hum. Genet.*, 26, 400
34. Sokal, R. R. and Rohlf, F. J. (1969). *Biometry. The Principles and Practice of Statistics in Biological Research*. (San Francisco: W. H. Freeman Co.)
35. Bliss, C. I. (1957). Some principles of bioassay. *Am. Sci.*, 45, 449
36. Finney, D. J. (1971). *Probit Analysis*. 3rd Ed. (London: Cambridge University Press)
37. Bliss, C. I. (1935a). The calculation of the dosage–mortality curve. *Ann. Appl. Biol.*, 22, 134
38. Bliss, C. I. (1935b). The comparison of dosage–mortality data. *Ann. Appl. Biol.*, 22, 307
39. Landauer, W. and Bliss, C. I. (1946). Insulin-induced rumplessness of chickens. III. The relationship of dosage and of developmental state at time of injection to response. *J. Exp. Zool.*, 102, 1
40. Biddle, F. G. and Fraser, F. C. (1976). Genetics of cortisone-induced cleft palate in the mouse – embryonic and maternal effects. *Genetics*, 84, 743
41. National Center for Toxicological Research. (1975). *Scientific Report*. (Arkansas: Jefferson)
42. National Center for Toxicological Research. (1977). *5th Anniversary Report*. (Arkansas: Jefferson)
43. Snedecor, G. W. and Cochran, W. G. (1967). *Statistical Methods*. (Ames: Iowa State University Press)
44. Litchfield, J. T. and Wilcoxon, F. (1949). A simplified method of evaluating dose–effect experiments. *J. Pharmacol. Exp. Ther.*, 96, 99
45. Williams, D. A. (1975). The analysis of binary responses from toxicological experiments involving reproduction and teratogenicity. *Biometrics*, 31, 949
46. Lin, F. O. and Haseman, J. K. (1976). A modified Jonckheere test against ordered alternatives when ties are present at single extreme value. *Biom. Z.*, 18, 623
47. Shirley, E. (1977). A non-parametric equivalent of Williams' test for contrasting increasing dose levels of a treatment. *Biometrics*, 33, 386
48. Zimmerman, E. F. and Bowen, D. (1972). Distribution and metabolism of triamcinolone in inbred mice with different cleft palate sensitivities. *Teratology*, 5, 335
49. Biddle, F. G. (1977b). 6-Aminonicotinamide-induced cleft palate in the mouse: the nature of the difference between the A/J and C57BL/6J strains in frequency of response and its genetic basis. *Teratology*, 16, 301
50. Biddle, F. G. (1977a). Can we discriminate between mechanisms of cleft palate induction? *Teratology*, 15, 21A (abstract)
51. Biddle, F. G. (1978a). Use of dose–response relationships to discriminate between the mechanisms of cleft-palate induction by different teratogens: an argument for discussion. *Teratology*, 18, 247
52. Biddle, F. G. (1975). Teratogenesis of acetazolamide in the CBA/J and SWV strains of mice. II. Genetic control of the teratogenic response. *Teratology*, 11, 37

53. Wright, S. (1968). *Evolution and the Genetics of Populations*. Vol. II. *Genetic and Biometric Foundations*. (Chicago: University of Chicago Press)
54. Bloom, J. L. and Falconer, D. S. (1964). A gene with major effect on susceptibility to induced lung tumors in mice. *J. Nat. Cancer Inst.*, 33, 607
55. Biddle, F. G. and Fraser, F. C. (1977a). Cortisone-induced cleft palate in the mouse. A search for the genetic control of the embryonic response trait. *Genetics*, 85, 289
56. Bonner, J. J. and Slavkin, H. C. (1975). Cleft palate susceptibility linked to histocompatibility-2 (*H-2*) in the mouse. *Immunogenetics*, 2, 213
57. Salomon, D. S. and Pratt, R. M. (1976). Glucocorticoid receptors in murine embryonic facial mesenchyme cells. *Nature (Lond.)*, 264, 174
58. Goldman, A. S., Katsumata, M., Jaffe, S. Y. and Gasser, D. L. (1977). Palatal cytosol cortisol-binding protein associated with cleft palate susceptibility and H-2 genotype. *Nature (Lond.)*, 265, 643
59. Kaplan, N. O., Goldin, A., Humphreys, S. R., Ciotti, M. M. and Venditti, J. M. (1954). Significance of enzymatically catalyzed exchange reactions in chemotherapy. *Science*, 120, 437
60. Biddle, F. G. and Fraser, F. C. (1978). Genetic independence of the embryonic reactivity difference to cortisone- and 6-aminonicotinamide-induced cleft palate in the mouse. *Teratology* (In press)
61. Siegel, S. (1956). *Nonparametric Statistics for the Behavioural Sciences*. (New York: McGraw-Hill)
62. Elston, R. C. and Stewart, J. (1970). A new test of association for continuous variables. *Biometrics*, 26, 305 and 860
63. Biddle, F. G. (1978b). Palate development in the mouse – revisited. (In preparation)
64. Vekemans, M. and Fraser, F. C. (1977b). Stage of palate closure as one indication of liability to cleft palate. In: J. W. Littlefield (ed.). *Fifth International Conference of Birth Defects*, p. 158A. (Amsterdam: Excerpta Medica)
65. Shire, J. G. M. (1974). Endocrine genetics of the adrenal gland. *J. Endocrinol.*, 62, 173
66. Taylor, B. A. (1972). Genetic relationships between inbred strains of mice. *J. Hered.*, 63, 83
67. Juriloff, D. M. (1978). Genetics of spontaneous and 6-aminonicotinamide-induced cleft lip in mice. Ph.D. Thesis. (Montreal: McGill University Library)
68. Layton, W. M. (1978). Personal communication. (In preparation)
69. Bailey, D. W. (1971). Recombinant-inbred strains. *Transplantation*, 11, 325
70. Klein, J. (1975). *Biology of the Mouse Histocompatibility-2 Complex*. (New York: Springer-Verlag)
71. Snell, G. D., Dausset, J. and Nathenson, S. (1976). *Histocompatibility*. (New York: Academic Press)
72. Snell, G. D. and Stimpfling, J. H. (1966). Genetics of tissue transplantation. In: E. L. Green (ed.). *Biology of the Laboratory Mouse*, pp. 457–491 (New York: McGraw-Hill Book Co.)
73. Boyse, E. A. (1977). The increasing value of congenic mice in biomedical research. *Lab. Animal Sci.*, 27, 771
74. Chernoff, G. F. (1977). The fetal alcohol syndrome in mice: an animal model. *Teratology*, 15, 223
75. Lambert, G. H. and Nebert, D. W. (1977). Genetically mediated induction of drug-metabolizing enzymes associated with congenital defects in the mouse. *Teratology*, 16, 147
76. Hewlett, P. S. (1969). Measurement of the potencies of drug mixtures. *Biometrics*, 25, 477
77. DeJongh, S. E. (1961). Isoboles. In: H. DeJonge (ed.). *Quantitative Methods in Pharmacology*, pp. 318–327. (Amsterdam: North-Holland Publishing Co.)
78. Tammes, P. M. L. (1964). Isoboles, a graphic representation of synergism in pesticides. *Neth. J. Plant Pathol.*, 70, 73
79. Aitchison, J. and Silvey, S. D. (1957). The generalization of probit analysis to the case of multiple responses. *Biometrika*, 44, 131
80. Ashford, J. R. (1959). An approach to the analysis of data for semi-quantal responses in biological assay. *Biometrics*, 15, 573
81. Mantel, N. (1966). Models for complex contingency tables and polychotomous dosage response curves. *Biometrics*, 22, 83

82. Ben-David, M., Heston, W. E. and Rodbard, D. (1969). Mammary tumor virus potentiation of endogenous prolactin effect on mammary gland differentiation. *J. Nat. Cancer Inst.*, 42, 207

83. Bross, I. D. J. (1958). How to use ridit analysis. *Biometrics*, 14, 18

84. Mead, R. and Pike, D. J. (1975). A review of response surface methodology from a biometric viewpoint. *Biometrics*, 31, 803

85. Warburton, D., Trasler, D. G., Naylor, A., Miller, J. R. and Fraser, F. C. (1962). Pitfalls in tests for teratogenicity. *Lancet*, ii, 1116

86. Miller, K. K. (1977). Commercial dietary influences on the frequency of cortisone-induced cleft palate in C57BL/6J mice. *Teratology*, 15, 249

87. Miller, T. J. (1972). Cleft palate formation: a role for pyridoxine in the closure of the secondary palate in mice. *Teratology*, 6, 351

88. Vekemans, M. and Fraser, F. C. (1978). Effects of two diets on the frequency of cortisone-induced cleft palate in mice. *Teratology*, 17, 24A (abstract)

89. Bedrick, A. D. and Ladda, R. L. (1978). Epidermal growth factor potentiates cortisone-induced cleft palate in the mouse. *Teratology*, 17, 13

90. Vekemans, M. and Fraser, F. C. (1977a). Characteristics of a new strain of mice with unusually high sensitivity to cortisone-induced cleft palate. *Teratology*, 15, 18A (abstract)

91. Wilson, J. G. (1973). *Environment and Birth Defects.* (New York: Academic Press)

92. Andrew, F. D. and Zimmerman, E. F. (1971). Glucocorticoid induction of cleft palate in mice: no correlation with inhibition of mucopolysaccharide synthesis. *Teratology*, 4, 31

93. Walker, B. E. (1971). Induction of cleft palate in rats with antiinflammatory drugs. *Teratology*, 4, 39

94. Hackman, R. M. and Brown, K. S. (1972). Corticosterone-induced isolated cleft palate in A/J mice. *Teratology*, 6, 313

95. Loevy, H. (1972). Lack of difference in sex ratio of mice with cortisone-induced cleft - palate. *J. Dent. Res.*, 51, 1010

96. Loevy, H. T. and Wade, M. A. (1972). Sex of mouse (CD1) offspring not a factor in hydrocortisone induced cleft palate. *Cleft Palate J.*, 9, 210

7

The pathogenesis of thalidomide embryopathy

W. G. McBRIDE

INTRODUCTION

Thalidomide (α-phthalimido glutarimide) was synthesized in West Germany in 1954 by Chemie-Gruenenthal Limited. It was marketed in West Germany in 1956 and distributed by Distillers Company (Biochemicals) Limited in England in 1958. Animal experiments were used in its preliminary pharmacological assessment[1,2] and subsequent clinical trials indicated that thalidomide was an effective sedative and hypnotic drug which even at extremely high doses failed to evoke significant toxic effects.

Early in 1960 isolated reports were received by Burley[3], of Distillers Company (Biochemicals) Limited in Great Britain suggesting that regular therapy with thalidomide for six months or more might provoke polyneuritis. Nevertheless, it was not until December of 1960 that the association of thalidomide and neuropathy was reported in the literature by Florence[4].

In September 1961, Wiedemann[5] drew attention to an alarming increase in the incidence within Germany of hypoplastic and aplastic malformations of the extremities. In November of the same year Pfeiffer and Kosenow[6] presented a paper referring to a number of similarly deformed babies born during the previous two years. In subsequent discussion, Lenz, guided by personal experience, raised the question of thalidomide ingestion by the mothers during pregnancy. At the same time, the teratogenic potential of thalidomide had been recognized in Australia by McBride who subsequently communicated his initial suspicions to the Distillers Company (Biochemicals) Limited in England and submitted a paper to the *Lancet* linking thalidomide with teratogenesis in June 1961.

Thalidomide was withdrawn from the market by Chemie Gruenenthal Limited, on November 27, 1961, and by Distillers Company (Biochemists) Limited on December 2, 1961. On December 9, 1961 McBride[7,8] published the first report in the literature on the possible association of thalidomide with congenital malformations. This was followed by confirmatory responses from Lenz[9,10] and Pfeiffer and Kosenow[11]. Epidemiological evidence was later collected to positively implicate the drug[12–14].

More subtle aspects of thalidomide embryopathy soon became apparent[15]. Malformations of the extremities affected the upper limb in the majority of the cases and although the involvement was generally bilateral, it was almost invariably asymmetrical. The pattern occurred in varying degrees of severity from amelia to minor thumb anomalies, although short phocomelic appendages were less frequent than the anomalies of the more distal segments. There was a definite tendency to reduction of the pre-axial bones in size and number. In some cases, the humerus was markedly reduced in size although portions of the forearm and the hand remained (proximal phocomelia). In such cases, the radius was shortened or absent and the pre-axial digits, the thumb and index finger, were not developed. The ulna and post-axial fingers or a monodactylous structure were the only elements preserved[16]. Together with the limb defects, a number of other anomalies have been reported (Table 7.1).

Table 7.1 Associated anomalies of the thalidomide syndrome

Organ or System	Abnormalities
Skull and associated structures	Cranial defects and hydrocephalus, microphthalmia and anophthalmia, deformities or absence of the pinna and atresia of the external canal, saddle nose, cleft palate, webbed neck, capillary haemangioma of upper lip, tip of nose, glabella and forehead
Respiratory system	Bilobed right lung
Cardiovascular system	Aortic hypoplasia, atrial and ventricular septal defects, transposition of great vessels, tetrology of Fallot, pulmonary stenosis, anomalous pulmonary veins
Gastrointestinal system	Pyloric stenosis, duodenal stenosis and atresia, malrotation, agenesis of appendix and caecum, imperforate anus, oesophageal atresia and tracheo-oesophageal fistulae, absence of gallbladder, atresia of common bile duct
Genitourinary system	Hypoplasia and agenesis of urinary tract and kidney, horseshoe kidney, double kidney, bicornuate uterus, atresia or absence of vagina, rectovaginal fistulae, undescended testis, hypospadias

McBride[17] indicated that the risk to the fetus exposed at critical stages of organogenesis was in the vicinity of 20%. Lenz[10], Lenz and Knapp[14] and Nowack[19] defined the critical period of gestation to be the 35th to 50th day from the last day of menstruation and suggested that the distribution of the organ systems involved, and the severity of the defects, depended on the specific period during the susceptible range, during which the drug was taken.

TERATOLOGICAL EXPERIMENTS WITH THALIDOMIDE

Somers[20] and later Spencer[21] were able to produce typical malformations in rabbits given thalidomide at appropriate doses during the organogenic period. Similar deformities were not consistently produced by thalidomide in rodents, the animals which had previously been most frequently used for experimental teratology[20, 22].

The consistency with which thalidomide induced congenital defects in the connective tissue component, not only in the developing limb but elsewhere in the body, suggested that this teratogen primarily assaulted the mesoderm of the fetal limb bud and its mesenchymal derivatives[7, 21, 23]. Therefore, as this belief became more generally accepted, experimental teratologists concentrated on investigation of its possible mechanisms of growth inhibition on the mesoderm.

A list of subsequent investigations into the specific processes of dysmorphogenesis related to thalidomide is shown in Table 7.2.

Table 7.2

Different theories discussed	Significant References
Vitamin B6 antimetabolite effect	Kemper[61], Leck and Millar[14], Tewes[53], Robertson[52], Hyssey[62], Buckle[63], Frank et al.[64], Everedd and Randall[65], Cuthbert et al.[66], Nystrom[67], Felisati[68], Staples[69], Toivanen[70], Naber[71]
Faulty glutamic acid metabolism and inhibition of nucleic acid synthesis	Faigle[72], Williams[73], Beckmann[74], Boylen et al.[75], Felisati[76], Fabro et al.[77], Schumacher et al.[78]
Acylation of natural aliphatic diomines e.g. putrexine, spermidine	Fabro et al.[79, 80]
Acylation of subcellular components	Schumacher[78, 81]
Deranged nucleic acid and protein synthesis and its effect on differentiation of tissues	Bakay and Nyhan[82]
Immunosuppression with survival of malformed fetus	Hellman et al.[83, 84], Locker[85]
Chromosomal abnormalities	Hughes et al.[86], Tsuda et al.[87], Hirsche[88], Benda and Baughmann[89], Villa[90], Jensen[91], Soukup[92], Roux[93]
Trophoblastic degeneration	Marin–Padilla[94]
Decreased mesonephric induction of chondrogenesis	Lash[95, 96]
Dystrophic action on already formed segments due to motor and sensory nerve damage	Cuthbert et al.[66], Gordon[97]
Axial limb artery degeneration	Jurand[98]
'More primitive paw pattern'	Vickers[29]
Faulty chrondrification and calcification processes	Nudleman and Travill[60]

In 1973, McBride suggested a quite different mechanism of the drug's action[24-26] suggesting a possible role of sensory nerves in normal limb development and postulating that this tissue represents the primary focus of attack in thalidomide embryopathy. McBride[25] examined the ganglionic hypoplasia accompanying thalidomide induced limb deformities in 29 rabbit fetuses. The limbs were variably reduced and this was correlated with degenerative changes in the ganglia. Abnormal neurons exhibiting cytoplasmic vacuolation and nuclear karyolysis were prominent in cervical ganglia of deformed fetuses, whereas normal neurons of control fetuses showed large

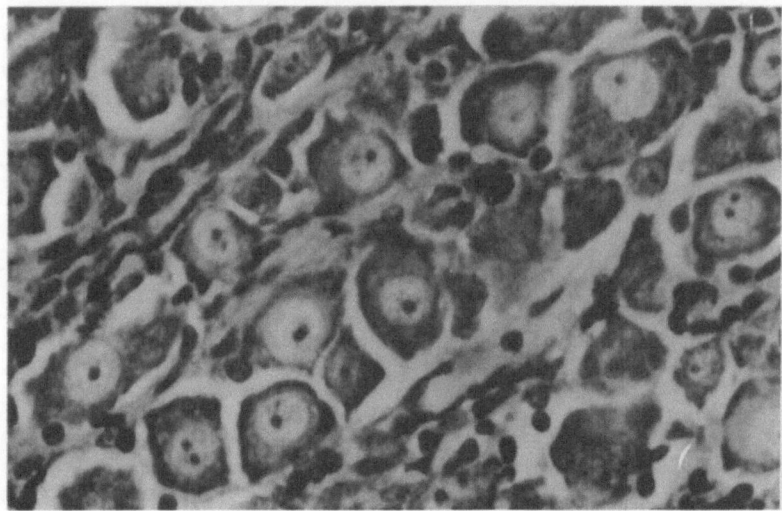

Figure 7.1 Control rabbit fetus. Sensory ganglion. CVl. Cresyl violet × 420

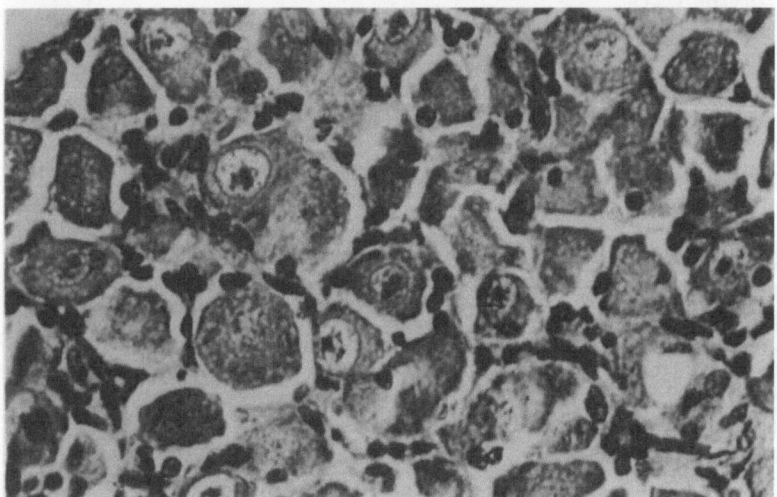

Figure 7.2 Thalidomide treated amelic fetus. Sensory ganglion CVl. Cresyl violet × 480

vesicular nuclei with prominent nucleoli. Some degree of hypoplasia with a small proportion of abnormal neurons was found where thalidomide had been administered without resultant deformity, but hypoplastic ganglia with a high proportion of abnormal neurons were present in severe degrees of reduction deformities (Figures 7.1 and 7.2). From this experiment it was postulated that it is the diminution in the number of sensory neurons that interferes with peripheral organ development. To support this hypothesis the ultrastructure of sensory neurons of the dorsal root ganglia of New Zealand White rabbit embryos and fetuses exposed to thalidomide was examined at early stages of

Table 7.3   Summary of ultrastructural changes found in the lower cervical dorsal root ganglia

	Neurons	*Axons*	*Presence of myelin whorls*
Control	Nuclear indentation, days 13, 15 only	Normal	Axons – associated with Schwann cells*
Experimental, normal	Nuclear indentation†	Loss of neurofilaments*	Present in axons*
	Cell shrinkage*	Membrane-bound vesicles*	Associated with Schwann cells†
Experimental, deformed	Nuclear indentation‡	Loss of neurofilaments‡	Present in axons‡
	Cell shrinkage†	Membrane-bound vesicles†	Associated with Schwann cells‡

* present
† marked
‡ prominent

development, before the limb was completely formed[27, 28]. The ultrastructural changes found in the neurons of the dorsal root ganglia consisted of nuclear indentation, cell shrinkage and increased ribosomal activity (Table 7.3),

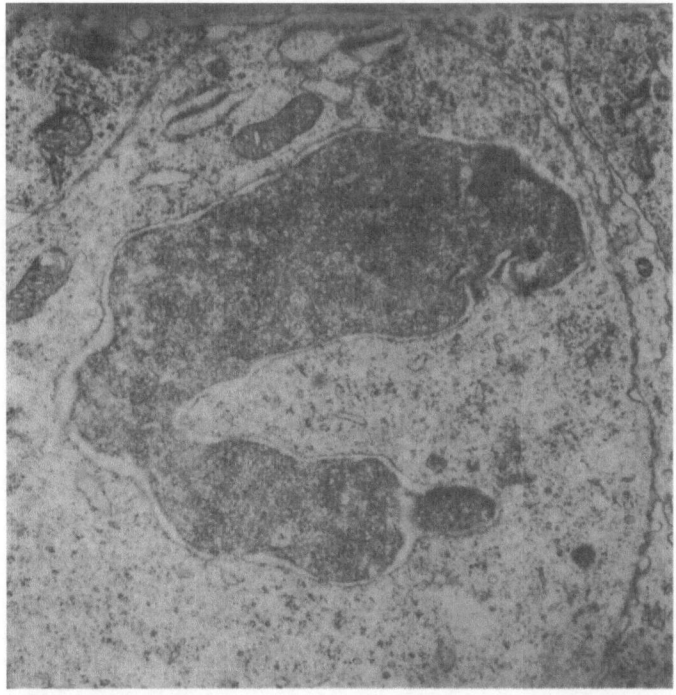

Figure 7.3   Day 13 thalidomide treated embryo. Neuron sensory ganglion. Nuclear condensation and loss of cytoplasmic detail, although some mitochondria are intact. Uranyl acetate and lead citrate. × 27 200

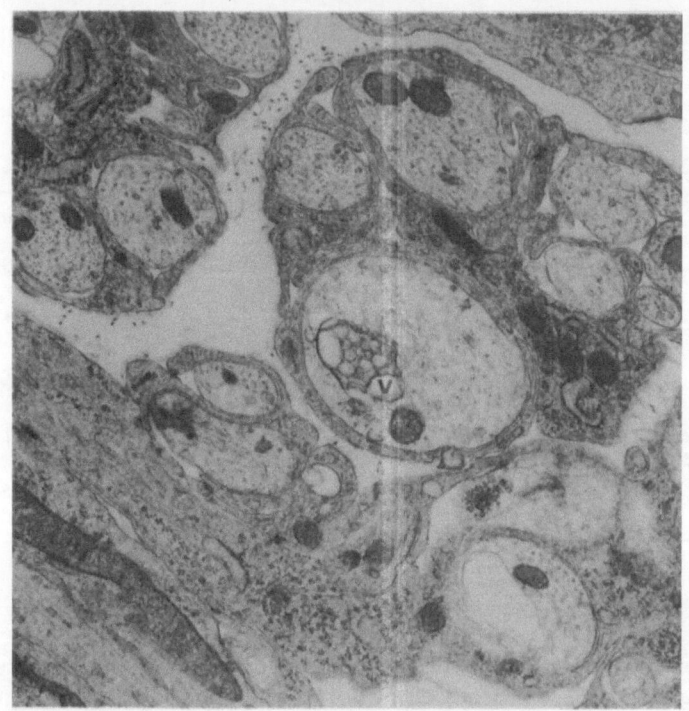

Figure 7.4 Day 13 experimental. The most frequent axonal lesion, loss of neurofilaments and neurotubules, and vacuolation of mitochondrion (v). Uranyl acetate and lead citrate. × 27 200

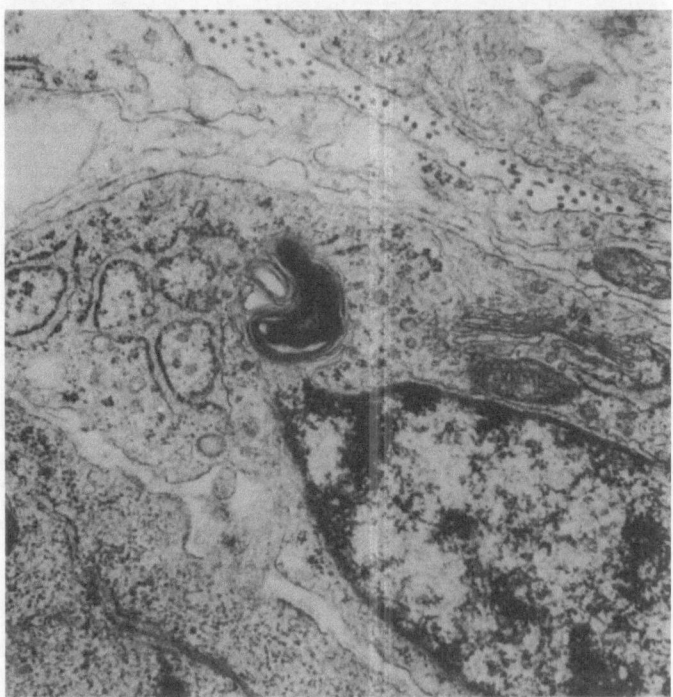

Figure 7.5 Day 21 experimental. Axons replaced by 'myelin-whorl' formation. Uranyl acetate and lead citrate. × 13 600

whilst the axons exhibited loss of microtubules and microfilaments, mitochondrial degeneration, the presence of membrane bound vesicles and the presence of myelin whorls (Figures 7.3, 7.4 and 7.5). As ultrastructural damage to the components of the dorsal root ganglia viz., neurons and axons could be found in day 13 experimental embryos, that is at least 16 h before the earliest signs of thalidomide dysmelia have been reported in rabbits[29], it is regarded as corroborative evidence of the importance of neurons in peripheral development.

EMBRYONIC INDUCTION PROCESSES

During the first two decades of this century, the organizer theory of embryonic induction was developed by Spemann and his co-workers[30–32]. Experiments on amphibian embryos revealed that the dorsal lip of the blastopore acted as a 'primary organizer'. When the blastoporal lip invaginates beneath the ectoderm during gastrulation, it induces differentiation of the overlying ectoderm into the different parts of the central nervous system; these sites then assume an organizer capability (the secondary organiser centres) and they themselves exert inductive influences on neighbouring plastic tissues which therefore undergo dependent differentiation, e.g. the optic cup, ear vesicles, nasal pits, etc.

By 1932, considerable evidence had accumulated indicating that the inductive processes revealed by Spemann were mediated through some form of chemical transmitter[33]. During the following years, evidence of more than one active inducing factor was found. With newer techniques for the separation of proteins, it has been shown that these initial assumptions were correct. Although not all their chemical properties are yet fully known, highly-purified low-molecular weight proteins and ribonucleoprotein fractions with a regionally specific inducing activity have been isolated from chick and amphibian embryos[34, 35].

LIMB MORPHOGENESIS

In recent years, certain factors underlying the morphological changes occurring during early stages of limb development have been elucidated by the study of chick and rodent embryos. As early as the gastrula stage, the presumptive limb mesoderm is already determined with respect to its further transformation into limb tissue, although this determination is limited to the formation of limb girdles[36–38].

The first demonstrable activity of definitive limb morphogenesis takes place in the dorsalsomatic mesoderm in the whole length of the trunk region; this proliferation does not originally involve ectodermal changes. The lengthwise swelling so formed is known as Wolff's crest[39]. A few hours later, the mesodermal growth proceeds actively at both extremities, giving rise to the anterior and posterior limb buds. Simultaneously, activity decreases in the intermediate portion which will later form the thoracic and abdominal walls.

Further development of the vertebrate limb depends on the formation of an apical ectodermal ridge; this originates from the early inductive influence of the young mesoderm on the overlying ectoderm, predominantly in the ventral surface of the bud[40]. A few hours later, the apical ectodermal ridge forms, beginning at the post-axial extremity of the marginal border, where the simple epithelium assumes a pseudostratified, columnar configuration. Subsequently, the ectodermal ridge demonstrates organizer capability, inducing definitive limb outgrowth. It communicates to the mesoderm a message stimulating its proliferation and the laying down of limb parts in normal proximo-distal order.

The ectoderm in mammals does not acquire a crest-like appearance before the mesoderm for the two proximal limb segments is laid down. Nevertheless, selective histochemical properties have been demonstrated in the area of the future apical ridge indicating that its morphogenetic properties develop very early, even before it has fully reacted to mesodermal induction[41]. Finally, there is a mesodermal feedback which maintains the ectoderm in a hyper-trophied/hyperplastic form enabling it to continue to induce outgrowth in the mesoderm[42].

The 'messages' involved in these reciprocal inductive influences between the ectoderm and mesoderm (apical ectoderm maintenance factor, mesoderm maintenance factor) have been shown to be diffusible substances capable of passing through millipore filters although their chemical nature is as yet unknown[43].

With histochemical analysis, ectodermal and mesodermal properties have been found to be asymmetrically distributed between the pre-axial and post-axial territories. It has been well established that the apical ectodermal ridge is thicker in the post-axial region and that the responsible mesodermal influence is stronger post-axially than pre-axially. The maintenance of the induced ectodermal ridge requires a continuous stimulation exerted by the same mesoderm which acted as its initial induction[40, 41, 44].

Milaire[41] demonstrated that later in development, at the stage of foot plate formation, there is a sudden cell degeneration in the pre-axial portion of the ectodermal ridge. It was suggested that the order of development of the five radiated blastemata in a caudocephalic direction is due to a consequent redistribution in metabolic activity of the apical ridge to predominantly a post-axial region. 'The shortness of the first toe may be considered the result of weak morphogenetic properties present in the pre-axial part of the apical ectoderm'[41]. Separation of the digits depends on the selective maintenance of the ectodermal ridge over the outgrowing digits; intervening areas lose enzymatic properties of the ectoderm and the underlying mesoderm demonstrates necrotic changes.

PERIPHERAL NERVOUS TISSUE AND LIMB BUD DEVELOPMENT IN THE HUMAN EMBRYO

In the early part of the fourth week of gestation (stage 10), the neural tube commences to close and this process is completed before the end of this period (Figure 7.6). The cells which lie in the line of fusion of the dorsal edges of the neural groove constitute the neural crest. When the folds meet in the median plane, the two neural crests fuse and form a wedge-shaped area along

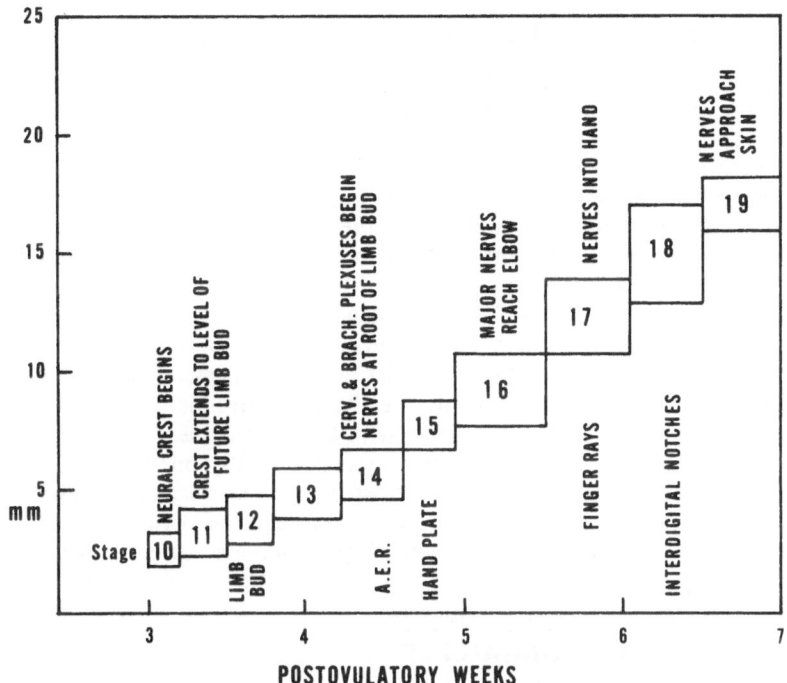

Figure 7.6   Diagram relating stage of upper limb development with development of the peripheral nerves (after Gardner and O'Rahilly[99])

the line of closure. Opposite the primitive segments these cells rapidly proliferate to form a series of oval-shaped masses, the primitive spinal ganglia, which then migrate for a short distance in a ventro-lateral direction. Towards the end of the fourth week, the central processes of the neuroblastic component of the spinal ganglia grow back into the wall of the neural tube, forming the fibres of the dorsal roots, while the peripheral processes grow forward to mingle with the ventral root in the spinal nerve.

Also, towards the end of the fourth week, the limbs appear as small elevations or buds from a slight lateral ridge (Wolff's crest) at either side of the trunk. 'The limb bud is small and a mere nodule of mesenchyme when the first nerves reach it close to the blood vessels that precede them. At least some of the primary connections may be assumed to be made then, although the definitive nerve endings, according to observations upon embryos of birds and mammals do not develop until later. Subsequently, the elongation of the bud takes place, and the individual muscles and cartilages segregate out of the original blastemata'[45].

The limb buds lengthen, and during the second half of the fifth week, divide into upper and lower parts. At the beginning of the sixth week, paddle-shaped hands and feet begin to be recognized at the extremities. Towards the end of this period, digital segmentation occurs, followed in the seventh week by digital separation.

121

PERIPHERAL NEURITIS INDUCED BY THALIDOMIDE

Florence[4] first reported peripheral neuritis in patients on long term therapy with thalidomide. This was soon confirmed by similar reports by Burley[3], Kuennsberg et al.[46], Fullerton[47], Meade and Robalski[48], and Kremer and Fullerton[49]. The neuropathy typically began 2–18 months after treatment had begun and was predominantly sensory in nature though some patients showed mild motor weakness, particularly in the proximal muscles, while others showed evidence of mild pyramidal tract damage. As with the embryopathies, the neuropathies were more prevalent in the upper extremities. Klinghardt[50] studying necropsy material from two patients with thalidomide neuropathy, found degeneration of axon cylinders and myelin sheaths in peripheral nerves, as well as changes in dorsal root ganglia and in the posterior columns of the spinal cord. Later, Fullerton[51] demonstrated a selective loss of large diameter fibres in nerve biopsies from patients with thalidomide neuropathy. Because of the similarity to the neuropathy associated with thiamine deficiency, therapy with the B-vitamins was instigated in some cases, but with rather limited success[52, 53]. Most patients showed evidence of permanent neuronal damage, with only minor improvement of symptoms even after many years.

PERIPHERAL NERVOUS TISSUE –
THE SECONDARY ORGANIZER

It has been established that at an early stage of embryogenesis, the somato-pleuric mesoderm in the region of the future limb buds possesses a definite though limited potentiality for future development. In experiments on normal limb outgrowth, when the limb bud is denuded of its apical ectodermal ridge, limb growth ceases abruptly and the degree to which proximal segments later develop depends on the stage at which the operation is performed[44, 54]. Therefore, it has become generally considered that the ectodermal ridge represents the main growth controlling centre for limb development.

However, this author suggests that the outgrowing elements of the peripheral nervous system are responsible for primary induction and maintenance of limb morphogenesis, establishing a link between the primary inductor tissue and the secondary limb induction systems already known.

The initial outgrowth of the somatopleuric mesoderm to form Wolff's crest may be considered the result of early inductive influences of parts of the central nervous system, in a manner similar to that which has been observed in other organs e.g. the optic cup, ear vesicles, etc. In further limb bud development, the outgrowing processes of the peripheral nervous system would then exert inductive influences on the mesodermal core of the limb bud (Figure 7.6). As there exists a synergistic relationship between the ectodermal cap and the mesoderm, it is a reasonable hypothesis that there is an analogous relationship between the outgrowing neural elements and the mesoderm, a neural stimulus which is essential for the maintenance of outgrowth and differentiation of the mesodermal core.

Therefore, the selective maintenance of the ectodermal ridge over the outgrowing digits in later development depends on the proliferation of nervous tissue elements into the five radiated blastemata and the simultaneous lack of a hypothetical mesodermal maintenance factor (of neural origin) in the intervening areas, where breakdown of mesoderm–ectoderm interaction and subsequent necrosis has been observed.

The role of nervous tissue as an inductor tissue has gained credence from work by Singer and co-workers. Regeneration of adult and larval limbs in the newt fails to occur if limbs are denervated at the time of amputation[55, 56]. Furthermore, infusion of agents which block the action of acetylcholine causes an immediate arrest of limb regeneration[57]. However, since infusion of acetylcholine into denervated limbs does not promote regeneration, Singer concludes that acetylcholine is not the only trophic stimulus from nerves which is essential for regeneration[58].

THALIDOMIDE – EMBRYONIC NEUROPATHY

The above assumptions provide a basis for elucidating the nature of thalidomide embryopathy. If thalidomide selectively damages nervous tissue elements, in particular the neural crest cells and/or their derivatives, the dorsal root ganglia, then the peripheral deformities might result from an arrest of development due to failure of neural induction.

Thalidomide is known to derange the adult peripheral nervous system and it would seem reasonable to assume that the primitive neural crest and posterior root ganglion cells would be equally and probably even more susceptible to the toxic effects of the drug. McCredie[59] noted that the defects seen in peripheral segments due to thalidomide corresponded to a specific nervous supply and therefore suggested a selective neural crest injury at the level of C6. However, it is proposed that thalidomide can have a toxic effect at all levels of the developing neural crest, particularly in the areas of rapid proliferation of nervous tissue into the segments of the anterior and posterior limbs, and that graded severities of this initial damage are then reflected by variable induction of outgrowth in the mesodermal core of the limb buds.

The more severe toxic effects completely eliminate neural inductive influence preventing limb growth and resulting in a total amelia. With less severe damage, mesodermal activity is incompletely maintained. It has been indicated earlier that mesodermal influence on the overlying ectoderm is much stronger post-axially than pre-axially and, as a result, the ectodermal cap is thicker in that region. It might be strongly suspected, therefore, that the pre-axial ectoderm is more sensitive to diminished mesodermal influences than the post-axial ridge. A significantly reduced neural induction of the limb bud mesoderm would be first reflected by maldevelopment of the pre-axial ectodermal ridge (and therefore an inhibition of differentiation and growth of pre-axial mesoderm – the thumb, radius, etc.) before comparable changes occurred in the post-axial region. Previous workers have shown that the limb defects due to thalidomide represent an arrest in differentiation of the mesoderm; these first appear in the pre-axial region and in more severe cases also affect the post-axial structures[29, 60].

The experimental evidence outlined above suggests that the primary lesion produced by thalidomide is destruction of the mitochondria in the neurons resulting in reduction in the number of neurons in the spinal ganglia and autonomic ganglia. The mesodermal arrest is secondary to an interference in inductive influences exerted by the outgrowing peripheral nervous system, and that the distribution of anomalies within affected limbs reflects the asymmetric distribution of mesodermal and ectodermal properties between the pre-axial and post-axial territories of the primitive limb bud.

References

1. Kunz, W., Keller, H. and Muckter, H. (1956). N-Phthalylglutaminsäure-Imid: experimentelle Untersuchungen an einem neuen synthetischen Produkt mit sedativer Eigenschaft. *Arzneimittel-Forsch.*, 6, 426
2. Somers, G. F. (1960). Pharmacological properties of thalidomide (α-phthalimido glutarimide): new sedative hypnotic drug. *Br. J. Pharmacol.*, 15, 111
3. Burley, D. (1961). Is thalidomide to blame? *Br. Med. J.*, 1, 130
4. Florence, A. L. (1960). Is thalidomide to blame? *Br. Med. J.*, 2, 1954
5. Wiedemann, H. R. (1961). Hinweis auf eine derzeitige Häufung hypo- und aplastischer Fehlbildungen der Gliedmassen. *Med. Welt*, 37, 1863
6. Pfeiffer, R. A. and Kosenow, W. (1962). Frage einer exogenen Verursachung von schweren Extremitätenmissbildungen. *Münch. Med. Wschr.*, 104, 68
7. McBride, W. G. (1961). Thalidomide and congenital abnormalities. *Lancet*, ii, 1358
8. McBride, W. G. (1961). Congenital abnormalities and thalidomide. *Med. J. Aust.*, 2, 1030
9. Lenz, W. (1961). Kindliche Missbildungen nach Medikament-einnahme während der Gravidität? *Dtsch. Med. Wschr.*, 86, 2555
10. Lenz, W. (1962). Thalidomide and congenital abnormalities. *Lancet*, i, 45
11. Pfeiffer, R. A. and Kosenow, W. (1962). Thalidomide and congenital abnormalities. *Lancet*, i, 45
12. Lenz, W. (1963). Das Thalidomid-Syndrom. *Fortschr. Med.*, 81, 148
13. Lenz, W. (1965). Epidemiologie von Missbildungen. *Pädiat. u. Pädol.*, 1, 38
14. Leck, J. M. and Millar, L. (1962). Incidence of malformations since the introduction of thalidomide. *Br. Med. J.*, 2, 16
15. Mellin, G. W. and Katzenstein, M. (1962). The saga of thalidomide: neuropathy to embryopathy, with cast reports of congenital anomalies. *N. Engl. J. Med.*, 267, 1184 and 1238
16. Smithells, R. W. (1966). Drugs and human malformations. In: D. H. M. Woollam (ed.) *Advances in Teratology* (New York: Academic Press)
17. McBride, W. G. (1963). Teratogenic action of drugs. *Med. J. Aust.*, 2, 689
18. Lenz, W. and Knapp, K. (1962). Die Thalidomid-Embryopathie. *Dtsch. Med. Wschr.*, 87, 1232
19. Nowack, E. (1965). Die sensible Phase bei der Thalidomid-Embryopathie. *Humangenetik*, 1, 516
20. Somers, G. F. (1962). Thalidomide and congenital abnormalities. *Lancet*, i, 912
21. Spencer, K. E. U. (1962). Thalidomide and congenital abnormalities. *Lancet*, ii, 100
22. Pliess, G. (1962). Thalidomide and congenital abnormalities. *Lancet*, i, 1128
23. Woollam, D. H. (1965). Principles of teratogenesis: mode of action of thalidomide. *Proc. R. Soc. Med.*, 58, 497
24. McBride, W. G. (1973). Foetal nerve cell degeneration produced by thalidomide in rabbits. *Int. Res. Commun. Syst.* (73–7) 5-5-4
25. McBride, W. G. (1974). Fetal nerve cell degeneration produced by thalidomide in rabbits. *Teratology*, 10, 283
26. McCredie, J. and McBride, W. G. (1973). Some congenital abnormalities; possibly due to embryonic peripheral neuropathy. *Clin. Radiol.*, 24, 204
27. Stokes, P. A., Lykke, A. W. and McBride, W. G. (1976). Ultrastructural changes in the dorsal root ganglia evoked by thalidomide preceding limb development. *Experientia*, 32, 597
28. McBride, W. G. (1976). Studies of the etiology of thalidomide dysmorphogenesis. *Teratology*, 14, 71

29. Vickers, T. H. (1967). Concerning the morphogenesis of thalidomide dysmelia in rabbits. *Br. J. Exp. Pathol.*, **48**, 579

30. Spemann, H. (1901). Entwicklungsphysiologische Studien am Tritonei I. *Arch. Entw. Mech. Org.*, **12**, 224

31. Spemann, H. (1912). Zur Entwicklung des Wirbeltierauges. *Zool. Jber. (Abt. Zool. Phys.)*, **32**, 1

32. Spemann, H. and Mangold, H. (1924). Über Induktion von Embryonalanlagen durch Implantation artfremder Organisatoren. *Arch. Entw. Mech. Org.*, **100**, 599

33. Bautzmann, H., Holtfreter, J., Spemann, H. and Mangold, O. (1932). Versuche zur Analyse der Induktionsmittel in der Embryonal-Entwicklung. *Naturwissenschaften*, **20**, 971

34. Saxen, L. and Toivanen, S. (1962). *Primary Embryonic Induction*. (London: Lagos Press)

35. Weber, R. (1967). *The Biochemistry of Animal Development*, Vol. II (New York: Academic Press)

36. Detwiler, S. R. (1933). On the time of determination of the antero-posterior axis of the forelimb in ambystoma. *J. Exp. Zool.*, **64**, 405

37. Chaube, S. (1959). On axiation and symmetry in transplanted wing of the chick. *J. Exp. Zool.*, **140**, 29

38. Kieny, M. (1960). Rôle inducteur du mésoderme dans la différentiation précoce du bourgeon de membre chez l'embryon de poulet. *J. Embryol. Exp. Morphol.*, **8**, 457

39. Hertwig, O. (1906). *Handbuch der vergleichenden und experimentellen Entwicklungslehre der Wirbeltiere.* Fischer, Jena, 166–338

40. Milaire, J. (1961). *Advances in Morphogenesis*, Vol. 2. Abercrombie, M. and Brachet, J. (eds.) (New York: Academic Press)

41. Milaire, J. (1967). The contribution of histochemistry to our understanding of limb morphogenesis and some of its congenital deviations. In: *Normal and Abnormal Embryological Development.* Frantz (ed.) (Washington: National Research Council)

42. Saunders, J. W. (1967). Control of growth patterns in limb development. In: *Normal and Abnormal Embryological Development.* Frantz (ed.) (Washington: National Research Council)

43. Saunders, J. W. and Gasseling, M. T. (1963). Trans-filter propagation of apical ectoderm maintenance factor in the chick embryo wing bud. *Dev. Biol.*, **7**, 64

44. Zwilling, E. (1961). *Advances in Morphogenesis*, Vol. I. Abercrombie, M. and Brachet, J. (eds.) (New York: Academic Press)

45. Harrison, R. G. (1969). *Organisation and Development of the Embryo.* Sally Wilens (ed.) (New Haven: Yale University Press)

46. Kuennsberg, E. V., Simpson, J. A. and Stanton, J. B. (1961). Is thalidomide to blame? *Br. Med. J.*, **1**, 291

47. Fullerton, P. M. and Kremer, M. (1961). Neuropathy after intake of thalidomide ('Distaval'). *Br. Med. J.*, **2**, 855

48. Meade, B. W. and Rosalki, S. B. (1961). Neuropathy after thalidomide ('Distaval'). *Br. Med. J.*, **2**, 1223

49. Kremer, M. and Fullerton, P. M. (1961). Neuropathy after thalidomide ('Distaval'). *Br. Med. J.*, **2**, 1498

50. Klinghardt, G. W. (1966). Experimentelle und human-pathologische Untersuchungen. *Proc. Vth. Int. Congr. Neuropath, Amsterdam*, p. 292

51. Fullerton, P. M. and O'Sullivan, D. J. (1968). Thalidomide neuropathy: a clinical, electrophysiological and histological follow-up study. *J. Neurol. Neurosurg. Psychiat.*, **31**, 543

52. Robertson, W. F. (1962). Thalidomide ('Distaval') and vitamin-B deficiency. *Br. Med. J.*, **1**, 792

53. Tewes, H. (1962). Vitamin therapy in neuritis due to thalidomide. *Munch. Med. Wschr.*, **104**, 269

54. Saunders, J. W. (1948). The proximal sequence of origin of the parts of the chick wing and the role of the ectoderm. *J. Exp. Zool.*, **108**, 363

55. Singer, M. (1952). The influence of the nerve in regeneration of the amphibian extremity. *Qu. Rev. Biol.*, **27**, 169

56. Singer, M. (1959). In: *Regeneration in Vertebrates*. Thornton, C. S. (ed.) pp. 59–80. (Chicago, Illinois: University of Chicago Press)
57. Singer, M. (1954). Apparatus for continuous infusion of microvolumes of solution into organs and tissues. *Proc. Soc. Exp. Biol. Med.*, 86, 378
58. Singer, M. (1960). In: *Developing Cell Systems and their Control*. Rudnick, D. (ed.) pp. 115–133. (New York: Ronald)
59. McCredie, J. (1974). A hypothesis of neural crest injury as the pathogenesis of congenital malformations. *Med. J. Aust.*, 1, 159
60. Nudelman, K. L. and Travill, A. A. (1971). A morphological and histochemical study of thalidomide-induced upper limb malfunctions in rabbit fetuses. *Teratology*, 4, 409
61. Kemper, F. (1962). Thalidomide and congenital abnormalities. *Lancet*, ii, 836
62. Hussey, L. M. (1962). Action of thalidomide. *N. Engl. J. Med.*, 268, 624
63. Buckle, R. M. (1963). Blood pyruvic acid in thalidomide neuropathy. *Br. Med. J.*, 2, 973
64. Frank, O., Baker, H., Ziffer, H., Aaronson, S., Hunter, S. A. and Leevy, C. M. (1963). Metabolic deficiencies in protozoa induced by thalidomide, *Science*, 139, 110
65. Everedd, D. F. and Randall, H. G. (1963). Thalidomide and the B vitamins. *Br. Med. J.*, 1, 610
66. Cuthbert, R. and Spiers, A. L. (1963). Thalidomide induced malformations – a radiological survey. *Clin. Radiol.*, 14, 163
67. Nystrom, C. (1963). Biochemical effects of thalidomide. *Scand. J. Clin. Lab. Invest.*, 15, 102
68. Felisati, D. (1964). Teratogenic action of thalidomide. *Lancet*, i, 724
69. Staples, R. E. (1963). Effects of parental thalidomide treatment on gestation and fetal development. *Exp. Mol. Pathol. (Suppl.)*, 2, 81
70. Toivanen, A., Markkanen, T., Mantyjarvi, R. and Toivanen, P. (1964). Microbiologically determined pantothenic acid and nicotinic acid content of chick embryos after treatment with thalidomide and of rat fetuses, newborns and placentas from mothers treated with thalidomide. *Biochem. Pharmacol.*, 13, 1489
71. Naber, E. C. and Largent, E. J. (1965). Thalidomide teratogenesis in the developing chick embryo and its relationship to vitamin metabolism. *Poult. Sci.*, 44, 1583
72. Faigle, J. W., Keberle, H., Riess, W. and Schmid, K. (1962). The metabolic fate of thalidomide. *Experientia*, 18, 389
73. Williams, R. T. (1963). Teratogenic effects of thalidomide and related substances. *Lancet*, 8, 723
74. Beckmann, R. (1963). Über das Verhalten von Thalidomid im Organismus. *Arzneimittel-Forsch.*, 13, 185
75. Boylen, J. B., Horne, H. H. and Johnson, W. J. (1963). Teratogenic effects of thalidomide and related substances. *Lancet*, i, 552
76. Felisati, D. (1964). Teratogenic action of thalidomide. *Lancet*, i, 724
77. Fabro, S., Schumacher, H., Smith, R. L., Stagg, R. B. L. and Williams, R. T. (1967). The fate of the hydrolysis products of thalidomide in the pregnant rabbit. *Biochem. J.*, 104, 570
78. Schumacher, H., Blacke, D. A. and Gillette, J. R. (1967). Acylation of subcellular components by phthalimidophthalidomide as a possible mode of embryotoxic action. *Fed. Proc.*, 26, 730
79. Fabro, S., Schumacher, H., Smith, R. L. and Williams, R. T. (1964). Identification of thalidomide in rabbit blastocyts. *Nature (Lond.)*, 201, 1125
80. Fabro, S., Smith, R. L. and Williams, R. T. (1965). Thalidomide as a possible biological acylating agent. *Nature (Lond.)*, 208, 1208
81. Schumacher, H., Blake, D. and Gillette, J. (1966). Thalidomide solutions. *Science*, 154, 1362
82. Bakay, B. and Nyhan, W. L. (1968). Binding of thalidomide by macromolecules in the fetal and maternal rat. *J. Pharmacol. Exp. Ther.*, 161, 348
83. Hellman, K., Duke, D. I. and Tucker, D. F. (1965). Prolongation of skin homograft survival by thalidomide. *Br. Med. J.*, 2, 687
84. Hellmann, K. (1966). Immunosuppression by thalidomide: implications for teratology. *Lancet*, i, 1136
85. Locker, D., Superstine, E. and Sulman, F. G. (1971). The mechanism of the push and pull

principle. VIII: endocrine effects of thalidomide and its analogues. *Arch. Int. Pharmacodyn. Ther.*, **194**, 39

86. Hughes, D. T., Delhanty, J. D. A., Chitham, R. G., Playfair, J. H. L. and Hopper, P. K. (1962). Chromosomes of thalidomide-deformed fetuses. *Lancet*, ii, 836
87. Tsuda, F., Abe, K., Kokubun, K., Hashimoto, T. and Nemoto, H. (1963). Chromosomes of newborn child with limb deformities. *Lancet*, i, 726
88. Hirsch, M. (1963). Chromosomal studies on so-called thalidomide embryopathy. *Med. Klin.*, **58**, 397
89. Benda, C. E. and Baughmann, F. A. (1963). An unusual case of developmental disorder – probably based on prenatal thalidomide damage, with chromosomal analysis. *Med. Welt*, **34**, 1661
90. Villa, L. and Eridani, S. (1963). Cytological effects of thalidomide. *Lancet*, i, 725
91. Jensen, M. K. (1965). Chromosome aberrations in human cells induced by thalidomide in vitro. *Acta Med. Scand.*, **177**, 783
92. Soukup, S., Takacs, E. and Warkany, J. (1967). Chromosome changes in embryos treated with various teratogens. *J. Embryol. Exp. Morphol.*, **18**, 215
93. Roux, C., Emerit, I. and Taillemite, J. L. (1971). Chromosomal breakage and teratogenesis. *Teratology*, **4**, 303
94. Marin-Padilla, M. and Benirschke, K. (1963). Thalidomide induced alterations in the blastocyst and placenta of the armadillo, *dasypus novemcinctus mexicanus*, including a choriocarcinoma. *Am. J. Pathol.*, **43**, 999
95. Lash, J. W. (1964). Normal embryology and teratogenesis. *Am. J. Obstet. Gynecol.*, **90**, 1193
96. Lash, J. W. and Saxen, L. (1971). Effect of thalidomide on human embryonic tissues. *Nature (Lond.)*, **232**, 634
97. Gordon, G. (1966). The mechanism of thalidomide deformities correlated with the pathogenic effects of prolonged dosage in adults. *Dev. Med. Child. Neurol.*, **8**, 761
98. Jurand, A. (1966). Early changes in limb buds of chick embryos after thalidomide treatment. *J. Embryol. Exp. Morphol.*, **16**, 289
99. Gardner, E. and O'Rahilly, R. (1976). Neural Crest, limb development and thalidomide embryopathy. *Lancet*, i, 635

8
The inductive influence of neurons in limb development

W. G. McBRIDE

INTRODUCTION

The inductive influence of neural tissue has been established in regard to the optic cup, ear vesicles and nasal pits[1]. From his transplant experiments Hamburger[2] concluded that morphogenesis is independent of innervation, and that atypical or completely abnormal morphogenetic development which occurred in a large percentage of transplants cannot be attributed to deficient nerve supply. However Hamburger and Waugh[3] again working with chick embryo limb transplants state that the nervous system has a trophic influence on the developing limb bud including the skeleton. Kieny and Fouvet[4] found that excision of a portion of the spinal cord in 2-day chick embryos, although it is made in front of the prospective leg level, will produce a high frequency of pre-axial hemimelia and a lower proportion of post-axial hemimelia, and concluded that the question of the influence of the nervous system in limb development is open for further investigation.

Cïhák[5] produced wing deformities in chick embryos by excision of portion of the brachial section of the neural tube at stages 13–15 of development[6]. Because of the conflicting evidence on the importance of neurons in limb development it was decided to study the effects on leg development of excision of portion of the lumbar segment of neural tube of chick embryos at stage 13–16 of development[6] (Figure 8.1).

EXPERIMENTAL MODEL

In order to ensure that any effects which may follow neuron damage were in fact due to damage of the neural tube, some embryos had the four lowest somites excised at stage 13–16 leaving the neural tube intact (Figure 8.2). Other embryos were allowed to develop to stage 20 then the tip of the limb bud was excised without damage to the neural tube in order to compare the type of deformity which results from excision of the apical ectodermal ridge.

The eggs used in this investigation were from a flock of White Leghorns

129

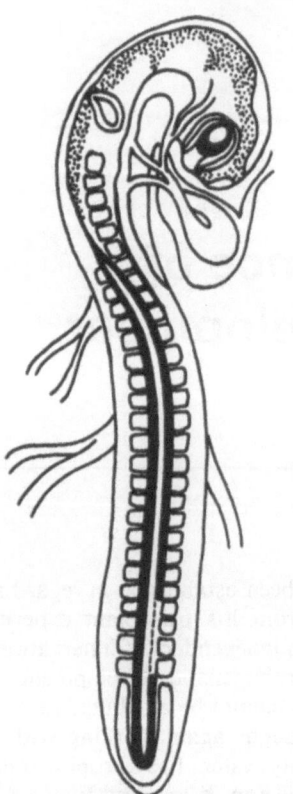 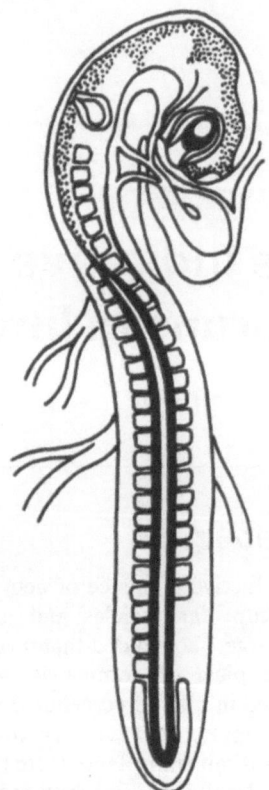

Figure 8.1 Diagrammatic representation of chick embryo H.H. stage 16. Excision of right half of neural tube opposite lower four somites

Figure 8.2 Chick embryo H.H. stage 16. Excision of somites 24–28, right side

(Hazletts Dunrobin Poultry Stud, Ingleburn, N.S.W.). Over a period of ten months 375 eggs have been incubated for periods varying from 52–60 h at 38°C ± 0.1°C and 70–80% relative humidity (in a Qualtex commercial incubator). The eggs were labelled with time and date of insertion on the upper surface. When removed from the incubator care was taken to keep the labelled surface uppermost thus eliminating the need to 'candle' the egg. Under aseptic conditions a window of approximately 1 sq. cm was sawed in the upper surface of the shell. The exposed membrane was moistened with 0.9% sodium chloride solution prior to excision. The vitelline membrane was torn back with fine watchmaker's forceps, exposing the embryo. Fifty-four eggs were unsuitable for further study because of damage whilst cutting the window or the situation of the embryo. A total of 148 eggs had a window cut but no further operative procedure, and these acted as controls for the study.

The stage of embryonic development at the time of the operations varied from 13–16, that is, at the 19–28 somite stage. Using a microsurgical technique the right side of the lumbar neural tube was excised from 104 embryos. The excision was made using a Castroviejo's or Bond's microsurgical forceps

130

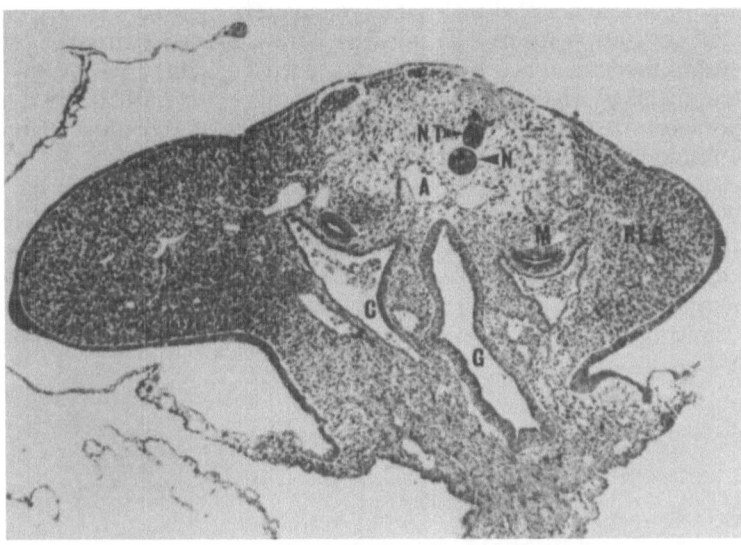

Figure 8.3 Chick embryo, stage 21, transverse section through leg buds. NT, neural tube, 24 hours after operation; N, notochord; A, dorsal aorta; G, hindgut; C, embryonic coelom; M, mesonephric duct; RLB, right limb bud. Haematoxylin and eosin. × 100

and a Beaver Blade 59S with a 3K handle. Utmost care was taken to avoid injury to the adjoining somites or blood vessels. It was considered that the operation was successful in 45 embryos. A sterile plastic cover was placed

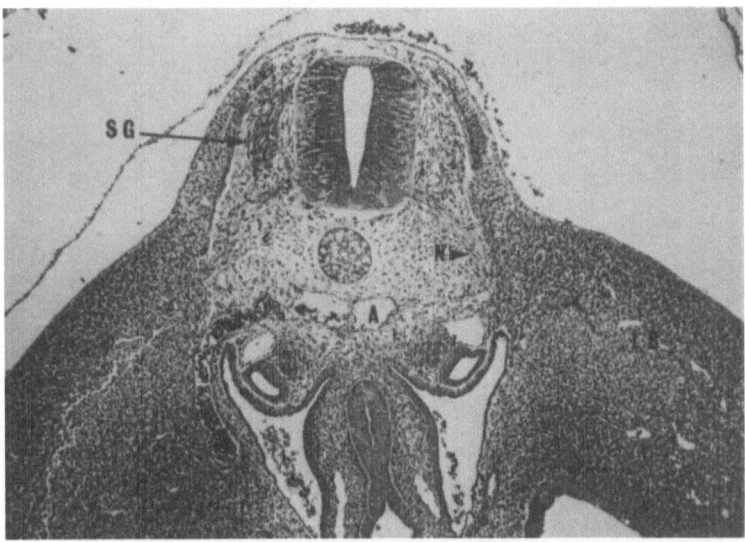

Figure 8.4 Chick embryo, stage 22, transverse section through limb buds. SG, spinal ganglion; N, nerve running into limb bud; LB, limb bud; A, dorsal aorta. Toluidine blue and eosin. × 100

over the window and sealed with adhesive tape. The eggs were opened on the 14th day or later. A further 20 embryos were subjected to excision of the right half of the lumbar neural tube, but were fixed one day after operation so that the histology of the lesions could be studied (Figure 8.3). Five other embryos were fixed and sectioned on the third or fourth day of incubation to act as histological controls. (Figure 8.4).

Seventeen of the embryos subjected to this operation survived to the 14th day of incubation, the others dying 1–7 days after the operation. Nine of the 17 chick embryos were found to have amelia (Figure 8.5). Five others had some degree of reduction in size of the right leg, although the morphology was normal in four, one having a reduction deformity of the right foot (Figure 8.6). Three had no appreciable difference in the appearance of the two legs. (Table 8.1).

Figure 8.5   Chick embryo with amelia of right leg following excision of right lumbar neural tube at stage 13

Figure 8.6 Chick embryo with reduction deformity of right foot following excision of portion of right lumbar neural tube at stage 14

Table 8.1 Outcome of 45 operations for excision of the Right Half of the Lumbar Neural Tube

| | H.H. stage of development at the time of operation | | | |
	13	14	15	16
No. of embryos operated	14	16	11	4
No. of embryos that developed to 14th day	5	5	5	2
Right amelia	5	4	—	—
Reduction deformity of right foot	—	1	—	—
Reduced size of right leg	—	—	3	1
Normal right leg	—	—	2	1

In an effort to insure that it was the destruction of the lumbar neural tube that produced the changes, two further procedures were devised. The first was to surgically excise the four lower somites leaving the neural tube intact. Thirty-four embryos were subjected to this operation, at stages 13–16 of development[6]. Ten of the operated embryos developed to the 14th day of incubation. These 10 embryos were smaller than the controls and in each case the right leg was smaller than the left (Figure 8.7).

In addition a further 10 embryos were allowed to develop until stage 20, then the tip of the right limb bud was excised in order to compare the deformities produced by excision of the apical ectodermal ridge. Ten other embryos exposed with a window, acted as controls. Six of the operated

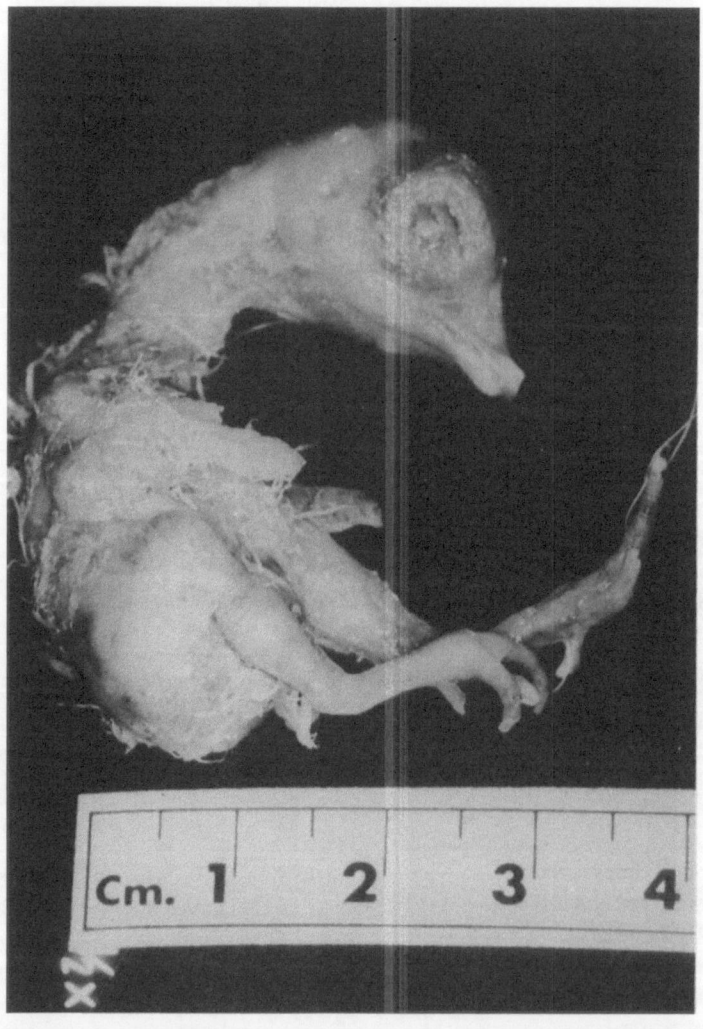

Figure 8.7   Chick embryo which has had lower right somites excised at stage 14

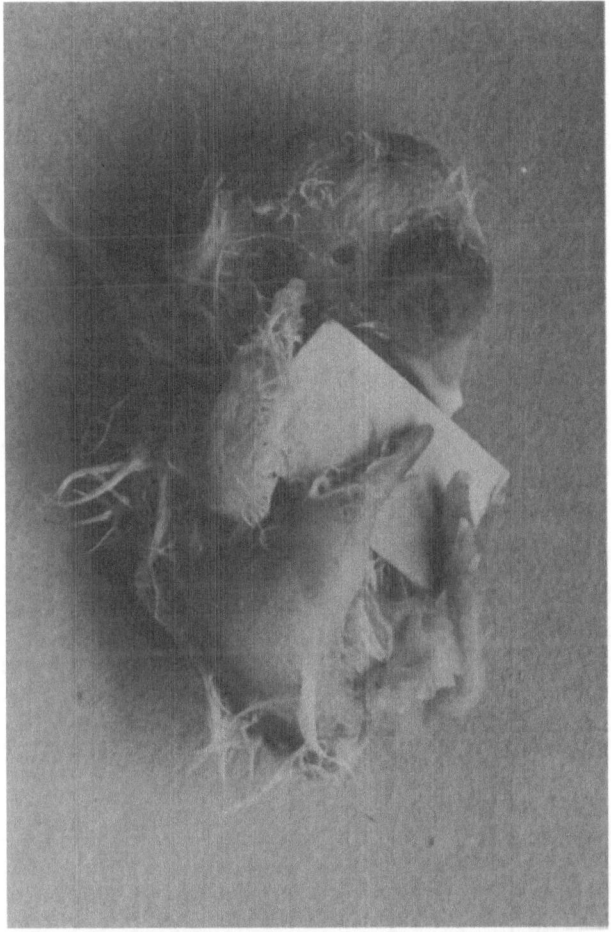

Figure 8.8 Chick embryo showing absence of right lower leg and foot following removal of the tip of right lower limb bud at stage 20

embryos developed until day 14 of incubation. All six exhibited a normal thigh but absence of the lower leg and foot (Figure 8.8).

All the operations were carried out on the right side of the embryo. Of the 148 control embryos not subjected to operative procedure 122 developed to or beyond the 14th day of incubation. None of the control embryos exhibited any deformity (Table 8.2).

In experiments on normal limb outgrowth, limb growth ceased abruptly when the limb bud was denuded of the apical ectodermal ridge, and the degree to which proximal segments later develop depends on the stage at which the operation is performed[7-9]. Therefore it is generally accepted that the ectodermal ridge represents the main controlling factor in limb development. Experiments with limbless mutant chicks showed that the underlying mesenchyme produced a 'maintenance factor' without which the ectodermal thick-

135

Table 8.2 Summary of the outcome of operation on 296 White Leghorn embryos

Operation performed	No. of embryos	No. developed to day 14 or beyond	Type of deformity
Excision of right half of lumbar neural tube stage 13–16	104 attempted (45 successful)	17 (38%)	Right amelia (9) (53%) Reduction deformity of foot (1) (5.9%) Reduced size of right leg (4) (23.5%) Normal (3) (17.6%)
Window cut in shell and membrane stage 13–16	138	122 (88%)	NIL
Destruction of right lower somites stage 13–16	34	10 (29%)	Reduction in size of right leg (10) (100%)
Excision of tip of right leg bud stage 20	10	6 (60%)	Absence of right foot (6) (100%)
Window cut in shell and membrane stage 20	10	10 (100%)	NIL

136

ening cannot survive in its normal form[10]. The 'messages' involved in these reciprocal inductive influences between the ectoderm and mesoderm (apical ectodermal maintenance factor, mesoderm maintenance factor) are diffusible substances capable of passing through millipore filters although their chemical nature is as yet unknown[11].

Although the apical ectodermal ridge in mammals, reptiles and birds forms on the distal extremity of the limb bud, it is more prominent on the post-axial part of the limb bud, ventral to its free edge where it underlies the largest portion of the marginal sinus. The widest part of the marginal venous system differentiates in the *post-axial* half of the distal mesoderm in close association with the apical ectodermal ridge. This topographic relation between these temporary limb structures was reported for human[12] and rat embryos[13]. The latter author suggested that there might be a causal relationship between them; the marginal sinus is probably induced by the ectodermal ridge.

In both humans[14] and rabbits[15] the limb deformities caused by thalidomide (α-phthalimido glutarimide) treatment in early gestation affects predominantly the pre-axial side of the limbs, that is, the area 'where the apical ectodermal ridge is less prominent'. McBride[16] examined the ganglionic hypoplasia accompanying thalidomide-induced limb deformities in 29 rabbit fetuses. The limbs were variably reduced and this was correlated with degenerative changes in the ganglia. Abnormal neurons exhibiting cytoplasmic vacuolation and nuclear karyolysis were prominent in cervical ganglia of deformed fetuses, whereas normal neurons of control fetuses showed large vesicular nuclei with prominent nucleoli. Some degree of hypoplasia with a small proportion of abnormal neurons was found where thalidomide had been administered without resultant deformity, but hypoplastic ganglia with a high proportion of abnormal neurons were present in severe degrees of reduction deformities. From this experiment it was postulated that it is the diminution in the number of sensory neurons that interferes with peripheral organ development. To support this hypothesis the ultrastructure of sensory neurons of the dorsal root ganglia of New Zealand White rabbit embryos and fetuses exposed to thalidomide were examined at early stages of development, before the limb was completely formed[17, 18]. The ultrastructural changes found in the neurons of the dorsal root ganglia consisted of nuclear indentation, cell shrinkage and increased ribosomal activity, whilst the axons exhibited loss of microtubules and microfilaments, mitochondrial degeneration, the presence of membrane bound vesicles and the presence of myelin whorls. As ultrastructural damage to the components of the dorsal root ganglia viz., neurons and axons could be found in day-13 experimental embryos, that is at least 16 h before the earliest signs of thalidomide dysmelia have been reported in rabbits[15], it is regarded as corroborative evidence of the importance of neurons in peripheral development.

According to Lillie[19] the hind limb buds develop lateral to somites 26–32. Seventeen of the 45 embryos where successful excision of the right half of the neural tube was judged to have been accomplished developed to the 14th day of incubation. Nine of these 17 embryos had amelia of the right leg. If the tube excision was performed at stage 13–14 it appeared more likely to produce amelia, if at stage 15–16 reduction in size with normal morphology or no apparent change seemed more probable. Ten of the 34 embryos, where the right lower somites were excised, developed to the 14th day of incubation. None of these embryos had amelia, but all exhibited a reduction in size of the right leg. The type of deformity produced by excision of the right half of the lumbar neural tube differed from the deformities produced by damage to the adjoining somites or to that produced by excision of the distal portion of the limb bud.

In the experiments described, neural tube excision was carried out at the 19–28 somite stage, whereas in Hamburger's transplant operations[2], the donor embryos varied from 24–45 somites. The difference in the stage of development could explain the difference in the results between this experiment and that of Hamburger[2]. It is interesting to note that Hamburger et al.[20], when carrying out a study of motility in the chick embryo in the absence of sensory input, observed that in 37 embryos where they excised the dorsal half of the lumbar spinal cord and carried out simultaneous extirpation of the entire spinal cord at the thoracic level to exclude sensory input from more rostral levels, in three of the embryos one leg was missing and, in several others, distal parts of the leg or toes were absent. These workers dismissed the deformities as 'probably nonspecific effects of the operation'.

This experiment has shown that excision of a portion of the neural tube in chick embryos will produce reduction deformities in the ipsilateral leg if performed at H.H. stage 13–14. Earlier experiments undertaken by this author have shown that chemical neuron damage in rabbit embryos is associated with reduction deformities. It is postulated that neural tissue exerts an inductive influence in the development of the limb. It is possible that limb development depends upon a combination of organizers, which are the neurons which will supply the limb and the apical ectodermal ridge. Destruction of the nerve supply of the early embryonic limb bud will deprive it of the neurotrophic influences, thus resulting in morphological deformities of the limb. It is possible that in transplant experiments where aneurogenic but otherwise morphologically normal limbs form, that the limb bud having developed prior to its excision is no longer dependent upon the neurotrophic influences.

References

1. Spemann, H. (1938). *Embryonic Development and Induction.* (New Haven: Yale Univ. Press)
2. Hamburger, V. (1939). The development and innervation of transplanted limb primordia of chick embryos. *J. Exp. Zool.*, **80**, 347
3. Hamburger, V. and Waugh, M. (1940). The primary development of the skeleton in nerveless and poorly innervated limb transplants of chick embryos. *Physiol. Zool.*, **13**, 367
4. Kieny, M. and Fouvet, B. (1974). Innervation et morphogenèse de la patte chez l'embryon de poulet. *Arch. Anat. Microsc.*, **63**, 281
5. Cihák, R. (1976). *Proceedings of International Symposium on Limb Development.* Prague (In press)
6. Hamburger, V. and Hamilton, H. L. (1951). A series of normal stages in the development of the chick embryo. *J. Morphol.*, **88**, 49
7. Saunders, J. W. Jr. (1948). The proximo-distal sequence of origin of parts of the chick wing and the role of ectoderm. *J. Exp. Zool.*, **108**, 363

8. Zwilling, E. (1961). Limb morphogenesis. *Adv. Morphogenet.*, 1, 300
9. Saunders, J. W. Jr., Cairns, J. M. and Gasseling, M. T. (1957). The role of the apical ridge of ectoderm in the differentiation of limb parts in the chick. *J. Morphol.*, 101, 57
10. Saunders, J. W. Jr. and Gasseling, M. T. (1963). Trans-filter propagation of apical ectoderm maintenance factor in chick embryo wing bud. *Dev. Biol.*, 7, 64
11. Saunders, J. W. (1969). The interplay of morphogenic factors. In: *Limb Development and Deformity: Problems of Evaluation and Rehabilitation.* C. A. Swinyard (ed.) pp. 84–100 (Springfield, Illinois: Thomas)
12. O'Rahilly, R., Gardner, E. and Gray, D. J. (1956). The ectodermal thickening and ridge in the limbs of stages human embryos. *J. Embryol. Exp. Morphol.*, 4, 254
13. Milaire, J. (1956). Morphological and cytochemical research on limb buds in the rat. *Arch. Biol.*, 67, 297
14. Lenz, W. (1962). Thalidomide and congenital abnormalities. *Lancet*, i, 271
15. Vickers, T. H. (1967). Concerning the morphogenesis of the thalidomide dysmelia in rabbits. *Br. J. Exp. Pathol.*, 48, 579
16. McBride, W. G. (1974). Fetal nerve cell degeneration produced by thalidomide in rabbits. *Teratology*, 10, 283
17. Stokes, P. A., Lykke, A. W. J. and McBride, W. G. (1976). Ultrastructural changes in the dorsal root ganglia evoked by thalidomide preceding limb development. *Experientia*, 32, 597
18. McBride, W. G. (1976). Studies of the etiology of thalidomide dysmorphogenesis. *Teratology*, 14, 71
19. Lillie, F. R. (1927). *The Development of the Chick.* 2nd Ed. (New York: Holt)
20. Hamburger, V., Wenger, E. and Oppenheim, R. (1966). Motility in the chick embryo in the absence of sensory input. *J. Exp. Zool.*, 162, 133

9
Mechanisms of limb teratogenesis: malformations in chick embryo induced by nitrogen mustard

BERTHE SALZGEBER

INTRODUCTION

Limb malformations occur frequently and have approximately the same features in the different vertebrate species. As with other malformations appearing spontaneously, they can be attributed either to hereditary, to environmental factors or to the association of both. It is well known that the teratogenic response may be different from one strain to another.

The vertebrate limb represents a very convenient model to study the genesis of malformations[1]. Limb abnormalities of different kinds may be produced by means of numerous chemical or physical agents. On the other hand, mutants are known which also present various types of limb defects. Finally, to understand the genesis of malformations, it is important to know the normal development of the organs. In this field of descriptive and experimental embryology, a great deal of work has been done on limb development.

The present report is concerned with studies on the chick embryo, which is a very useful laboratory model . Microsurgical operations are easy to perform on the embryos and for this reason, much experimental work has been carried out on limb morphogenesis[2].

Teratogenic agents may be applied to the embryo at a precise stage of development. Thus, specific types of malformations can be induced, allowing the study of the genesis of some limb abnormalities.

PRODUCTION AND DESCRIPTION OF LIMB DEFECTS IN CHICK

Limb malformations were induced by means of numerous teratogenic agents. Some were non-specific in their effects; others caused a high rate of limb deformities[3, 4]. Thus, X-rays and alkylating agents were used with success in the production of various limb defects.

The total irradiation of the chick embryo at a dose of 500 to 800 r and at various stages of development, from stage 8 to stage 35 H.H. (2 to 9 days of incubation) produced different kinds of limb malformations[5].

A regional sensitivity to X-rays was observed in a disto-proximal direction if the irradiation was performed on embryos, stages 12 to 17. But with increasing age, stage 19 to 35, this sensitivity showed progression in a proximo-distal direction. The abnormalities consisted in reduction in size, fusion or absence of some bones[5].

With the technique of localized irradiation[6], it was possible to submit to X-rays either a whole or a part of the limb bud[7–9]. Specific malformations were thus produced. The results of these investigations showed that the digits were always inhibited in the same order. In the case of the leg, digit I disappeared first, followed by digits II, IV and III. Such a predominance of one area on another has also been observed in the case of the radius and the ulna, the tibia and the fibula. The irradiation of a leg bud determined the disappearance of the fibula, whereas the tibia persisted. It was suggested that, after injury of the bud, there exists a competition for mesenchymal cells between some areas of the bud. The tibia develops at the expense of the tissues which should form the fibula in the normal embryo. If the X-rays were applied to the presumptive areas of the tibia only, the fibula developed and was of the same size as the tibia. The competitive effect was suppressed. Such differential sensitivity of some limb bud primordia was also observed after treatment of the embryos with teratogenic substances.

Nitrogen mustard (methyl-bis-β-chloroethyl amine) coded HN2 is a strong mutagenic and carcinogenic substance which interferes with mitosis and DNA synthesis[10]. In addition, this drug is teratogenic in a variety of species

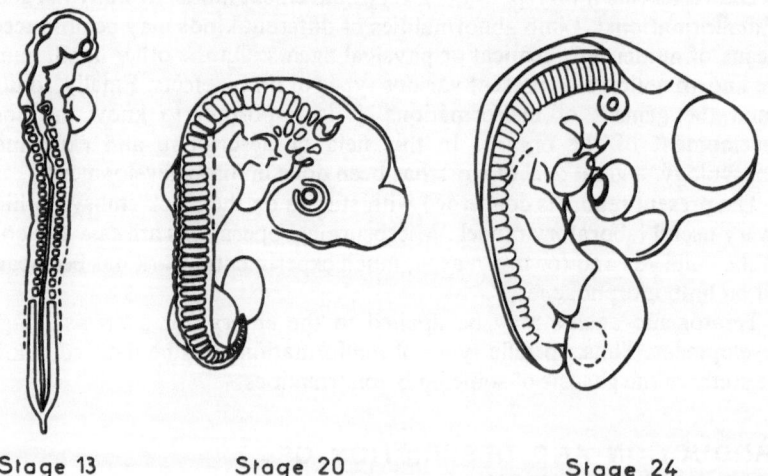

Stage 13   Stage 20   Stage 24

Figure 9.1 Nitrogen mustard (HN2) was applied onto chick embryos at various stages of development. For example: stage 13 = 19 somites (50 h); stage 20 = 3.5 days; stage 24 = 4.5 days

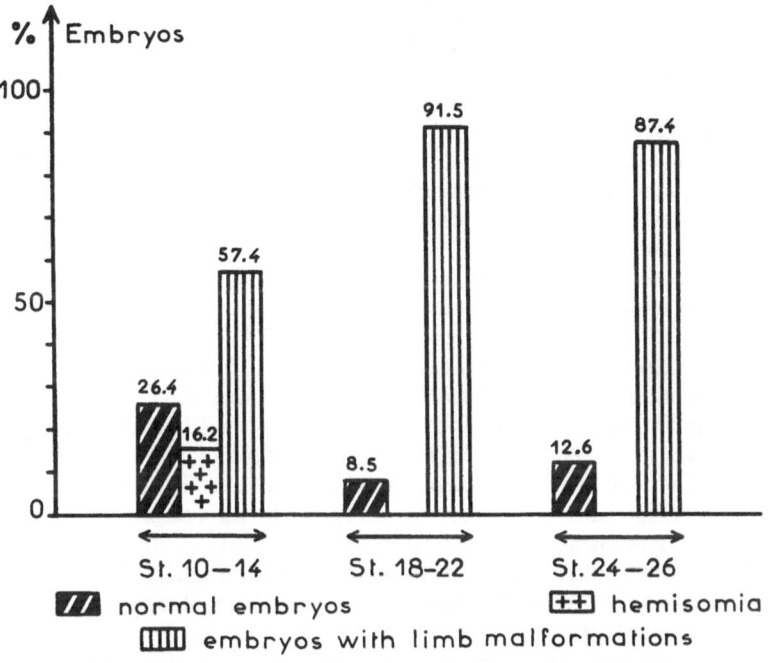

Figure 9.2 Incidence of limb defects after treatment of the embryos at different developmental stages

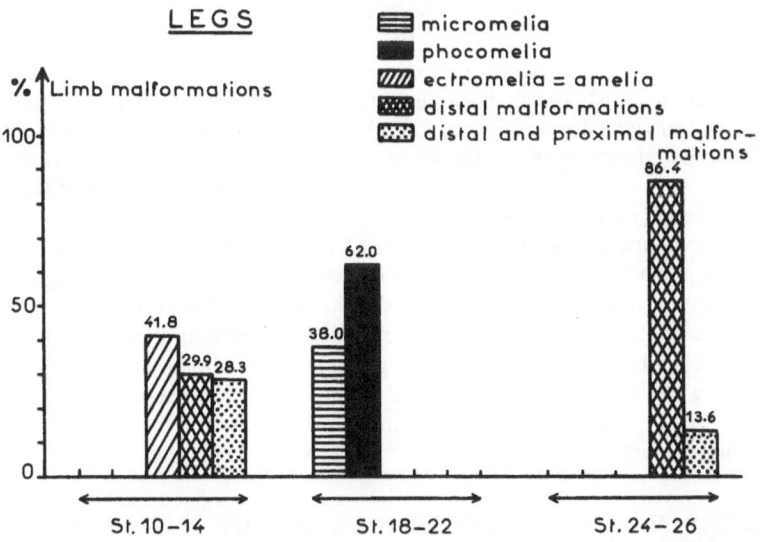

Figure 9.3 Distribution of the various leg malformations in chicks treated with HN2 and which survived the 8th day of incubation. The values are expressed in percentage of the number of embryos which showed limb malformations. Stages 10–14 (Hamburger and Hamilton), 10 to 23 pairs of somites, dose: 0.6–0.9 μg/egg. Stages 18–22 (H.H.) 3–3.5 days, dose: 1.2–1.6 μg/egg. Stages 24–26 (H.H.) 4.5 days, dose: 1.5–3 μg/egg

including amphibians, birds and mammals[11]. Among numerous malformations, limb defects appeared frequently.

We investigated the effects of nitrogen mustard (mustine hydrochloride; Boots) on the development of chick and quail embryos.

The substance diluted in Tyrode solution or distilled water, was applied to chick or quail embryos from 2 to 5 days of incubation. In most cases, it was inoculated over the embryo through a small window cut in the shell. At the time of treatment, the embryo stages were noted according to the seriation of Hamburger and Hamilton[12].

The type of malformations observed 8–15 days later depended on the dose of the teratogenic drug, the method of administration and the stage of the embryo at the time of treatment (Figures 9.1, 9.2 and 9.3).

Specific digit malformations (adactylia) occurred after administration of the drug to embryos, stages 24 to 26 H.H. (4.5 to 5 days)[13].

Micromelia (brachymelia) and phocomelia appeared when the embryos were treated at stages 18 to 21 H.H. (3 to 4 days of incubation)[14]. These malformations are characterized by an important reduction or absence of the long bones (Figures 9.4, 9.5, 9.9 and 9.11). In the case of phocomelia, distal parts, entirely or partly preserved, were linked directly to the hip (legs) or to the shoulder (wings) (Figure 9.9). In the case of micromelia, the reduction of some bones (femur, tibia, metatarsals) was less severe. But even in this case, the fibula was absent.

It should be mentioned that some results concerning the reduction in the number of digits and the action on the fibula were similar to those observed after treatment of the embryos with X-rays.

Distal abnormalities and long bone deficiencies were observed after treatment of younger embryos, with 10 to 23 pairs of somites[13, 15]. The foot or the hand or both structures were missing (Figure 9.7), the limb often being reduced to a short stump (complete hemimelia) (Figure 9.12). In extreme cases, the limb was absent (Figure 9.6). Amelia (ectromelia) appeared in 41.8% of the affected embryos. It should be noted that some embryos treated at stage 15 (24 to 27 pairs of somites) showed a reduction in the size of the tibia and the fibula (Figure 9.10).

The incidence of the different kinds of malformation after administration of nitrogen mustard at various stages of development is shown in Figure 9.3. Among these limb abnormalities, we should mention the appearance of some other type of malformations such as eye, beak defects and lordosis. Furthermore, the embryos showed general growth retardation. The limb malformations were very characteristic and occurred in a constant way. For this reason, we decided to use nitrogen mustard for the study of the genesis of some limb defects.

GENESIS OF LIMB DEFECTS INDUCED BY NITROGEN MUSTARD

The normal development of the limbs depends on a sequence of embryological processes. I should briefly recall that the limb bud of a 3-day-old chick embryo is made up of two tissues: a mesenchyme forming the central part of

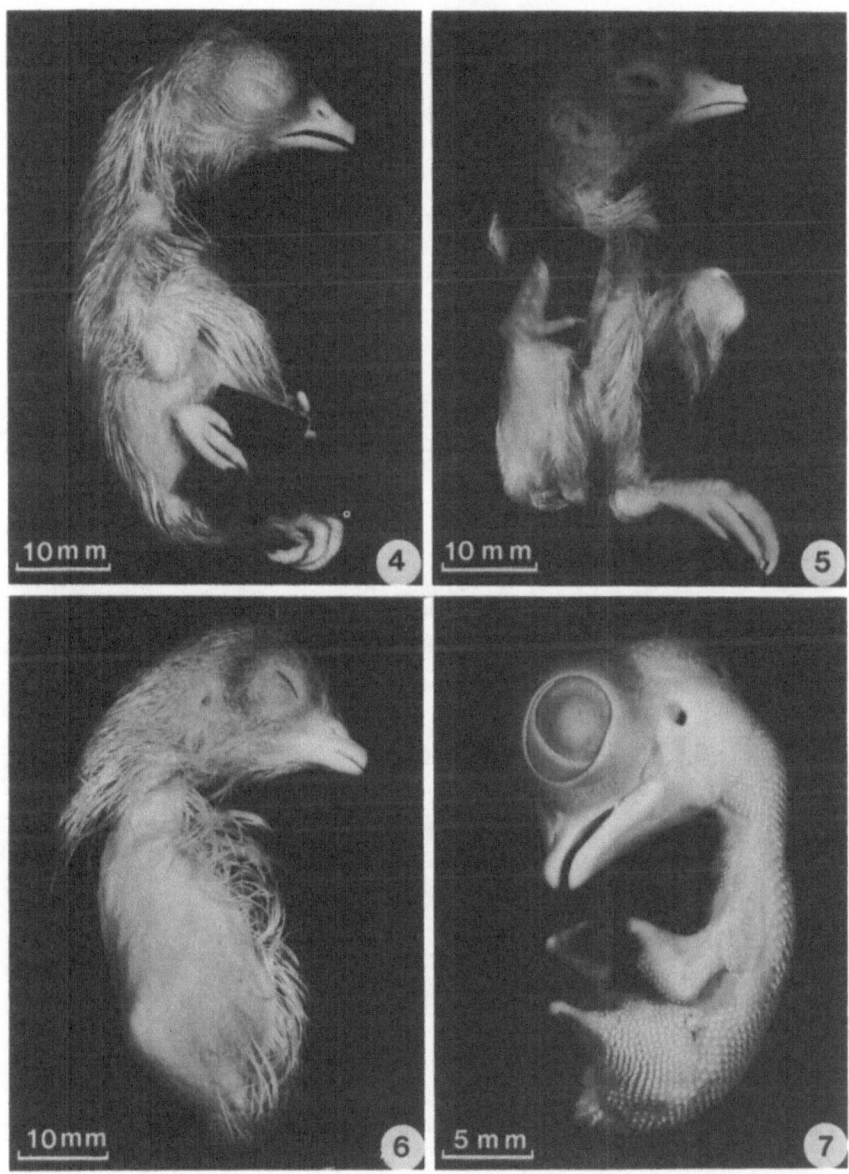

Figure 9.4   Limb malformation (phocomelia) after HN2 treatment on day 3.5 (stage 19 H.H.). The foot is linked directly to the body of the 17-day-old embryo

Figure 9.5   Another type of a phocomelia limb (right leg). Two digits are fused (syndactyly). The leg on the left side is only slightly reduced

Figure 9.6   Amelia (ectromelia) malformation. Note the absence of all four limbs. This embryo was treated at stage 12 with HN2. Feathers have been removed

Figure 9.7   An 11-day-old embryo which had been exposed to HN2 at stage 13 H.H. (19 pairs of somites). Note the stumpy left leg. The right leg shows a single digit

145

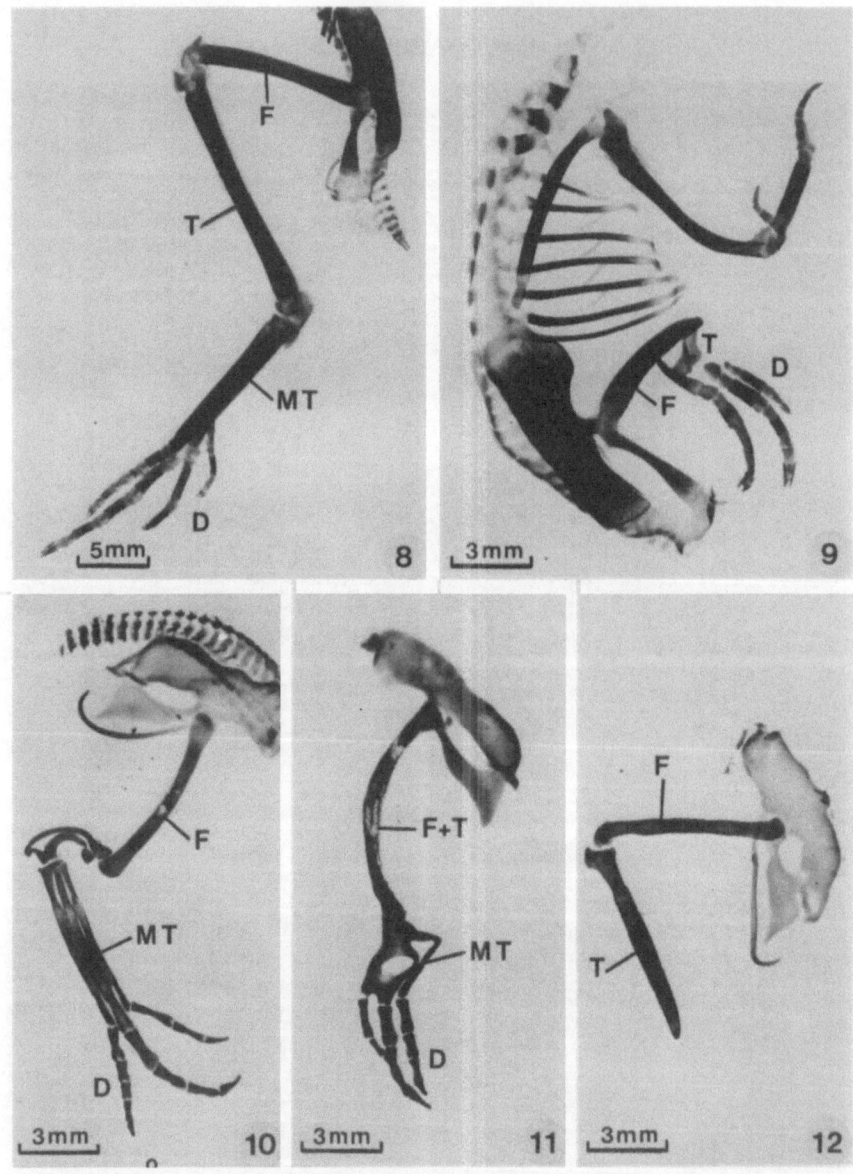

Figure 9.8 Normal leg skeleton of a chick embryo (age 13 days). Cleared alcian blue and alizarine red stained preparations. F = femur, T = tibia, MT = metatarsals, D = digits

Figure 9.9 Leg and wing of a HN2 treated embryo. Note phocomelia of right hindlimb. The tibia (T) and the metatarsals are reduced in size; three digits are well developed. Absence of the first digit

Figure 9.10 Hindlimb skeleton of an embryo treated at stage 15. Note the presence of fibula and tibia, both reduced in size. The metatarsals (MT) are normal. Absence of the first digit

Figure 9.11 Hindlimb skeleton of a chick embryo treated at stage 19. Fusion of femur and tibia. Metatarsals abnormal and reduced in size

Figure 9.12 Leg reduced to a stumpy limb, observed in a 13-day-old chick embryo treated at stage 13. Note the absence of a distal part

the bud and an ectodermal unicellular covering which is thickened at its distal end to a pluricellular epithelium called the apical ectodermal ridge (AER). The removal of the AER at various developmental stages inhibited the growth of the limb bud and resulted in limb defects characterized by the absence of the distal part[16, 17]. The normal development of the limb is determined by a succession of interactions between both tissue components of the limb bud, thus ensuring its growth, differentiation and polarity. (For a recent review on this subject see ref. 2.)

In the course of the present study, we shall see that these tissues may be selectively sensitive to a teratogenic substance, at a precise stage of development. The failure of either tissue affects the inductive processes leading thus to typical malformations.

We investigated the ectodermal–mesodermal interactions in limb buds treated with nitrogen mustard. These experiments were performed on 3.5 and 2.5-day-old embryos in order to study the genesis either of phocomelia and micromelia or hemimelia and amelia.

Proximal part defects: phocomelia

These limb malformations were observed after treatment of 3- to 4-day-old chick embryos, stages at which the limb buds are evident. Most of the surviving embryos, up to 8 days, showed a high frequency of phocomelia malformations.

The first detectable morphological lesions could be distinguished between 24 and 30 h after administration of the drug. Growth was inhibited. After 48 h, haematomas of different sizes were seen in the central part or at the base of the bud. The examination at later stages revealed no more oedema or haemorrhagic cysts. The limbs were reduced in size, but they looked healthy.

Histological studies

Histological study[18] of the limb buds was carried out within the first hours or days of treatment. No significant alterations were observed during the first hours. The examination, 7 h after exposure, revealed a depletion of mitosis in the mesodermal part of the bud. The changes were much more pronounced 15–24 h after treatment. Numerous abnormal chromosomal figures and enlarged nuclei were found in the mesenchyme whereas the ectodermal apical cap develops normally in most cases.

The Feulgen–Rossenbeck reaction showed clearly the difference in the sensitivity of the two tissue components. The reaction was much lower in the nuclei of the mesodermal tissue than in the apical cap nuclei (Figures 9.13 and 9.14). These results indicate a reduction in DNA content in the cells of the mesodermal tissue.

An autoradiographic study of tritiated thymidine incorporation ([^{3}H]TdR; spec. act. 5–11 Ci/mM, dose 30 μCi/egg) into limb buds has been carried out in order to examine the effects of nitrogen mustard on the synthesis of DNA in limb buds *in vivo*[19]. Four hours after the treatment of the embryos with nitrogen mustard, a marked inhibition (40–60%) of the incorporation of the labelled thymidine was observed in some areas of the mesenchyme and

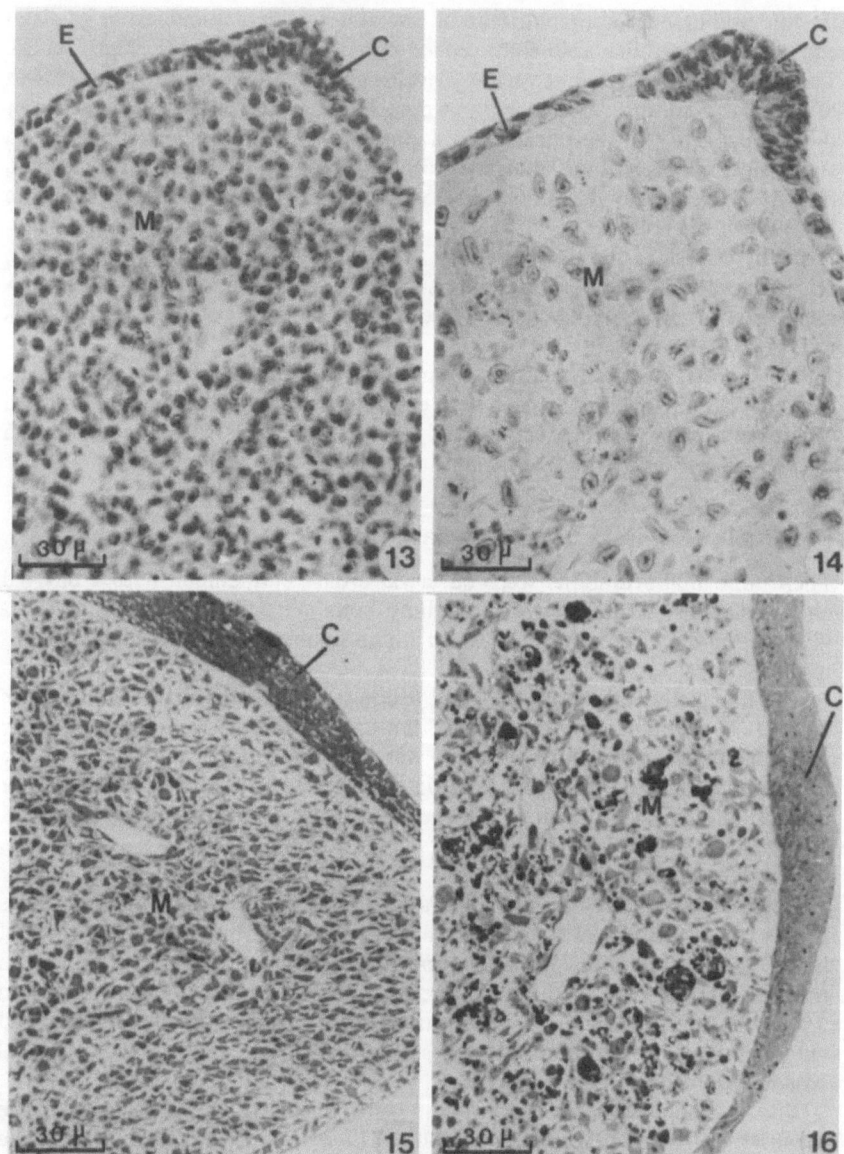

Figure 9.13 Section through a control leg bud removed from a 4.5 day-old embryo. Feulgen–Rossenbeck reaction. E = ectodermal layer, C = apical crest, M = mesenchyme

Figure 9.14 Section through a limb bud fixed 24 h after treatment of a 3.5-day-old embryo (stage 19 H.H.). Note the enlarged nuclei in the mesenchyme. The Feulgen reaction is less intense in these nuclei than in those of the untreated tissue

Figure 9.15 Thin epon section (1–2 μ) stained with toluidine blue through a control leg bud

Figure 9.16 Thin epon section through a bud, 30 h after treatment with HN2. In the mesenchyme, note numerous necrotic areas. The structure of the apical cap is not modified

the apical cap of the leg bud. However, later on, 18–24 h after treatment, there was still a significant difference between control and treated buds in the case of the mesenchyme, whereas the incorporation values were identical to those of the controls in the case of the ectodermal cap (Figure 9.17). It may be that there is some correlation between the degree of inhibition of DNA synthesis and the teratogenic response.

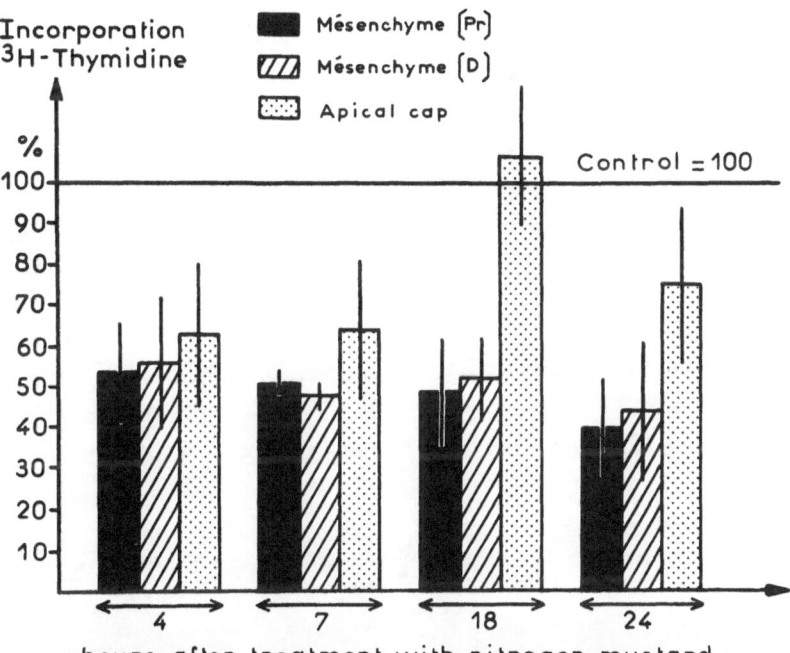

Figure 9.17 Tritiated thymidine incorporation in limb buds after treatment with HN2. The line represents the control values. The amount of incorporation is expressed as a percentage of the controls. The values are the mean of 10 to 14 determinations and they are expressed as the percentage of the same number of control values. Standard error of the mean ($p \geqslant 95\%$)

Thirty to 48 h after treatment, the mesodermal tissue showed a loosened structure in which vacuoles, lacunae, pycnotic nuclei and necrotic areas can be recognized.

These areas of necrosis (Figures 9.15 and 9.16) were essentially detected either by staining the limb bud sections with methyl green pyronine[18] or by staining the embryo with vital dyes such as Nile blue sulphate or neutral red according to the techniques of Saunders and Fallon[20] and Hinchliffe and Ede[21]. Numerous deeply stained masses were distributed throughout the whole mesenchyme of the limb buds, whereas in controls, cell death occurred in very limited areas called the anterior and posterior necrotic zones (ANZ and PNZ). Histochemical studies showed that strong acid phosphatase activity was localized in these stained regions.

These results indicate that there is a strong phagocytic activity which could be attributed to macrophages.

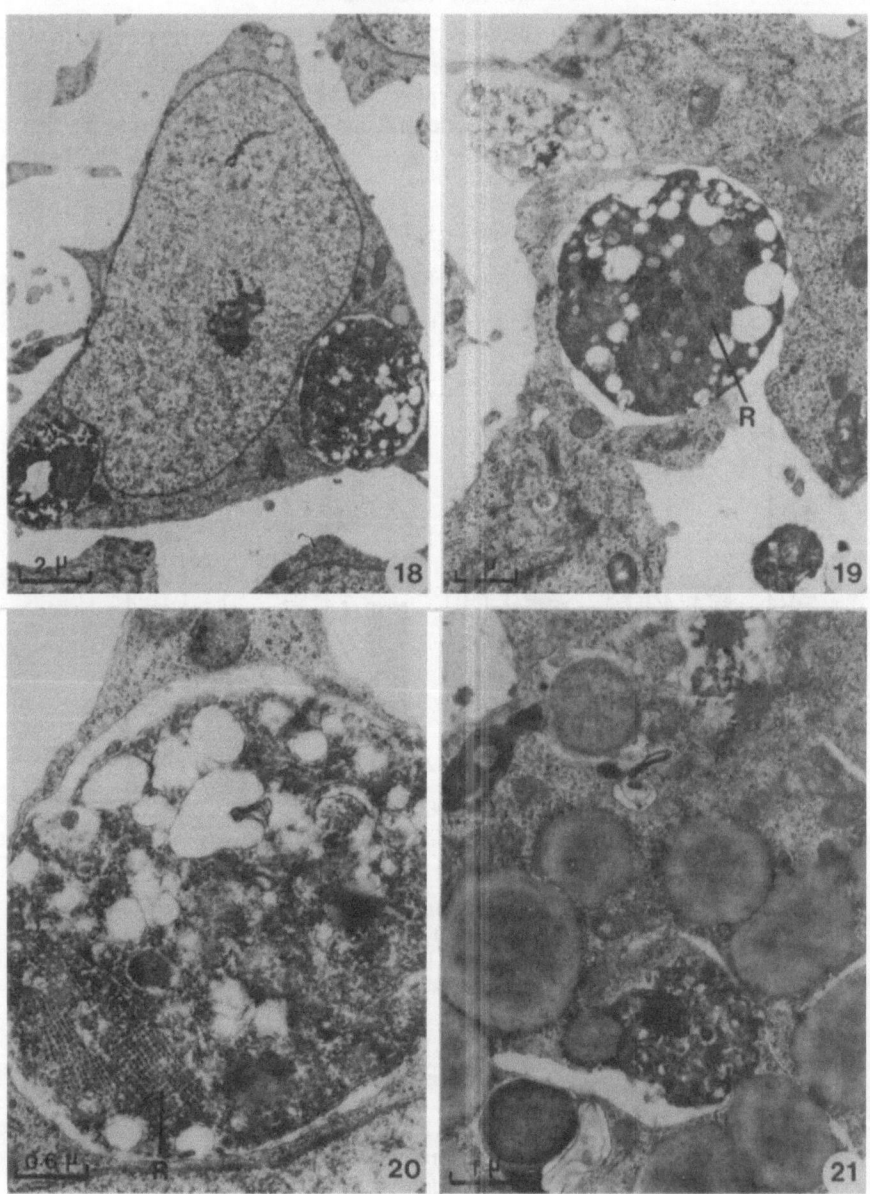

Figure 9.18   Electronmicrograph of the mesenchyme of a limb bud, 24 h after treatment with HN2. Note the presence of necrotic areas in the cytoplasm of the cell

Figure 9.19   Vacuolated mass of cytoplasm containing banded granules (R) and degenerating mitochondria (M)

Figure 9.20   Phagocytosed mass of cytoplasm with numerous vacuoles, altered mitochondria and banded granules (R)

Figure 9.21   Electronmicrograph of a bud fixed 48 h after treatment with HN2. Numerous lipid droplets are present in the degenerating cells

Electron microscopical studies were carried out to study the nature of these intensely stained cells which are morphologically identical to those observed by several authors in the necrotic areas of normal buds[22]. These observations showed that from 15–48 h after treatment, numerous kinds of alterations appeared in the mesenchyme: enlargement and distortion of the nuclei, dense bodies, numerous vacuoles, phagolysosomes (Figures 9.18, 9.19, 9.20 and 9.21). A most characteristic feature was the presence of banded granules. Important lipid inclusions and myelin figures were found in these degenerated cells, essentially 48 h after treatment.

As a result of this ultrastructural analysis, the localization of acid phosphatase could be observed in some cell structures: autophagic vacuoles, lysosomes, Golgi apparatus. Seventy-two hours after treatment, the pycnotic areas disappeared; most cells were healthy and divided actively. But wide spaces which often contain cellular debris or haemorrhagic zones were found in different areas of the limb bud. At this stage of development, the control limb showed a cellular condensation which preceded the formation of the cartilage. In the treated buds, chondrogenesis was absent or delayed. Four to five days after treatment, the limb, although reduced, showed a typical wing or leg structure. The number of subsisting cartilaginous pieces depended on the degree of degeneration of the bud.

Most of the defects observed in the mesodermal tissue of the treated limb buds resulted in the appearance of phocomelic or micromelic types of malformation. In the case of more severe mesodermal injuries, the apical cap was also affected; the buds having thus degenerated, the absence of limbs (amelia) was noticed a few days later.

These observations suggest that of the two limb components, it is the mesodermal tissue which is most affected. In contrast, the ectodermal cap appeared healthy.

Tissue interaction studies
In order to study the interactions of both mesodermal and ectodermal tissues, the following experiments were carried out. Firstly, we examined the effect of nitrogen mustard on isolated whole limb buds, which were then grafted onto a host embryo, according to the technique of Hampé[17]. Wing buds of a host embryo were removed and replaced by the treated buds. These grafts were kept in place with small platinum needles. The host embryos were sacrificed 7 or 10 days after the operation.

The results were similar to those observed in *in vivo* studies. Depending on the doses and on the time of exposure, the teratogenic substance affected the buds in varying degrees resulting in the appearance of the three typical malformations: brachymelia, phocomelia and amelia. This type of experiment showed clearly that nitrogen mustard acts directly on the limb bud itself[23].

Secondly, the next step in these studies was to determine what tissue was responsible for the appearance of the various malformations[24].

The HN2 or the Tyrode treated limb buds were dissociated either with trypsin or EDTA into their two components according to Zwilling's method[25].

The two isolated tissues were then recombined in the following manner. The mesoderm of a nitrogen mustard treated limb bud was associated with

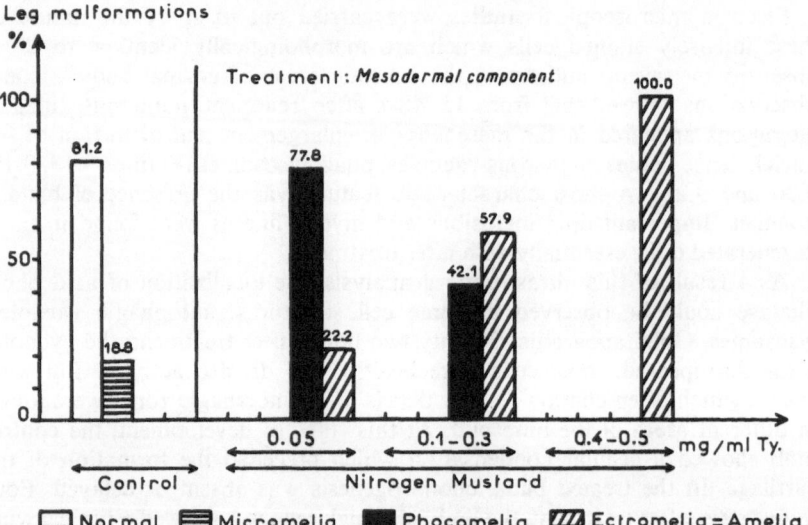

Figure 9.22 Interchange of limb bud components. The mesodermal component removed from buds treated with various doses of HN2 (0.05–0.5 mg/ml Tyrode) was recombined to a non-treated ectodermal part. This bud was grafted on to a host embryo, which was sacrificed at 10 days of incubation. At lower doses, phocomelia appeared; with increasing doses, the buds did not develop (amelia)

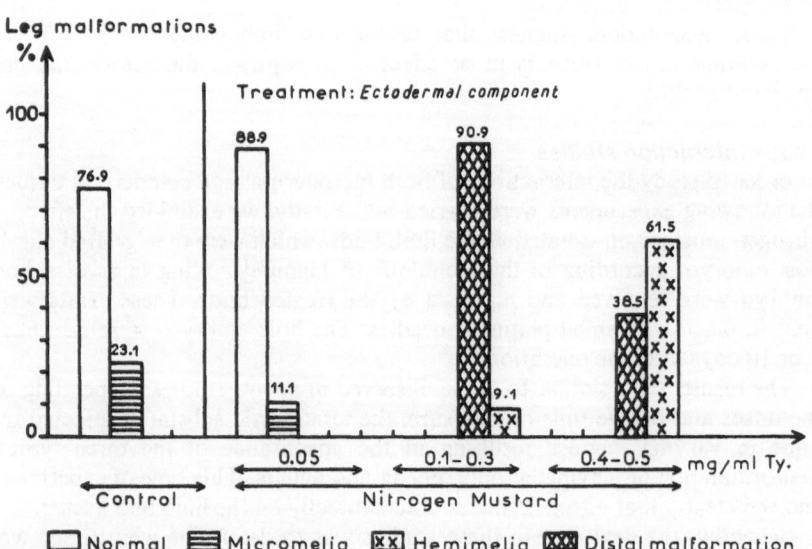

Figure 9.23 Interchange of limb bud components. The ectodermal component came from buds treated with HN2 (0.05–0.5 mg/ml Tyrode) and was associated to a normal, non-treated mesodermal component. The development of the grafted bud was normal when lower doses (0.05 mg/ml Tyrode) were applied. With increasing doses, distal malformations and stumpy legs appeared

an ectodermal cap of a healthy bud. On the other hand, the mesoderm of a normal limb bud was recombined with an ectodermal part of a treated bud.

The restored limb buds were then grafted onto host embryos (stages 18 to 20) as in the experiments of grafting whole limb buds described above. The embryos were examined 24–48 h later to verify the success of the graft, and were sacrificed at 8 to 10 days of incubation for external morphology and skeleton studies.

It should be noted that in most cases, the two tissues which were recombined came either from wing or from leg buds. But in some cases, chimaeric associations were noted between a wing component and a leg component.

The results of these investigations (Figures 9.22 and 9.23) showed clearly that in the case of a recombination of a nitrogen mustard treated mesoderm with an untreated ectodermal cap, the limb malformations which appeared were similar to those obtained after exposure of whole buds, namely micromelia, phocomelia and amelia. The treatment of the mesodermal component with a solution of 0.05 mg HN2/ml Tyrode produced micromelic–phocomelic malformations (Figures 9.24 and 9.25). With increasing doses of HN2, phocomelic limbs with only one digit appeared (Figure 9.31) and in extreme cases no limbs developed (amelia). The treatment of the ectodermal part produced quite a different picture. A bud resulting from the association of a treated ectodermal tissue (0.05 mg HN2/ml Tyrode) with a healthy mesoderm developed normally (Figures 9.26 and 9.27). But with increasing doses (0.1 to 0.5 mg HN2/ml Tyrode) the distal part of the limb was affected. With lower doses, only slight abnormalities were observed. The distal phalanges were reduced in size or fused (Figures 9.28 and 9.29). When higher concentrations of the drug (0.5 mg HN2/ml Tyrode) were used, a stumpy limb appeared, in which case, the distal and a part of the proximal segment were missing (Figure 9.30).

These experiments are illustrated in the table below which gives the results concerning the leg bud (Figure 9.32). But similar effects have been found in the case of the wing bud.

Chimaeric associations[26] between leg and wing bud components showed that the malformations were similar to those one may observe after homologous recombination of limb bud components. The nature of the limb was not modified. It developed into leg or wing depending on the origin of the mesoderm.

The results establish clearly that the induced limb malformations in these experiments were due to a specific effect on the bud's mesodermal component. These observations allowed us to give an embryological explanation for the genesis of the malformations.

The studies on normal limb development showed that the growth of the limb bud occurs in a proximo-distal sequence; the primordia of proximal segments differentiate first whereas those of distal structures differentiate later. A topographical map of the presumptive cartilaginous areas has been established by means of carbon particle markers[16, 17, 27], radioactive thymidine labelling[28] and growth rate studies[29].

At the stage the embryos were treated with HN2, the proximal prospective bone areas were proliferating regions and thus were especially sensitive to the

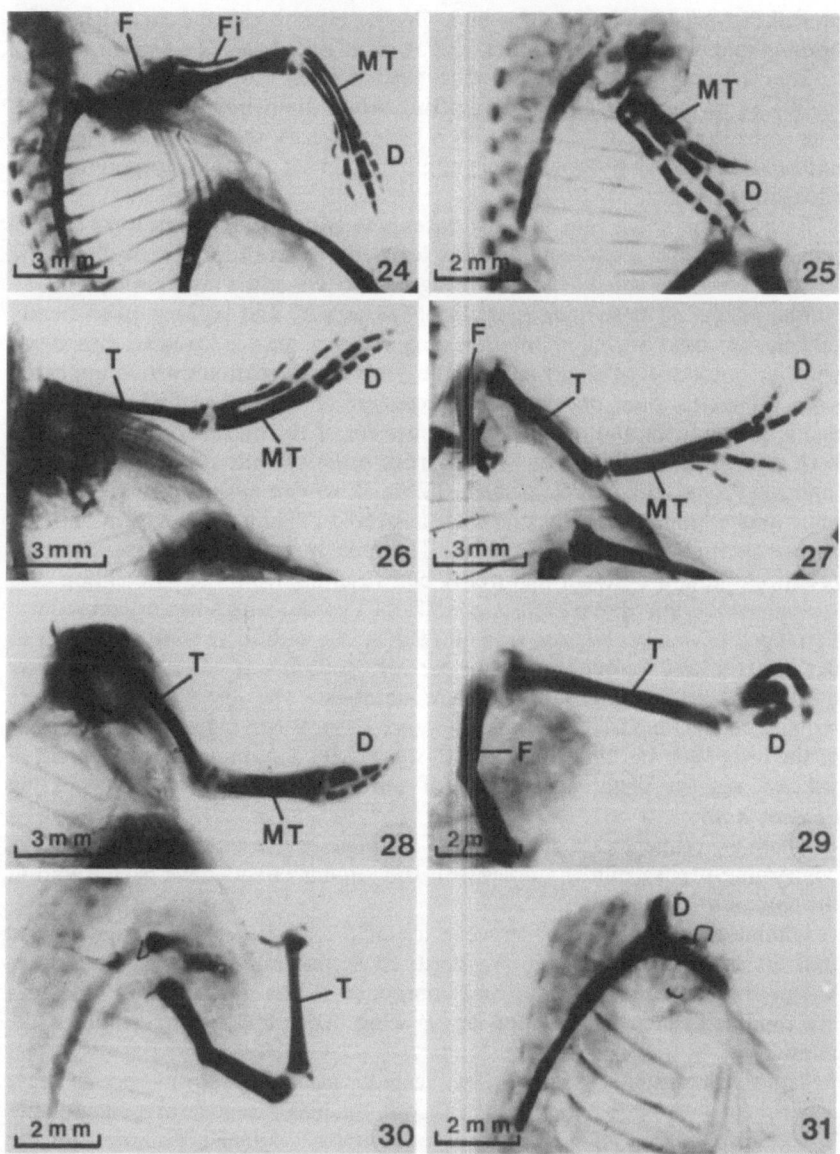

Figure 9.24 Control leg developed from an ectoderm–mesoderm exchange between two leg buds of the same stage (stage 19). Treatment of the mesodermal component with Tyrode. The recombined limb bud was grafted in place of the excised wing bud of the host. F = femur, Fi = fibula, MT = metatarsals, D = digits

Figure 9.25 Phocomelic leg developed from a limb bud composed of a normal ectoderm and a treated mesoderm (8 min *in vitro*, HN2 0.05 mg/ml Tyrode. Control leg Figure 9.24)

Figure 9.26 Control leg. Treatment of the ectodermal component with Tyrode

Figure 9.27 The leg developed from a limb bud composed of a normal mesoderm and a treated ectoderm (8 min *in vitro*, HN2 0.05 mg/ml Tyrode). The leg is normal

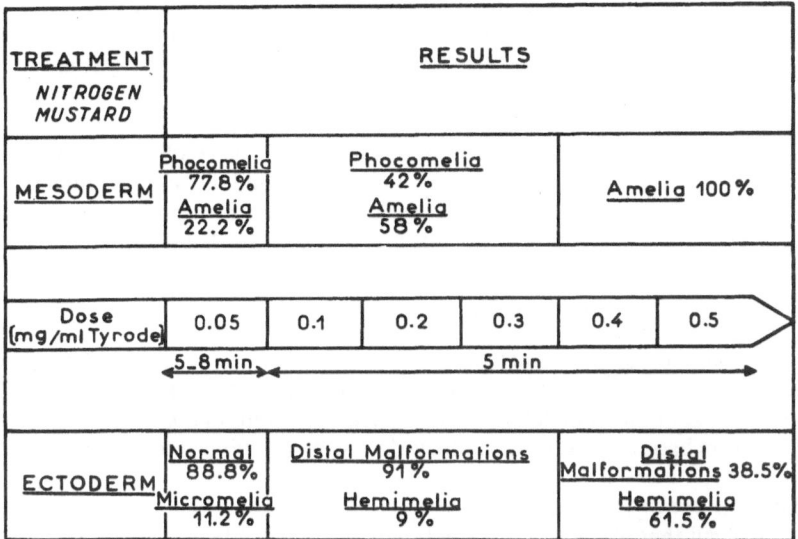

TREATMENT *NITROGEN MUSTARD*	RESULTS		
MESODERM	Phocomelia 77.8% Amelia 22.2%	Phocomelia 42% Amelia 58%	Amelia 100%

Dose (mg/ml Tyrode)	0.05	0.1	0.2	0.3	0.4	0.5
	◄ 5–8 min ►	◄	5 min			►

ECTODERM	Normal 88.8% Micromelia 11.2%	Distal Malformations 91% Hemimelia 9%	Distal Malformations 38.5% Hemimelia 61.5%

Figure 9.32 Production of different types of limb defects after treatment, dissociation and recombination of the leg bud components (stage 18 to 20). 1. Treatment mesoderm: HN2 treated mesodermal component was associated with a normal ectodermal part. 2. Treatment ectoderm: HN2 treated ectodermal component was recombined with a normal mesoderm

drug. At this stage, the AER induced these mesodermal areas to differentiate into the proximal limb structures. After treatment, numerous mesenchymal cells degenerated; a part of the tissue was destroyed and lost, whereas the apical cap looked healthy. The affected areas, at that time in a state of differentiation were unable to respond to the inductive action of the apical crest. But later on, 72 h after treatment, cell death was discontinued and there was a recovery of the tissue. This latter and above all the more distal regions, which at this stage of development were destined to produce the distal parts of the limb, responded to the induction of the non-affected ectodermal ridge. It should be noted that some experiments support this hypothesis.

If nitrogen mustard was applied *in vivo* at precise stages of development, it produced specific abnormalities. If given to 4.5–5-day-old embryos, at the

Figure 9.28 The leg developed from a recombined bud composed of a healthy mesodermal part and a treated ectodermal part (5 min *in vitro*, HN2 0.2 mg/ml Tyrode). Note the abnormal digits

Figure 9.29 The leg developed from a bud made of a normal mesoderm and a treated ectoderm (5 min *in vitro*, HN2 0.2 mg/ml Tyrode). Note the shortening of the digits (D), and fusion of phalanges

Figure 9.30 The leg developed from a bud composed of a normal mesoderm and a HN2 treated ectoderm (5 min *in vitro*, HN2 0.5 mg/ml Tyrode). Note the absence of the distal part (hemimelia)

Figure 9.31 A leg rudiment (a single digit) developed from a limb bud composed of a normal ectoderm and a treated mesoderm (5 min *in vitro*, HN2 0.1 mg/ml Tyrode)

stage when the distal parts of the limb were forming, the latter would be affected. Proximal long bones (femur, tibia-fibula and metatarsals) developed but the foot was missing. At this stage, the distal area was the most sensitive to the drug and it may be that the apical cap could not induce this area to develop the distal parts of the limb.

Another possible explanation is that after the recovery of the mesodermal tissue the ectodermal cap had aged and thus could no longer induce the proximal part of the bud. At this stage of development it acts on the underlying distal area of the bud which is destined to produce hand and foot.

Distal part defects: hemimelia

After treatment of the embryos at earlier stages of development (10 to 14 H.H.; 2 to 2.5 days of incubation) many types of malformations were observed. Most of the limb deformities consisted of reduction or absence of hand and foot; in extreme cases, the limbs were absent or reduced to stumps.

Development of the bud

Let us recall that at the stage of treatment, 10 to 14 H.H. the limb bud cannot be distinguished. However, the wing and leg buds exist as presumptive areas.

The development of the buds was followed in the egg, a few hours to several days after the treatment of the embryos. Forty hours after the administration of the drug, the buds had developed; most of them (43 cases out of 51) had an opaque aspect with an abnormal shape (Figures 9.33, 9.34, 9.35 and 9.36). They became elongated cones, with an apical cap at the distal end. Three to four days after treatment, the growth of some buds ceased; at autopsy, the limbs were absent. The site of the limb bud was indicated by a small swelling. Histological studies carried out 24 to 96 h after exposure of the embryo to the drug revealed that an apical crest was present, but both components, the mesenchyme and the apical crest, were altered (Figures 9.37 and 9.38). Numerous enlarged nuclei and abnormal mitotic figures were first observed in the mesenchyme, but 40 to 48 h after treatment, lesions could be recognized in the apical cap (Figure 9.39). This tissue degenerated, and 60–72 h after treatment, 10 wing buds out of 17 (58.8%) were devoid of an apical crest, whereas the other 7 buds showed this structure (Figures 9.40 and 9.41). In three cases, however, the apical crest invaginated into the mesenchyme instead of being prominent. Some buds did not develop. On removing the embryo from the egg at 10 to 12 days of incubation, the limbs were absent (amelia). These experiments showed that the malformation amelia was not the result of agenesis but of a degeneration of the buds soon after their appearance. The growth of other buds was less inhibited, but after several days no distal paddle was observed, and the bud became pointed.

Ectodermal–mesodermal interactions

Embryological studies showed that the mesodermal tissue is the primary inductor of the limb[30, 31]. Thus prospective leg mesoderm, devoid of its ectodermal covering, grafted into the flank of 48 h-old embryos developed a supernumerary limb. To explore the effects of nitrogen mustard on either

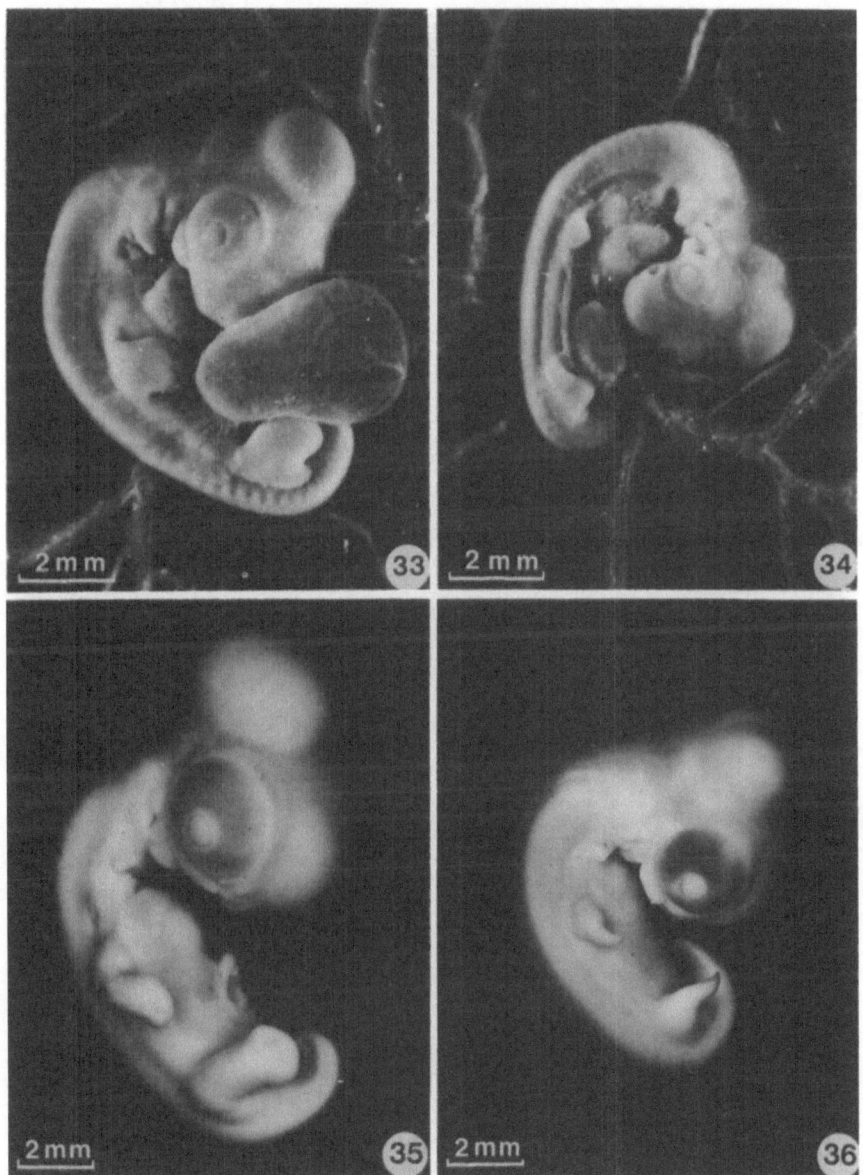

Figure 9.33 Normal embryo, stage 24 (H.H.), 4.5 days

Figure 9.34 Embryo treated with HN2 at stage 14 (H.H.), 21 pairs of somites and fixed 50 h later. Note the small size of the embryo and the cone-shaped limb buds

Figure 9.35 Normal embryo, stage 26 (H.H.), 5 days

Figure 9.36 Embryo treated with HN2 at stage 14 (H.H.) and fixed 72 h later. Note the absence of the distal plate. Both limb buds, wing and leg are reduced to stumps

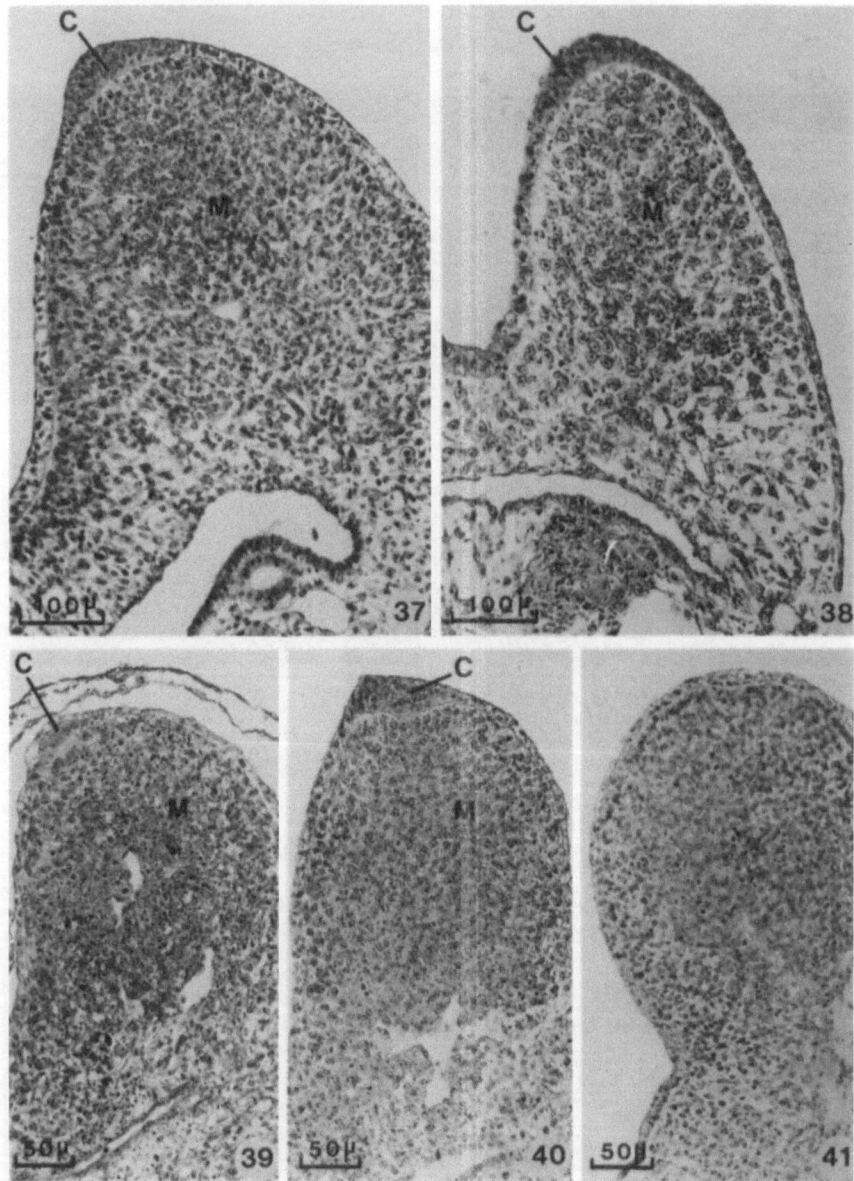

Figure 9.37 Section through a normal limb bud from a 3.5-day-old embryo. C = apical cap, M = mesenchyme

Figure 9.38 Limb bud from an embryo treated with HN2 at stage 14 (21 pairs of somites) and fixed 28 h later. Note the enlarged nuclei in the mesenchyme

Figure 9.39 Treatment of the embryo at stage 14. Fixation of the limb bud, 42 h after treatment. Degeneration in both components, the mesenchyme (M) and the apical cap (C)

Figure 9.40 Embryo treated with HN2 at stage 14 (21 pairs of somites). Fixation 64 h after treatment. The apical cap (C) is abnormal in shape

Figure 9.41 Limb bud from an embryo treated with HN2 at stage 14. Note the absence of an apical cap

component of these presumptive areas, the following experiments were carried out[32].

(i) The mesodermal tissue of a presumptive leg area was taken from a HN2 treated embryo (24 to 26 pairs of somites), and was then grafted into the flank of a normal embryo (stage 15 to 24 pairs of somites).

(ii) On the other hand, the mesoderm of a normal limb area was implanted under the flank of a treated embryo.

The exposure time to HN2 (0.03 mg/ml Tyrode) varied from 4–6 h. The host embryos were removed from the eggs at 12 days of incubation.

The results of these experiments showed that the injury of the mesodermal component (experiment (i)) caused the reduction of the proximal segments of the limb; the distal parts were present in 58% of the recovered grafts. In contrast (experiment (ii)), after the treatment of the ectodermal part (host's flank), stumpy limbs appeared (42%). According to these results, the absence of the limb's distal parts observed after treatment of young embryos, may be due to a deficiency of the epidermal tissue. The malformation amelia (ectromelia) appeared in both experimental groups, after treatment either of the mesodermal or the ectodermal component.

CONCLUSIONS

Through research in experimental embryology, knowledge has been gained not only about the normal development but also about the genesis of abnormalities. On the other hand, studies on the effects of various teratogenic factors have contributed to the understanding of limb morphogenesis.

The stage of treatment of the embryo, the dose and the nature of the teratogenic agent are of considerable importance in the production of a specific limb malformation. Thus phocomelia was observed after treatment of 3- to 4-day-old embryos (18–21 H.H.). Hemimelia and adactylia appeared after treatment of the embryo at an early stage (10–14 H.H.) when the apical cap was developing, or at a later one (24–26 H.H.) when the distal part of the bud differentiated. Ectromelia was produced at various stages.

The type of limb defects observed in chick embryos following nitrogen mustard treatment approximates that caused by some mutants. Heterozygote creeper fowl show a shortening of the long bones (achondroplasia) of micromelic aspect. But this defect is more marked in homozygotes whose four limb buds are of phocomelic type[33].

In wingless mutants, the wings are absent or reduced to a stump. Zwilling[34] reported that in this mutant, the buds developed until the third or fourth day of incubation and then regressed. The apical ectodermal ridge (AER) degenerated.

Other mutants are affected with distal limb defects, e.g. syndactylism, brachydactylism and polydactylism. To our knowledge, phenotype-specific polydactyly, with one or more extra toes at the extremities has not yet been produced experimentally.

Abnormal development may result from interference with various embryological processes, such as morphogenetic movements, regulation, growth, degeneration and tissue interactions[35].

One of the mechanisms involved in the genesis of limb defects is the failure or disturbance of tissue interactive processes at a precise stage of development.

Our experiments concerning ectoderm—mesoderm interchange between nitrogen mustard treated and normal tissues showed that different types of malformations were produced according to the component, either ectodermal or mesodermal, which is affected.

Reciprocal recombinations of normal and mutant limb bud tissues were carried out with several mutant embryos, for example, talpid[36], eudiplopodia[37], duplicate polydactyly[38], wingless[39]. These experiments showed that the genetic disease was localized, either in the mesodermal component (talpid) or in the ectodermal part of the bud (eudiplopodia).

Cell death and degeneration of tissues occurred in most cases after treatment of the embryos with teratogenic substances. The resulting malformation may depend on an extension of the necrotic areas. Thus, some tissues recover from the effects of the treatment, whereas other tissues are destroyed and totally lost. Abnormal degeneration and cellular death were also observed in case of genetic defects. In the wingless mutant, there is an extension of areas of naturally occurring cell death[21] whereas in the mutant talpid, the normal cell death is inhibited[40].

Most of the studies on teratogenesis which have been reported in the literature reveal the diversity of limb malformations which may be obtained by means of a chemical substance. On the other hand, various chemical compounds produce the same type of malformation. However, the mechanisms by which a limb defect arises in a particular case, may be different.

The mode of action of most of the teratogenic drugs is actually unknown. Some substances act in a direct manner by injuring specific tissues at a precise stage of development, other substances act in an indirect manner by disturbing either the vascular system of the organ or the metabolism of the embryo[4].

Evidence gained from studies of embryological mechanisms may be directed towards the elucidation of the genesis of limb malformation. In addition, the explanation of limb malformation in biochemical terms, together with an understanding of the mode of action of teratogenic agents, are two important aspects which require further study before the goal of prevention can be achieved.

ACKNOWLEDGEMENTS

I wish to express my thanks for technical assistance to B. Henri for the photographic illustration, to A. Placenti for the line drawings and E. Bourson for typing the manuscript.

References

1. Bergsma, D. and Lenz, W. (1977). *Morphogenesis and malformation of the Limb*. Birth Defects: Original Article Series, Vol. 13, p. 364 (New York: Alan R. Liss, Inc.)
2. Ede, D. A., Hinchliffe, J. R. and Balls, M. (1977). *Vertebrate Limb and Somite Morphogenesis*, p. 498 (Cambridge University Press)
3. Ancel, P. (1950). *La Chimiotératogenèse chez les Vertébrés*, p. 397. (Paris: G. Doin)
4. Landauer, W. (1969). Dynamic aspects of hereditary and induced limb malformations. In: C. A. Swinyard (ed.). *Limb Development and Deformity: Problems of Evaluation and Rehabilitation*, pp. 120–137 (Springfield: Charles C. Thomas)
5. Goff, R. A. (1962). The relation of developmental status of limb formation to X-radiation sensitivity in chick embryos. I. Gross study. *J. Exp. Zool.*, 151, 177
6. Wolff, Ét. (1936). Les bases de la tératogenèse expérimentale des Vértebrés amniotes, d'après les résultats de méthodes directes. *Arch. Anat. Hist. Embryol.*, 22, 1
7. Schué, M. (1951). L'action des rayons X sur le développement des ébauches de pattes chez l'embryon de poulet. *C.R. Soc. Biol.*, 145, 752
8. Wolff, Ét. and Kieny, M. (1962). Mise en évidence par l'irradiation aux rayons X d'un phénomène de compétition entre les ébauches du tibia et du péroné chez l'embryon de poulet. *Dev. Biol.*, 4, 197
9. Wolff, Ét. (1958). Le principe de compétition. *Bull. Soc. Zool. Fr.*, 83, 13
10. Wheeler, G. P. (1962). Studies related to mechanism of action of cytotoxic alkylating agents – a review. *Cancer Res.*, 22, 651
11. Chaube, S. and Murphy, M. L. (1968). The teratogenic effects of the recent drugs active in cancer chemotherapy. In: D. H. M. Woollam (ed.) *Advances in Teratology*, Vol. III, pp. 181–237 (London: Logos Press)
12. Hamburger, V. and Hamilton, H. L. (1951). A series of normal stages in the development of the chick embryo. *J. Morphol.*, 88, 49
13. Salzgeber, B. (1972). Production expérimentale de malformations distales des membres chez l'embryon de poulet et de caille après traitement par l'ypérite azotée. *Ann. Embryol. Morphol.*, 5, 145
14. Salzgeber, B. (1963). Production élective de la phocomélie sous l'influence d'ypérite azotée. *J. Embryol. Exp. Morphol.*, 11, 413
15. Salzgeber, B. (1972). Limb malformation production in chick embryo treated with nitrogen mustard at 48 hours of incubation. In: M. A. Klingberg, A. Abramovici and J. Chemke (eds.). *Drugs and Fetal Development*, pp. 175–187 (New York: Plenum Press)
16. Saunders, J. W. (1948). The proximo-distal sequence of origin of the parts of the chick wing and the role of the ectoderm. *J. Exp. Zool.*, 108, 363
17. Hampé, A. (1959). Contribution à l'étude du développement et de la régulation des déficiences et des excédents dans la patte de l'embryon de poulet. *Arch. Anat. Micr. Morph. Exp.*, 48, 345
18. Salzgeber, B. (1966). Production élective de la phocomélie sous l'influence d'ypérite azotée, chez l'embryon de poulet. II. Étude histologique des bourgeons de membres au cours du développement. *J. Embryol. Exp. Morphol.*, 16, 339
19. Salzgeber, B. (1975). Action de l'ypérite azotée sur l'incorporation de thymidine tritiée dans les bourgeons de membres de l'embryon de poulet. Étude histoautoradiographique. *C.R. Acad. Sci.*, 280, 911
20. Saunders, J. W. and Fallon, J. F. (1967). Cell death in morphogenesis. In: M. Locke. *Major Problems in Developmental Biology*, pp. 289–314 (New York: Academic Press)
21. Hinchliffe, J. R. and Ede, D. A. (1973). Cell death and the development of limb form and skeletal pattern in normal and wingless (ws) chick embryos. *J. Embryol. Exp. Morphol.*, 30, 753
22. Dawd, D. S. and Hinchliffe, J. R. (1971). Cell death in the 'opaque patch' in the central mesenchyme of the developing chick limb: a cytological cytochemical and electron microscopic analysis. *J. Embryol. Exp. Morphol.*, 26, 401
23. Salzgeber, B. and Pinot, M. (1968). Étude sur la genèse de malformations expérimentales des membres chez l'embryon de poulet: micromélie, phocomélie, ectromélie. I. Traitement à l'ypérite azotée de bourgeons de membres entiers. *Ann. Embryol. Morphol.*, 1, 83
24. Salzgeber, B. (1969). Étude sur la genèse de malformations expérimentales des membres

chez l'embryon de poulet: micromélie, phocomélie, ectromélie. II. Expériences de dissociation et de réassociation des deux constituants du bourgeon de membre après traitement par l'ypérite azotée. *Ann. Embryol. Morphol.*, 1, 313

25. Zwilling, E. (1955). Ectoderm–mesoderm relationship in the development of the chick embryo limb bud. *J. Exp. Zool.*, 128, 423
26. Salzgeber, B. (1969). Étude comparative des effets de l'ypérite azotée sur les constituants, mésodermique et ectodermique, des bourgeons de l'embryon de poulet. *J. Embryol. Exp. Morphol.*, 22, 373
27. Amprino, R. and Camosso, M. (1958). Analisi sperimentale dello sviluppo dell'ala nell'embrione di pollo. *Wilhelm Roux Arch. Entw. Mech. Org.*, 150, 509
28. Stark, R. J. and Searls, R. L. (1973). A description of chick wing bud development and a model of limb morphogenesis. *Dev. Biol.*, 33, 138
29. Lewis, J. H. (1975). Fate maps and the pattern of cell division: a calculation for the chick wing-bud. *J. Embryol. Exp. Morphol.*, 33, 419
30. Kieny, M. (1960). Rôle inducteur du mésoderme dans la différentiation precoce du bourgeon de membre chez l'embryon de poulet. *J. Embryol. Exp. Morphol.*, 8, 457
31. Dhouailly, D. and Kieny, M. (1972). The capacity of the flank somatic mesoderm of early bird embryos to participate in limb development. *Dev. Biol.*, 28, 162
32. Salzgeber, B. (1976). Étude sur la genèse des malformations des parties distales des membres chez l'embryon de poulet (Adactylie, Hémimélie). *C.R. Acad. Sci.*, 283, 1241
33. Landauer, W. (1932). Studies on the Creeper fowl. III. The early development and lethal expression of homozygous Creeper embryos. *J. Genet.*, 25, 367
34. Zwilling, E. (1949). The role of epithelial components in the developmental origin of the 'Wingless' syndrome of chick embryos. *J. Exp. Zool.*, 111, 175
35. Saxen, L. and Rapola, J. (1969). *Congenital Defects*, p. 247 (New York: Holt, Rinehart and Winston, Inc.)
36. Goetinck, P. F. and Abbott, U. K. (1964). Studies on limb morphogenesis. I. Experiments with the polydactylous mutant, Talpid. *J. Exp. Zool.*, 155, 161
37. Goetinck, P. F. (1964). Studies on limb morphogenesis. II. Experiments with the polydactylous mutant Eudiplopodia. *Dev. Biol.*, 10, 71
38. Zwilling, E. and Hansborough, L. A. (1956). Interaction between limb bud ectoderm and mesoderm in the chick embryo. III. Experiments with polydactylous limbs. *J. Exp. Zool.*, 132, 219
39. Zwilling, E. (1956). Interaction between limb bud ectoderm and mesoderm in the chick embryo, IV. Experiments with a wingless mutant. *J. Exp. Zool.*, 132, 241
40. Hinchliffe, J. R. and Thorogood, P. V. (1974). Genetic inhibition of mesenchymal cell death and development of form and skeletal pattern in limbs of talpid 3 (ta$_3$) mutant chick embryos. *J. Embryol. Exp. Morphol.*, 31, 747

10
Experimental amniocentesis and teratogenesis: clinical implications

L. A. KENNEDY AND T. V. N. PERSAUD

INTRODUCTION

Chemicals and drugs are often injected into the amniotic cavity of laboratory mammals in order to study their direct effects on the development of the embryo. In doing so the physiological fetal–placental–maternal unit remains intact and the mediating influence of the placenta and the maternal organism are eliminated[1]. The intra-amniotic route of administration has also been used in the chick embryo[2]. Evidence is accumulating, however, which suggests that the results of such experiments cannot be attributed exculsively to the test substance, but that the method itself is at least partially responsible[3, 4]. In rodents, the term amniocentesis dysmelia syndrome[5] has been used to describe the congenital anomalies which characteristically result from puncture of the amniotic sac. A comparable 'straight jacket syndrome' has been described in chicks[2].

Amniocentesis has also been used in obstetric practice for more than a century. In recent years it has become a routine procedure for the management of pregnancies complicated by rhesus or ABO isoimmunization disease and for the prenatal diagnosis of genetic and other disorders of the fetus. The possibility of risk to the fetus was first suggested by Dewhurst in 1956[6], and recently several cases of fetal injury following amniocentesis have been reported[7, 8].

Extensive clinical follow-up studies of the offspring of women whose pregnancies were monitored with amniocentesis have failed to demonstrate effects similar to those induced following experimental amniotic sac puncture[9, 10]. A problem exists, therefore, of how to reconcile the apparently conclusive yet contradictory reports from laboratory and clinical investigations regarding the safety of amniocentesis.

Recent investigations have linked fetal lung hypoplasia with withdrawal of amniotic fluid in rats and with oligohydramnios resulting from prolonged leakage of amniotic fluid in human pregnancy[11]. Furthermore, the features of the experimental amniocentesis syndrome bear a striking resemblance to those described for the oligohydramnios or Potter's syndrome of human infants. Both syndromes are associated with a premature loss of amniotic fluid and intra-uterine compression.

The following series of experiments describe various aspects of the teratogenicity of amniotic sac puncture in rats, attempts which were made to eliminate or ameliorate the observed deleterious effects, and the pathogenesis of the characteristic head and limb defects. Based on the results of these experiments, a working hypothesis of the aetiology is discussed which integrates the ostensibly contradictory results of clinical and experimental investigations regarding the safety of amniocentesis.

AMNIOTIC SAC PUNCTURE AND TERATOGENESIS

It is unclear whether the results of teratological studies using the intra-amniotic route of administration can be attributed solely to the test substance or if, in fact, the experimental procedure itself is not at least partially responsible. Previous reports have provided contradictory evidence regarding the safety of the method[1, 12–14]. Therefore, experiments were designed to assess the teratogenicity of various aspects of the procedure[3]. Group 1 animals received no treatment at all and served as untreated controls. In the four experimental treatment groups the animals were anaesthetized with sodium pentobarbital, laparotomy was performed, and both uterine horns were exteriorized. The fetuses of the right uterine horn were otherwise untouched and served as treatment controls. The experimental fetuses in the left uterine horn were treated as follows:

Group 2 – The amniotic sacs were not punctured.

Group 3 – A 26 gauge needle was introduced into the amniotic sac through the antimesometrial surface of the uterus, taking care to avoid vascular damage, pressure on the amniotic sac, and forceful entry and exit of the needle.

Group 4 – Following the insertion of the needle, 50 μl of sterile distilled water was injected.

Group 5 – 100 μl of sterile distilled water was injected. Each treatment group consisted of three pregnant Sprague-Dawley rats (10 to 25 fetuses) and each treatment was administered to different groups on one of days 14, 15 or 16 of gestation. On day 20 the fetuses were recovered and their growth and development were evaluated using the following parameters: fetal mortality; crown–rump length (CRL), weight (Wt), CRL/Wt ratio, placental weight (P.wt), placental index (Wt/P.wt), and gross external malformations in the survivors. These measurements were analysed using mixed analysis of variance, t-test, and percentage histograms. For more detail, see Kennedy and Persaud[3].

The interpretation of the quantitative results relating to the surviving fetuses was hampered by the high fetal mortality rates in some treatment groups and the loss of many fetuses from the analysis. However, it was possible to make the following conclusions:

(a) Since there were no differences between the fetuses of groups 1 and 2 it was concluded that the experimental procedure alone (i.e. anaesthesia, laparotomy, and the exteriorization of the uterine horns) had no deleterious effects on the growth and development of the offspring. This is consistent with previous findings[15] and suggests that the high mortality rates in the treated fetuses in groups 3, 4 and 5 was the result of puncturing the fetal membranes (Figure 10.1).

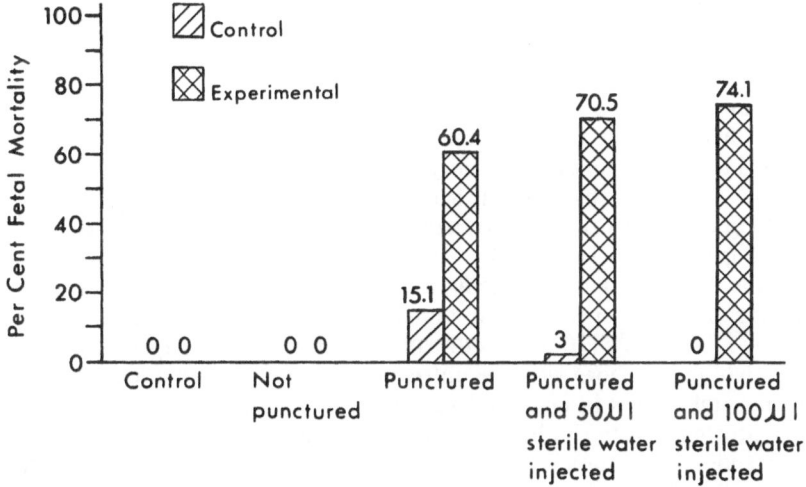

Figure 10.1   Fetal mortality rates (all treatment days combined) following the administration of the different treatments

(b) Amniocentesis on days 14, 15 and 16 resulted in a high incidence of mortality (Figure 10.2) and a significant reduction in the weight and CRL of surviving fetuses. Furthermore, the survivors generally displayed signs of severe compression, postural moulding and stunting. These observations are consistent with previous reports[13, 16, 17].

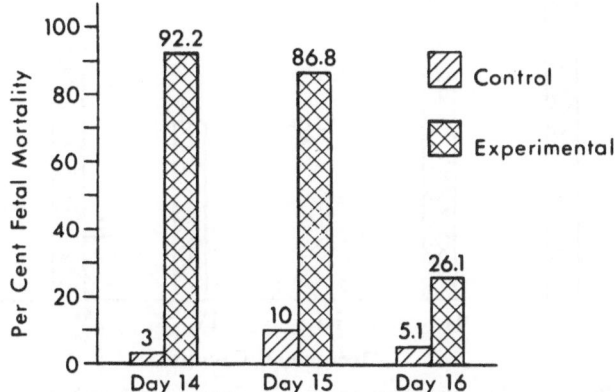

Figure 10.2   Fetal mortality rates (all treatments combined) on day 14, 15 or 16 of gestation

(c) With increasing gestational age at the time of treatment, the incidence of fetal mortality decreased (Figure 10.2). This probably reflects the increased ability of the fetus to resist damage with advancing maturity or a reduced fetal plasticity.

(d) By reducing the volume of solvent injected into the amniotic cavity, the fetal mortality rate was slightly reduced (Figure 10.1). It appears that by minimizing the volume of solvent used as a vehicle, the risk to the fetus can be reduced.

CONTROL OF TERATOGENICITY

The administration of a test substance using the intra-amniotic route has many advantages in experimental teratology. Therefore experiments were designed to investigate the possibility of eliminating or ameliorating the threat presented to the fetus by amniotic sac puncture[4].

Taking day 16 as the most resistant age, an attempt was made to reduce the leakage of amniotic fluid by reducing the calibre of the needle used to make the puncture, and by placing a square of surgical haemostatic sponge over the puncture site. Each treatment group consisted of five pregnant Sprague-Dawley rats (20 to 35 fetuses). In treatment groups 1 to 4 the fetuses of the right uterine horn were the untreated controls. The amniotic sacs of the experimental fetuses in Groups 1 to 3 were punctured with a 26, 30 and 34 gauge needle respectively. In Group 4 the puncture was made with a 26 gauge needle and a haemostatic sponge was placed over the puncture site. In Groups 5 and 6, the effects of puncturing alone (fetuses of the right horn) were compared with the effects of puncturing plus sealing (fetuses of the left horn) following the insertion of a 26 gauge needle (Group 5) or a 34 gauge needle (Group 6). For further details see Kennedy and Persaud[4]. The following conclusions were made:

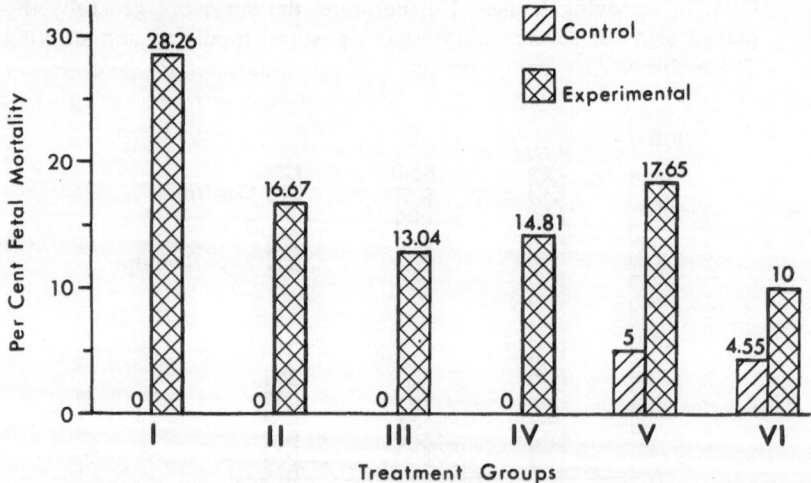

Figure 10.3   Fetal mortality rates following the administration of the different treatments on day 16 of gestation

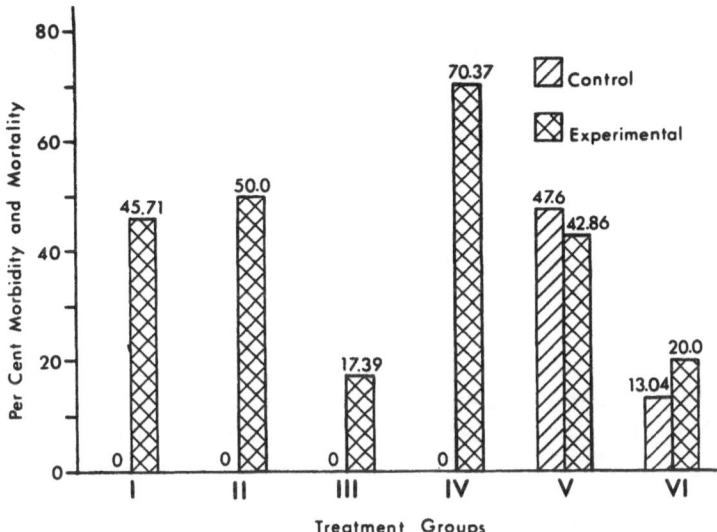

EXPERIMENTAL AMNIOCENTESIS

Figure 10.4   Combined fetal mortality and morbidity following the administration of different treatments on day 16 of gestation

(a) The incidence of fetal mortality, but not the combined incidence of fetal mortality and morbidity, was reduced as the calibre of the needle inserted into the amniotic sac (Figures 10.3 and 10.4) was decreased.

(b) The deleterious effects of amniocentesis on the weights, CRLs and CRL/wt ratios of the surviving fetuses were not ameliorated either by reducing the calibre of the needle or by attempting to seal the puncture site. It appeared that a compression mechanism, following the loss of amniotic fluid through the puncture site, was still in operation. This effect was present to variable degrees not only between different litters, but also between individual fetuses within each litter.

(c) The placement of the haemostatic sponge over the puncture site (group 4) reduced fetal mortality following puncture of the amniotic sac with a 26 gauge needle to the level induced by puncture with a 34 gauge needle (group 3). However, in groups 5 and 6, the application of the sponge was associated with a two- to three-fold increase in fetal mortality, compared with the puncture alone. When mortality and morbidity were combined, there was no large difference between puncture alone and puncture plus the attempt to prevent leakage. This conflicting and inconclusive evidence regarding the ability of the haemostatic sponge to reduce the leakage and thus ameliorate the effects of amniocentesis, is probably the result of the inability of the sponge to adhere to the fetal membranes.

Comment: From these first two experiments it can be seen that by increasing the gestational age at the time of puncture, by reducing the gauge of the needle used to make the puncture and by minimizing the volume of vehicle injected, fetal mortality (but not fetal mortality combined with morbidity) can be reduced. It appears that leakage of amniotic fluid from the puncture site is the primary aetiological factor. Attempts to prevent this leakage with a haemostatic sponge were unsatisfactory. However, Blackburn (personal communication) has reported considerable success in this regard by cauterizing the puncture site in the fetal membranes. With further improvements the intra-amniotic approach to the fetus may yet be a useful experimental model for teratological studies.

PATHOGENESIS OF AMNIOCENTESIS-INDUCED DEVELOPMENTAL DEFECTS

Previous reports have described the well-defined gross abnormalities characteristically found four to six days after amniotic sac puncture[5, 13, 14, 16–19]. These defects relate primarily to the head, limbs, and abdominal walls of affected fetuses. Cleft palate, microstomia, micrognathia, exencephaly/ exencephalocele have been described in the head region. Anomalies of the limbs have included meromelia, adactyly, syndactyly, malrotations, club foot, constriction rings and amputations. Short umbilical cords and abnormalities of the abdominal wall, short thick trunks and stunting, as well as haemorrhagic lesions in the extremities, have also been consistently reported. Love and Vickers[5] examined fetal rat limbs recovered six days after amniocentesis for microscopic changes. They found numerous subcutaneous haemorrhages which sometimes took the form of blebs, as well as impaired endosteal, periosteal and endochondral osteogenesis. Kino[20] studied the morphogenesis of amniocentesis-induced amputations and reductions in rats, and Singh and Singh[17], using the dissecting microscope, observed haemorrhagic lesions between the digits of fetal rats as early as five minutes after amniocentesis and the withdrawal of amniotic fluid. In the interval between the early haemorrhages and the well-defined defects observed in near-term fetuses up to six days later, however, many as yet undefined processes of damage and repair must occur from which evolve the final anomalies.

Most of the hypotheses regarding the aetiology of the amniocentesis-induced anomalies implicate intra-uterine compression or oligohydramnios subsequent to the leakage of amniotic fluid as the causal mechanism[13, 18, 21]. Although this direct compression of the fetus could explain some of the anomalies, it is apparent that a second more subtle or less direct mechanism must be sought to account for others. Therefore, experiments were designed to study the morphogenesis of the head and limb defects, with the intention of gaining greater insight into the possible causal mechanism(s)[22].

On day 16 of gestation, a 26 gauge needle was introduced into the amniotic sacs of fetuses contained in the left uterine horn; those in the right horn were untouched. The fetuses of five rats were recovered for each time interval of 15, 30 and 60 minutes, and 12, 24, 36 and 48 h after amniotic sac puncture. They were placed in Bouin's fixative and subsequently examined under a dissecting microscope for gross external malformations. Serial sections of the head regions and limbs of control fetuses and of the defective head regions and limbs of experimental fetuses were made for each time interval. They were stained routinely with haematoxylin and eosin and examined for microscopic changes.

The gross morphological changes observed immediately after amniocentesis, as well as the later defects, were consistent with previous reports. In addition, microscopic lesions were detected in the intervening period; these were present in varying degrees of frequency and severity and evolved into the final anomalies. The defects related to the head region and limbs were the most consistent and clearly defined. Where malformations were present, they were typically multiple. Furthermore, both the frequency and severity of the lesions increased with increasing time after treatment.

Typical changes in gross morphology suggested vascular damage and compression. Evidence of vascular damage included subcutaneous haemorrhages on the parts of the body likely to be compressed by the contracting uterus such as the dorsum of the hindlimbs, head, and spinal column, massive intracranial haematomas, general body oedema, and superficial blebs on the dorsum of distal limb segments which were filled with either blood or clear fluid. Evidence of intra-uterine compression included postural moulding such as limb malrotations and severe deviations of spinal curvature, many banding or constriction deformities, micrognathia, microstomia, and the pooling of blood in the umbilical cord or the absence of the normal herniation of the gut into the cord.

Microscopically, massive engorgement, with or without disruption, of both deep and peripheral blood vessels of both the head region and the limbs were observed as early as 15 min after amniotic sac puncture. This was accompanied by varying degrees of interstitial haemorrhage and/or interstitial oedema. By one hour after treatment these areas of vascular damage were associated with foci of cells displaying signs of cloudy swelling. In the limbs the initial changes were accompanied by areas of deep tissue cavitation. These early signs of tissue degeneration were morphologically different from the morphogenetic or physiological necrosis which was observed in the interdigital mesenchyme of control fetuses and which is an embryologically normal sculpturing process. Further degenerative changes were observed in the defective limbs of fetuses recovered from 12 to 48 h after treatment. Regenerative or repair processes were also detected. The final deformity appeared to be the product of the simultaneous occurrence of both degenerative and regenerative processes. For example, a typical reduction deformity (meromelia, adactyly) would evolve as follows: the initial hypervolaemia produces vascular damage and interstitial haemorrhage and oedema. As a result of the excessive increase in the extracellular fluid volume persisting superficial blebs and deep tissue cavities are formed. Necrotic changes begin in these areas and soon afterward the plastic fetal tissues begin regenerating periderm around the damaged areas. Eventually the necrotic portions of the limb are isolated by the reforming periderm, producing the limb reductions, amputations or constrictions which are characteristic of amniocentesis dysmelia. The deformities of the preskeleton of the fetal limbs, such as the longitudinal splitting of the phalanges reported by Love and Vickers[5] result from the necrotic invasion of the mesenchymal phalangeal condensations, the cartilaginous carpal preskeleton, and the distal epiphyses of the long bones. Severe bowing of the long bones was also detected and was accompanied by marked degenerative changes in both the chondrial and perichondrial cytoarchitecture[22].

Deformities in the head region of treated fetuses followed a similar course and involved both nervous and connective tissues. The initial pattern of hypervolaemia, vascular damage and excessive accumulation of fluid resulted in superficial cranial blebs, intracranial haematomas and separation of the intracranial connective tissue layers. The nervous tissue proper also displayed areas of haemorrhage, cavitation and degenerative changes. A reduction of the mitotic figures was detected in the ependyma adjacent to the necrotic tissues. Distortions were observed in the morphology of the ventricular system and in the normal hindbrain–forebrain relationship. These, plus the protrusions of cerebral tissue through the membranous fetal cranium (cranioschisis) were probably due to an increase in intracranial pressure resulting from a combination of direct pressure by the contracting uterus and the excessive accumulation of intraventricular, intravascular, and/or interstitial fluid volumes. From these changes the characteristic deformities of the head region evolved.

AN INTEGRATING HYPOTHESIS

Based on the reports of other investigators and the interpretation of the results of the preceding experiments the following hypothesis regarding the aetiology of amniocentesis-induced fetal deformities is proposed (see Figure 10.5):

(a) The initial lesion is a vascular one. The hypervolaemia and haemorrhage represent a state of severe central venous congestion which can be produced by compression of the umbilical cord between the fetus and the placenta, and/or by the compression of the placenta itself. Either of these would result from the contraction of the myometrium subsequent to the leakage of amniotic fluid from the puncture site. Purely circumstantial factors such as the spatial orientation of the fetus at the time of puncture and the volume of amniotic fluid lost would determine the degree of fetal compression and immobilization, and thus the degree to which the syndrome would be expressed.

(b) The subcutaneous haemorrhages seen in relation to the spinal column, the snout and the outer surfaces of the limbs suggest localized venous congestion and subsequent rupture of the fragile peripheral blood vessels. This congestion is probably caused by direct pressure from the contracting uterus and would explain the higher frequency of such lesions on the hindlimbs than on the forelimbs because the latter would be protected by the large overhanging fetal head.

(c) Whereas only vascular damage and the accumulation of interstitial fluid are evident for the first 30 minutes, by 60 minutes after amniotic sac puncture foci of interstitial oedema and cloudy swelling are detected. Although these degenerative changes could be explained by tissue hypoxia secondary to local vascular damage, they are frequently found in tissues with no evidence of haemorrhage. Therefore a second mechanism, embryonic oxygen deficiency, is proposed to be acting by this time. This would be due to the obstruction of the feto-maternal circulation following the constriction of the placenta and/or the compression of the umbilical cord. Grabowski[23], and Leist and Grauwiler[24] have discussed embryonic oxygen deficiency as a

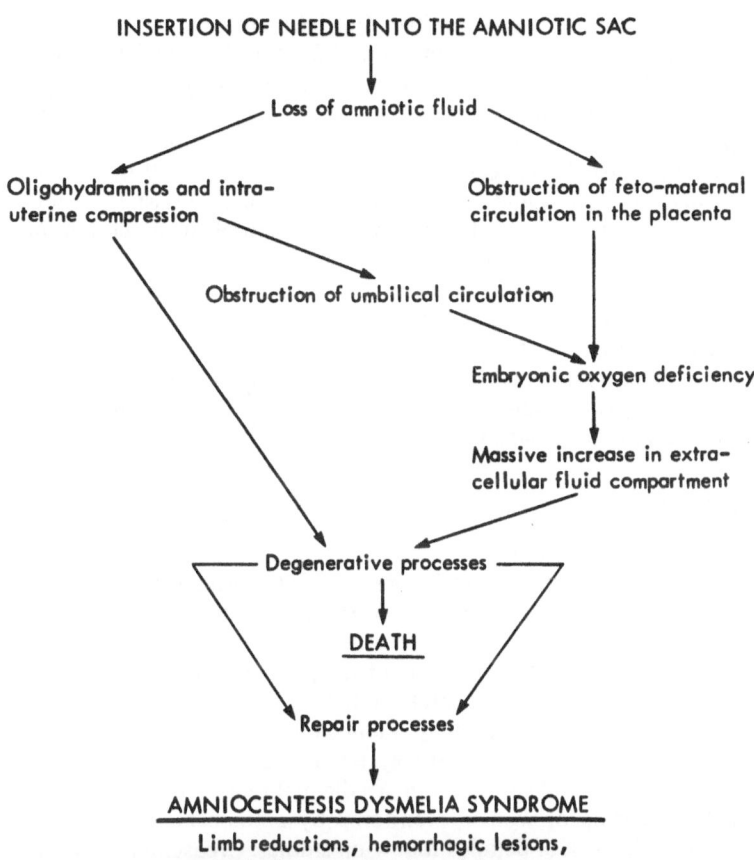

EXPERIMENTAL AMNIOCENTESIS

INSERTION OF NEEDLE INTO THE AMNIOTIC SAC

Loss of amniotic fluid

Oligohydramnios and intra-
uterine compression

Obstruction of feto-maternal
circulation in the placenta

Obstruction of umbilical circulation

Embryonic oxygen deficiency

Massive increase in extracellular fluid compartment

Degenerative processes

DEATH

Repair processes

AMNIOCENTESIS DYSMELIA SYNDROME

Limb reductions, hemorrhagic lesions,
cranioschisis, postural moulding and
compression defects, growth stunting,
etc.

Figure 10.5   Aetiology of amniocentesis-induced congenital anomalies – an integrating hypothesis

non-specific teratogen and have described congenital malformations associated with this state which parallel those observed in this study.

(d) On the basis of the observations in these and other experiments,
embryonic oxygen deficiency is implicated as the primary teratogenic insult
in the production of the major congenital anomalies induced by amniocentesis.
Whether this hypoxia is initiated peripherally by direct pressure on exposed
areas or centrally by obstruction of the feto-maternal circulation, the primary
aetiological factor would be the oligohydramnios and uterine contraction
subsequent to amniotic sac puncture.

Moderate hypoxia has been reported to be associated with 40–60% increase in blood plasma as early as 30 min after exposure. Both moderate and
acute hypoxia have been reported to result in the excessive accumulation of
extracellular fluid in the neural and mesenchymal tissues of the head and

171

mesenchymal tissues of the extremities. Furthermore, acute hypoxia has been implicated in the excessive accumulation of fluid in the ventricular system of the central nervous system[23]. This volume increase could cause the fragile fetal blood vessels to rupture or could lead to the leakage of plasma through the hypoxia-damaged circulatory endothelium. It would be manifested morphologically as vascular engorgement and disruption, interstitial oedema and haemorrhage, cavitation and separation of mesenchymal tissues, bleb formation, distortion of the ventricular system, herniation of the cerebrum through the membranous cranium, and changes in the normal intracranial relationship due to the increase of ventricular and intracranial pressure. The ensuing degenerative changes would lead either to fetal death, or if repair processes were initiated, to recovery. The degenerative changes and the simultaneous progress of regenerative processes, in combination with the effects of direct uterine compression, result in the persisting congenital deformities characteristic of the amniocentesis dysmelia syndrome.

CLINICAL IMPLICATIONS

The insertion of a needle into the amniotic sac for diagnostic purposes is an established procedure in obstetric practice. The principle underlying this technique has been exploited by the experimental teratologist as a direct route for the treatment of the embryo. Considerable concern has been expressed regarding the possible adverse effects of the procedure on the offspring both of humans[6, 25], and of experimental animals[16, 26]. Despite the many studies describing malformations typically induced by amniotic sac puncture in experimental animals[1-5, 11-20], relatively few cases of such defects have been reported in humans. Valenti (cited in Aladjem[25]) reported three cases of club foot in the offspring of women monitored by amniocentesis and expressed concern about the lack of detailed follow-up studies. Creaseman et al.[27] alerted those using amniocentesis to the possible complications involved. In a survey of the available case reports he noted that severe fetal trauma (haemothorax, lacerated myocardium, retained catheter tip, peritonitis, liver damage, sepsis and needle injuries to the brain) had been reported in as high as 11% of the offspring of treated mothers. Trauma to the placenta with increased iso-immunization and severe fetal anaemia and shock leading to intra-uterine death were discussed and sometimes the incidence of feto-maternal haemorrhage was reported to be as high as 10%. Hanid[28] reported a case of pneumothorax and surgical emphysema in a newborn baby which was caused by amniocentesis.

Thus, the initial human reports dealt primarily with fetal death or morbidity resulting from needle trauma, there being only one mention of the postural moulding suggestive of oligohydramnios or intra-uterine compression which is consistently observed in experimental animals.

Careful follow-up studies have shown that amniocentesis in humans, when carefully carried out so as to avoid mechanical trauma to the fetus and fetal membranes, has negligible risk and a risk of an entirely different nature than has been reported for the offspring of experimental animals[9, 10]. Indeed, in humans the experimental amniocentesis syndrome is paralleled more closely in mechanical complications of pregnancy such as oligohydramnios[29], premature rupture of the fetal membranes[30-32] or prolonged leakage of amniotic fluid[9].

The observations in the experiments and reports previously discussed offer valuable insight into the dilemma of how to reconcile the two contradictory sets of evidence regarding the safety of amniocentesis. In fact, they point to the loss of amniotic fluid as the source not only of the complications of experimental amniotic sac puncture in rats but also of the difference in risk between experimental and clinical applications. Except in cases of repeated attempts to obtain fluid, fluid loss following amniocentesis in human occurs and oligohydramnios is never persistent where it does occur (Hirschhorn, K., 1978, personal communication).

In examining some of the differences between the experimental and clinical procedures, possible explanations for this disparity in risk to the rat model and human fetus become more obvious. Poswillo[26], for example, investigated the effect of experimental first-trimester amniocentesis on non-human primates (*Macacus irus*) and failed to produce the congenital defects that have been reported in lower mammalian species. The factor protecting *M. irus* appeared to be the ability of the extra-embryonic coelomic fluid to coagulate within seconds, thus providing a self-sealing mechanism and minimizing the loss of amniotic fluid. In the experiments conducted in our laboratory using the rat, however, the leakage of amniotic fluid was observed not only as the needle was withdrawn but also as a vaginal discharge during the postoperative period. It is conceivable that the different relationship of the fetal membranes to the uterine wall and the different nature and contents of the extra-embryonic coelom play a significant role in the presence or absence and in the effectiveness of this self-sealing mechanism. Poswillo[26] concluded that on the basis of this more realistic, non-human primate model, the hazard of amniocentesis in humans may be less than had been predicted on the basis of animal experiments.

A second factor which may account for the disparity between the observed effects of clinical and experimental amniocentesis is the difference in the total volume of amniotic fluid contained within the fetal sacs of the various species. Gulienetti *et al.*[33], using a radioisotope dilution method, found that the mean volume of amniotic fluid for day 20 Wistar rats was 0.56 ± 0.16 ml, for day 17 DBA/1J mice was 0.16 ± 0.01 ml, and these values would be less during early gestation. In comparison, the volume of amniotic fluid surrounding a 17-week human fetus (which is within the recommended time span during which amniocentesis can be performed) is 225 ml; Poswillo[26] aspirated between 2.2 to 3.5 ml from first trimester *M. irus*. It is evident that the loss of even a small volume of fluid from the amniotic sac of the laboratory rodent would represent a very large proportion of its total volume and would suffice to produce severe and persistent oligohydramnios. On the other hand, the aspiration of 10 ml of amniotic fluid from the human subject for diagnostic purposes, even if followed by a certain amount of leakage, would be unlikely to induce an oligohydramnios of significant consequence to a small fetus surrounded by relatively abundant (225 ml) fluid.

A third factor to consider is the postoperative state and posture of the rat following amniocentesis. Having a bicornuate uterus with many offspring, a rodent must be subjected to general anaesthesia, laparotomy, and exteriorization of the uterine horns before a needle can be inserted into the amniotic sac. This procedure alone has been shown to have no deleterious effects on the growth and development of the rat fetus[3]. What may be of consequence, however, is the fact that the rat is laid on its abdomen in a comatose state for the period of recovery. In this position the contents of the uterine horns would be supporting a considerable portion of the maternal body weight. It is unlikely that a human patient would be allowed to engage in postures or activities as conducive as this to the leakage of amniotic fluid following puncture of the amniotic sac.

References

1. Persaud, T. V. N. (1972). Effect of intra-amniotic administration of hypoglycin B on foetal development in the rat. *Exp. Pathol.*, 6, 55
2. Jelinek, R., Doskocil, M. and Losticky, C. (1976). The 'straight jacket' syndrome in chicks. I. A morphological analysis. *Folia Morphol.*, 24, 98
3. Kennedy, L. A. and Persaud, T. V. N. (1976). Experimental amniocentesis and teratogenesis. 1. Evaluation of the intra-amniotic route of treatment in teratological studies. *Anat. Anz.*, 140, 267
4. Kennedy, L. A. and Persaud, T. V. N. (1976b). Experimental amniocentesis and teratogenesis. 2. Control of teratogenicity. *Anat. Anz.*, 140, 275
5. Love, A. M. and Vickers, T. H. (1972). Amniocentesis dysmelia in rats. *Br. J. Exp. Pathol.*, 53, 435
6. Dewhurst, C. J. (1956). Diagnosis of sex before birth. *Lancet*, i, 471
7. Broome, D. L., Wilson, M. G., Weiss, B. and Kellogg, B. (1976). Needle puncture of fetus: a complication of second trimester amniocentesis. *Am. J. Obstet. Gynecol.*, 126, 247
8. Karp, L. E. and Hayden, P. W. (1976). Fetal puncture during midtrimester amniocentesis. *Obstet. Gynecol.*, 49, 115
9. Simpson, N. E., Dallaire, L., Miller, J. R., Siminovich, L. and Hamerton, J. L. (1976). Prenatal diagnosis of genetic disease in Canada. Report of a collaborative study. *Can. Med. Assoc. J.*, 115, 739
10. Lowe, C. U. and Alexander, D. (1976). Safety and accuracy of mid-trimester amniocentesis for prenatal diagnosis. National Institute of Child Health and Human Development (Bethesda, Maryland: National Institute of Health)
11. Blanc, W. *et al.* (1978). Cited in Kastner, I., Amniocentesis connected to infants lung hypoplasia. *Med. Post*, 14 (6), 1
12. Dostal, M. (1973). Effect of some non-specific factors accompanying intra-amniotic injection in mouse foetus. *Acta Morphol.*, 21, 97
13. Persaud, T. V. N. (1973). Meromelia and other developmental abnormalities in experimental oligohydramnios. *Anat. Anz.*, 133 (Suppl.), 499
14. Singh, S., Mathur, M. M. and Singh, G. (1974). Congenital anomalies in rat foetuses induced by amniocentesis. *Ind. J. Med. Res.*, 62, 394
15. Nicholas, J. S. (1962). Experimental methods and rat embryos. In: E. J. Farris and J. Q. Griffith (eds.). *The Rat in Laboratory Investigation*, 2nd Ed., pp. 51–67. (New York: Hafner Publishing Co.)
16. Trasler, D. G., Walker, B. E. and Fraser, F. C. (1956). Congenital malformations produced by amniotic sac puncture. *Science*, 124, 439
17. Singh, S. and Singh, G. (1973). Hemorrhages in the limbs of fetal rats after amniocentesis and their role in limb malformations. *Teratology*, 8, 11
18. De Meyer, W. and Baird, I. (1969). Mortality and skeletal malformations from amniocentesis and oligohydramnios in rats: cleft palate, club foot, microstomia and adactyly. *Teratology*, 2, 33
19. Kendrick, F. J. and Feild, L. E. (1967). Congenital anomalies induced in normal and adrenalectomized rats by amniocentesis. *Anat. Rec.*, 159, 353

20. Kino, Y. (1972). Reduction malformations of the limbs in the rat fetus following amniocentesis. *Cong. Anom.*, 12 (1), 35
21. Singh, S. and Singh, G. (1974). The role of uterine contraction in producing hemorrhages in rat foetuses after amniocentesis. *Teratology*, 10, 145
22. Kennedy, L. A. and Persaud, T. V. N. (1977). Pathogenesis of developmental defects induced in the rat by amniotic sac puncture. *Acta Anat.*, 97, 23
23. Grabowski, C. T. (1970). Embryonic oxygen deficiency – a physiological approach to analysis of teratological mechanisms. In: D. M. H. Woollam (ed.). *Advances in Teratology*, 4, pp. 123–167. (London: Logos Press)
24. Leist, K. H. and Grauwiler, J. (1974). Fetal pathology in rats following uterine-vessel clamping on day 14 of gestation. *Teratology*, 10, 55
25. Aladjem, S. (1969). Artificially induced variations of amniotic fluid volume in early pregnancy. In: C. T. Lund and J. W. Choate (eds.). Transcript from *5th Rochester Trophoblast Conference*. (Unpublished)
26. Poswillo,D. (1972). Experimental first-trimester amniocentesis in non-human primates. *Teratology*, 6, 227
27. Creaseman, W. T., Lawrence, R. A. and Thiede, H. A. (1968). Fetal complications of amniocentesis. *J. Am. Med. Assoc.*, 204, 949
28. Hanid, T. K. (1975). Pneumothorax and surgical emphysema in a newborn baby caused by amniocentesis. *Br. J. Obstet. Gynaecol.*, 82, 170
29. Jeffcoate, T. N. A. and Scott, J. S. (1959). Polyhydramnios and oligohydramnios. *Can. Med. Assoc. J.*, 80, 77
30. Torpin, R. (1965). Amniochorionic mesoblastic fibrous strings and amnionic bands: Associated constricting fetal malformations or fetal death. *Am. J. Obstet. Gynecol.*, 91, 65
31. Torpin, R. (1968). *Fetal Malformations Caused by Amnion Rupture During Gestation*. p. 129. (Springfield: Charles C. Thomas)
32. Dunn, P. M. (1971). Congenital deformation following premature rupture of the membranes. *Teratology*, 4, 487
33. Gulienetti, R., Kalter, H. and Davis, N. C. (1962). Amniotic fluid volume and experimentally-induced congenital malformations. *Biol. Neonat.*, 4, 300

11
A mechanistic approach to teratogenesis: analysis of caudal dysplasia syndrome (sacral agenesis)

S. KAPLAN

INTRODUCTION
The mechanistic approach

Experimental teratology must ultimately concern the study of mechanisms; by mechanisms I mean the temporal sequence of events between the initial teratogenic insult and the final morphological alteration which is recognized as a congenital malformation. Figure 11.1 shows schematically an embryo exposed to a teratogenic insult, a drug for example, and the hypothetical sequence which follows. Presumably the first effect of the drug is at the

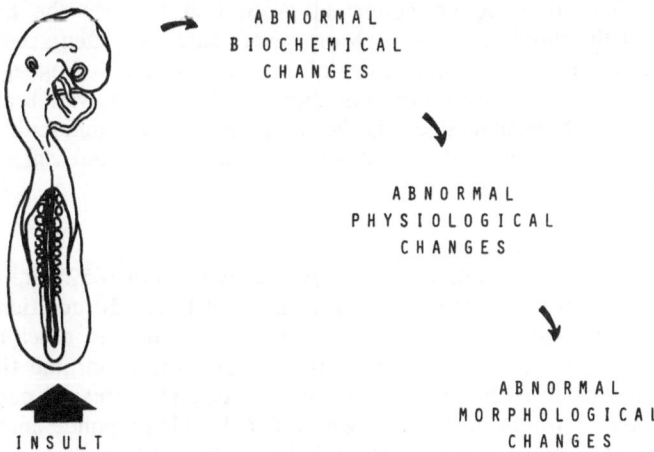

Figure 11.1   Hypothetical sequence of events following exposure of an embryo to a teratogen

biochemical level, perhaps the inhibition of an enzyme. This might lead either directly, or after a cascade of biochemical effects, to alterations at the physiological level. More and more we are finding that true physiological changes do occur in the embryo – alterations in the acid–base balance of the blood, changes in the osmolarity of the plasma, or changes in the blood pressure itself. These changes lead, ultimately, to the changes in morphology that we see with the naked eye or the dissecting microscope.

To date, no studies have revealed the entire sequence of events outlined above and shown schematically in Figure 11.1. Indeed, I know of no studies in embryos which have both characterized the biochemical perterbations and which have shown how these perterbations can lead to physiological abnormalities, and there are very few studies which have shown the transition between physiology and morphology. This chapter will show a systematic approach to the elucidation of this sequence for a specific malformation, caudal dysplasia syndrome or, in the chick, rumplessness. The entire sequence has not been revealed; nevertheless, we have made significant progress in understanding the genesis of this malformation following alterations at the level of physiology. It is this portion of the sequence that is outlined herein.

Human caudal dysplasia syndrome

Syndrome characterization
This syndrome in humans is also known as caudal regression, caudal agenesis and sacral or sacrococcygeal agenesis. These terms describe a spectrum of defects and a spectrum of severity. Varying degrees of developmental failure or regression may be present, involving the lower lumbar, sacral or coccygeal vertebrae. The corresponding spinal cord segments are often involved, leading to neurological defects ranging from mild lack of control of bladder function to total motor and sensory loss below the level of the deficit[1].

Externally, sacral agenesis is manifested by flattening and dimpling of the buttocks and shortening of the intergluteal cleft. If the lumbar spine is also involved, there may be an abnormal proportion between the transverse diameter of the thorax and pelvis. More severe cases are often accompanied by abduction and flexion deformities of the hips, causing a frog-leg appearance. Generally, there are no defects above the level of the vertebral lesion[2]. However, it has been suggested that the oesophageal atresia and renal agenesis sometimes found combined with caudal dysplasia may represent a distinct syndrome[3].

Incidence
Caudal dysplasia syndrome was first reported by Hohl in 1852[4] and a second case was described in 1857[5]. Since that time, at least 150 additional cases have been reported[6–8]. However, this number of published cases does not reflect the incidence. It has been estimated that the syndrome (including sirenomelia, which is the most severe form of caudal dysplasia) occurs at a rate of about 1 : 60000 live births, with a 2.7 : 1 male preponderance[9]. However, the defect may occur as often as 1 : 100 in live children born to diabetic mothers, as will be discussed shortly.

Sirenomelia

Sirenomelia (sympodia, monopodia, mermaid) represents the most severe form of caudal dysplasia syndrome[9]. It consists of fusion of the lower limbs combined with abnormalities of the bony pelvis. The spectrum of lower limb malformations ranges from limited fusion of soft tissue elements with bony elements present and normal, to complete fusion of bony and soft tissues affecting everything from the pelvis to the toes. These infants classically exhibit renal hypoplasia or agenesis, absence of external genitalia, bladder agenesis, a single umbilical artery, imperforate anus and Potter's face.

Association with other defects

The term caudal dysplasia syndrome then, encompasses a broad spectrum of defects of the caudal axis with a wide range of severity. That these defects are strongly associated can be seen by looking at them in another way. For example, the risk for other anomalies in an individual born with imperforate anus is: lower vertebral defects, 40%; urological defects other than fistulas, 19%; genital anomalies, 17%; and lower limb anomalies, 10%[9]. Statistics such as these, along with the constellation of defects seen in caudal dysplasia syndrome, can be interpreted to mean that a common mechanism is operating in the production of these defects. Indeed, Smith[9] has suggested that the cause is a deficit in a wedge-shaped portion of the caudal axial mesoderm. Absence or hypoplasia of this medial mass during the early, morphogenetic stages would permit fusion of the limb buds at their proximal ends, with absence or hypoplasia of intervening caudal structures. It is interesting to note that two medial structures composed largely of mesoderm, the bladder and external genitalia, are almost universally absent in sirenomelic infants.

Association with maternal diabetes

It had been suspected for many years that infants born of diabetic mothers exhibited a higher incidence of congenital malformations than infants born of normal mothers,[10, 11] but this was not confirmed until Pederson et al. concluded from their comprehensive survey that diabetic women do, in fact, produce about three times as many defective infants as do non-diabetics[12]. Although these authors did not feel their series of cases sufficiently large to permit a study of the differences in distribution of malformations when compared with controls, they did state their clinical impression that '... severe bony malformations, especially in the limb ...' seemed relatively common. Rusnak and Driscoll[7] noted that seven of the 133 cases of caudal agenesis syndrome reported in the world literature to that date were offspring of diabetic mothers. These authors described three cases of their own, among 1150 infants of diabetic mothers born in the Boston Lying-in Hospital between 1952 and 1964, and thus made a case for association of caudal dysplasia syndrome with maternal diabetes. Passarge and Lenz[13] suggested that caudal dysplasia syndrome occurs in about 1% of all children born to diabetic mothers, and this figure is the generally accepted upper limit today. Kucera[14] conducted a worldwide survey of 7101 fetuses from diabetic women and compared these with a control group of 431 764 cases obtained from the World Health Organization. The control group exhibited 0.65% malformations, whereas 4.8% of the fetuses from diabetic women were malformed. A comparison of the ratio of incidence of particular malformations showed that caudal dysplasia syndrome was 212 times more common in infants born of diabetic mothers, when compared with the control group.

Thus, there seems little doubt that caudal dysplasia syndrome is associated with maternal diabetes. The question then arises as to whether this association is due to the insulin taken by many of these women, or whether it is due to some deleterious effect of the disease itself. Although it will be shown later that insulin treatment can cause caudal dysplasia in animal embryos, current consensus is that neither the increased neonatal death rate nor the increased malformation rate seen in infants born of diabetic mothers is due to exogenous insulin. For example, Pederson et al.[12] presented convincing evidence that insulin reactions and diabetic coma in the first trimester were not the cause of malformations in such infants. Indeed, insulin treatment is probably advantageous for the infants, not deleterious. Karlsson and Kjellmer[15] showed that an expected perinatal mortality rate of 24% in a series of high risk diabetic pregnancies (greater than 150 mg per 100 ml mean blood sugar level) was reduced to 4% when insulin was used to depress the sugar below 100 mg per 100 ml of blood. Other similar reports show the advantages of insulin control of diabetes to both the fetus and the mother[16, 17].

We are left with the hypothesis that the high blood sugar level somehow causes deleterious effects in the embryo. Indeed, Hughes et al.[18] report that treatment of early chick embryos with sugars including glucose leads to a wide range of embryonic malformations. Elevated blood glucose levels are known to damage *adult* vascular endothelium; perhaps parallel effects on embryonic vascular endothelium play a role in the development of caudal dysplasia syndrome. The concept that lesions of the embryonic vascular system can lead to absence or reduction of caudal tissues can best be developed by considering experiments using animal models.

Animal models

Rumplessness

Rumplessness is a syndrome in the domestic fowl which appears to be identical to human caudal dysplasia syndrome. It consists of absence or reduction in the sacral and caudal, skeletal and soft tissue elements. The spontaneous defect has been known since the turn of the century[19], and the first detailed report of the associated anatomical changes was published in 1913[20]. The spontaneous rate of occurrence is about 1%[21], but it has long been known that physical agents such as abnormal incubation temperatures[22] or mechanical shaking prior to incubation[21] can markedly increase the incidence. A dominant gene for the condition has been described[23, 24], as well as a recessive gene[25].

The first experimental *chemical* induction of rumplessness was by Ancel and Lallemand in 1941[26], who produced 'an arrest in development of the caudal bud' by depositing several chemical agents on the early chick embryo (25–27 somites, later known as Hamburger–Hamilton[27] Stage 17, 52–64 h of development). Among the agents tested was the azo dye, trypan blue. This first use of trypan blue as a tool in experimental teratology was to be followed by a long history of its use in many experimental animals.

Insulin, the chief exogenous drug associated with diabetic pregnancies, is itself teratogenic in the chick and causes a high incidence of rumplessness. Landauer[28] reported that 2 units of insulin injected into the yolk sac prior to incubation resulted in rumplessness among 42% of the embryos which survived to the 17th day of incubation. Later, Landauer and Bliss[29] injected insulin at different times after incubation and found that the subsequent incidence of rumplessness was greatest when the injections were made at about 31 h of incubation. The rumplessness-inducing effects approached zero at about 72 h of injection. The general period of maximum susceptibility to rumplessness in the chick was then first defined, at least for this agent.

Many other drugs and chemical agents have since been shown to cause rumplessness in the chick embryo. Among these are inorganic salts, antibiotics, pesticides, solvents, volatile chemicals and other miscellaneous agents[30]. However, among all these agents, the most notable is trypan blue, which has consistently been shown to cause a high incidence of rumplessness and a relatively low incidence of embryonic death. For example, Beaudoin and Wilson found that the dye produced rumplessness in 88% of all experimental embryos surviving to day 10 or 12 of incubation[31]. In a later series, Beaudoin found rumplessness in 54% to 69% of the embryos treated, the percentage varying with the route of administration[32].

Thus, we have seen that there are several genetic conditions and, in otherwise genetically normal eggs, a wide spectrum of exogenous physical and chemical agents which can induce rumplessness. It seems clear that there exists a peculiar sensitivity of caudal tissues in the chick to these conditions and agents. Among the chemical agents, trypan blue has an unusual propensity for inducing this defect, but not only in the chick embryo. Trypan blue is also notorious for inducing caudal defects in other species, including several mammals.

Caudal defects in mammals due to treatment with trypan blue

Gillman et al. found that 18% of the pups born to rats treated with trypan blue exhibited shortened or absent tails[33]. Hamburgh[34, 35] also found kinky, shortened, or absent tails in mice born to mothers who had been treated with trypan blue. He recorded the presence of many blood-filled blisters (haematomas) in the tails of these embryos, a finding confirmed by Waddington and Carter[36] in a different strain of mice. Gunberg[37] found caudal defects in 19% of the abnormal young born to rat mothers treated with trypan blue. The severity of these defects varied from slight malformations of the tail to complete absence of the spinal cord and vertebral column caudal to the 12th thoracic vertebrae. Wilson, Beaudoin and Free[38] found absence or shortening of the tail or trunk in up to 40% of the fetuses surviving to day 20 after maternal treatment with trypan blue. We can conclude from these studies that trypan blue causes caudal defects in mammals, though perhaps not to the same extent or severity as in the chick embryo.

Caudal haematomas and blisters

Blisters (blebs), either filled with clear fluid or with blood (haematomas), have been repeatedly described in connection with trypan blue induced caudal defects in both bird and mammalian embryos, *provided the investigator examined the embryos at an early stage of development*. While Ancel and Lallemand[26] first discovered that rumplessness can be induced by trypan blue treatment, they examined their chick embryos near the end of the incubation period and could only conclude that the tail tissues exhibited a 'selective sensitivity' to the action of the dye. But later, presumably after studying the early stages in the development of rumplessness, Ancel[39] suggested that the rumplessness seen at later developmental stages might be associated with caudal haematomas observed at earlier stages. However, Ancel's proposal has received little attention from others in the field. Hamburgh's discovery[34, 35] of caudal haematomas in mouse embryos after maternal treatment with trypan blue has already been mentioned. Hamburgh quite clearly suggested that 'these haematomata and the invasion of clumps of blood cells into somite and mesenchymal tissue might have a destructive effect on these structures and their derivatives, leading through degenerative changes to shortening of the tail or to its bending or curling'[35]. Mulherkar[40] found clear blisters in very early chick embryos cultures with trypan blue. No haematomas were found, but this was to be expected since most of the embryos were cultured and examined at very early stages, before the establishment of a complete circulation. Stephan and Sutter[41] found that 24% of the chick embryos treated in their series on day 2 of incubation developed haematomas of the caudal region, while another 9% developed clear blisters.

Trypan blue is not the only teratogen known to cause caudal haematomas and clear blisters. In 1967, Ferm and Carpenter showed that lead salts consistently cause tail and sacral malformations in the golden hamster[42]. They more recently reported the discovery of caudal clear blisters and haematomas during the early genesis of these defects[43]. Recent experiments in my laboratory have shown that insulin, dextroamphetamine ('pep pills') and methamphetamine ('speed') can produce caudal haematomas in chicks which are indistinguishable from those produced by trypan blue[44].

Summary and rationale

It is hoped that this brief literature review has established the following:

1. Caudal dysplasia syndrome encompasses a number of specific defects of the caudal region, of varying degrees of severity. Maternal diabetes predisposes the birth of infants with this defect.

2. Rumplessness in chicken embryos and caudal defects in mammalian animal embryos represent caudal dysplasia syndrome in subhuman species.

3. The syndrome can be produced in animal embryos by a wide variety of chemical and physical agents. Among these are: (a) insulin (notable because many pregnant diabetic women are themselves exposed to this drug) and (b) trypan blue, (notable because this agent produces such a high incidence of the defect in several animal species).

4. Clear blisters or blood-filled blisters (haematomas) have been found

repeatedly in animal embryos treated to produce caudal defects, provided the embryos were examined early in the genesis of the caudal defect.

My own approach to the problem was to select a model – some combination of animal and teratogen that worked well – and to study it in some detail with the ultimate goal of understanding the sequential steps comprising the mechanism of the defect. Trypan blue was selected as the teratogen because it causes a relatively high percentage of caudal defects and a relatively low percentage of embryonic death. The chick embryo was used because this species is extraordinarily susceptible to caudal malformation with trypan blue. The aim was to study this defect in the sequence shown in Figure 11.1, beginning at the morphological level. What could one see in the time between administration of trypan blue during early embryogenesis and the caudal agenesis (rumplessness) that was evident later? Did the caudal region of the embryo really fail to form, as suggested by the term agenesis? Or did it form and later degenerate? What were the mechanisms involved?

EXPERIMENTAL ANALYSIS

Incubation and injection procedures

Fertile White Leghorn eggs were used throughout these studies. They were stored for no longer than one week prior to incubation (at 10°C) after receipt from the supplier.

After incubation at 38°C for 48 h, the eggs were removed from the incubator, swabbed with 70% ethanol, and a window was cut in the shell with a flame-sterilized hacksaw blade or motor driven microsaw. Trypan blue solution (0.1%, dissolved in aqueous 0.85% NaCl) was sterilized by filtration and a calibrated dose of 20 μl was drawn into a glass microneedle, made by

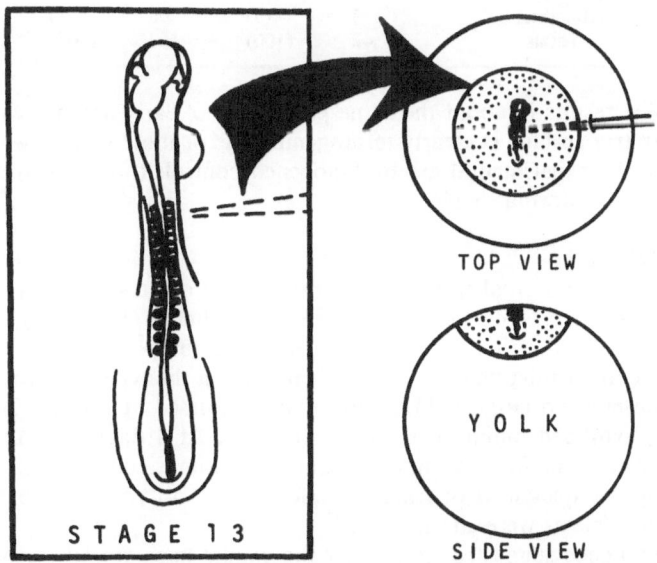

Figure 11.2 Injection technique. See text for description

183

flame-drawing capillary tubing to a fine point. The needle was inserted through the yolk sac at a point peripheral to the marginal vein of the area vasculosa (Figure 11.2). The needle tip was moved to a site just under the lateral body wall of the embryo, and the dye was released into the interface between the yolk sac and the underlying yolk. After the injection, the opening in the shell was sealed with Parafilm or a cover glass and melted paraffin, and the eggs were returned to the incubator.

Saline injected controls were treated identically, except that they were injected with sterile saline (0.85% NaCl, w/v). Unopened controls were removed from the incubator, handled and swabbed with alcohol, but were not treated further.

Gross morphology

Mortality and frequency of malformations

The embryos were examined daily between 1 and 8 days after treatment. Table 11.1 shows that opening and injecting the egg with either saline or

Table 11.1 Mortality and frequency of malformation in chicks treated at 48 h of incubation and examined 1 to 8 days later

	Saline-injected controls		Trypan blue treated	
	No.	(%)	No.	(%)
Died within one day of treatment	11	(26)	41	(25)
Normal	28	(67)	30	(17)
Abnormal	3	(7)	96	(58)
Totals	42	(100)	167	(100)

trypan blue results in about the same percentage of early deaths, while treatment with trypan blue is clearly teratogenic. One or more defects were found in 58% of the experimental group. Unopened controls (not shown) exhibited no deaths or malformations[45].

Types of defects

Among the 96 abnormal embryos, several kinds of defects involving a variety of embryonic structures were found. However, almost 60% of the abnormal embryos exhibited either caudal haematomas or rumplessness (Table 11.2). Even more interesting was the result of correlating the type of defect with the age of the embryo (Figure 11.3). Embryos examined at 3 days (24 h after injection) exhibited mainly caudal haematomas. At 4 days, few had caudal haematomas alone; 88% exhibited caudal haematomas coupled with incipient (beginning) rumplessness or frank rumplessness. Embryos older than 4 days with caudal defects were all rumpless[45].

This sequence suggested to me that the caudal haematomas seen at early developmental stages led to the rumplessness seen at later stages. To test this hypothesis, a series of embryos were treated as usual with trypan blue, but a glass coverslip was used to seal the shell. In this way, each egg could be removed from the incubator every few hours, and the presence of haematomas or rumplessness noted. When this was done over a period of several days, the

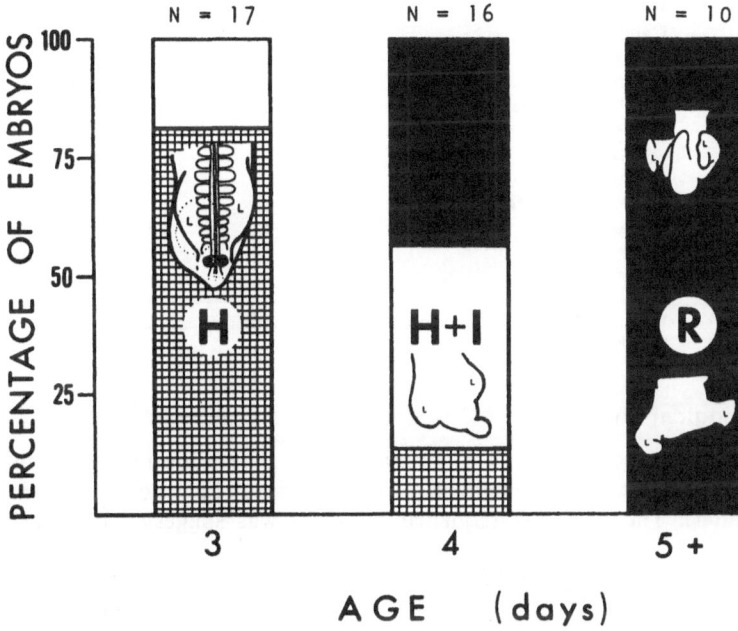

Figure 11.3   Correlation of the type of caudal defects found in 96 abnormal embryos with the age of the embryo. Ordinate, percentage of embryos with caudal defects; abscissa, age at autopsy. Among the 3-day embryos with caudal defects, most (82%) exhibited caudal haematomas alone (hatched portion of bar), while the remainder exhibited caudal haematomas coupled with incipient (beginning) rumplessness. None were rumpless. The drawing shows the caudal region of a 3-day embryo with a haematoma (black) just caudal to the last somite pair. L = hind limb bud. Among the 4-day embryos, 44% exhibited caudal haematomas coupled with incipient rumplessness (white portion of the bar), another 44% were rumpless (solid black) and the remainder exhibited only caudal haematomas. The drawing shows the caudal external contours of a 4-day embryo with an abnormal tail constriction. This embryo exhibited twin caudal haematomas at 3 days (it was examined while living) resulting in bilateral resorption of tissue leading to the constriction. Such embryos were classified as exhibiting haematomas + incipient rumplessness because remnants of the haematomas were still visible and resorption of tail tissue was under way. L = hind limb bud. Among the 5-day and older embryos, all were rumpless (solid black bar). The upper drawing shows the caudal external contour of a normal 5-day embryo. The lower drawing depicts complete rumplessness. L = hind limbs

results were unequivocal: every embryo which developed a caudal haematoma went on to become rumpless whereas every treated embryo which failed to develop a caudal haematoma went on to develop a normal caudal area.

Haematomas, then, are clearly associated with rumplessness. The histological evidence presented below adds credence to the hypothesis that they are a cause.

Table 11.2 Malformations found in 96 abnormal chick embryos treated with trypan blue at 48 hours of incubation and examined 1 to 8 days later

Malformation	Number*	Per cent
Rumplessness	27	30
Caudal haematomas	26	29
Oedema	20	22
Retarded growth	17	19
Head haematomas	16	17
Haematomas of extraembryonic membranes	11	11
Microphthalmia (uni- or bilateral)	5	5
Microcephaly	5	5
Clear blisters	3	3
Other	2	2

* Many embryos exhibited more than one malformation; each is tabulated here

Histological findings

Embryos were sectioned at various stages, both during haematoma formation and later during the development of rumplessness (12 h post-treatment and thereafter). The following sequence of events was suggested by this histo-

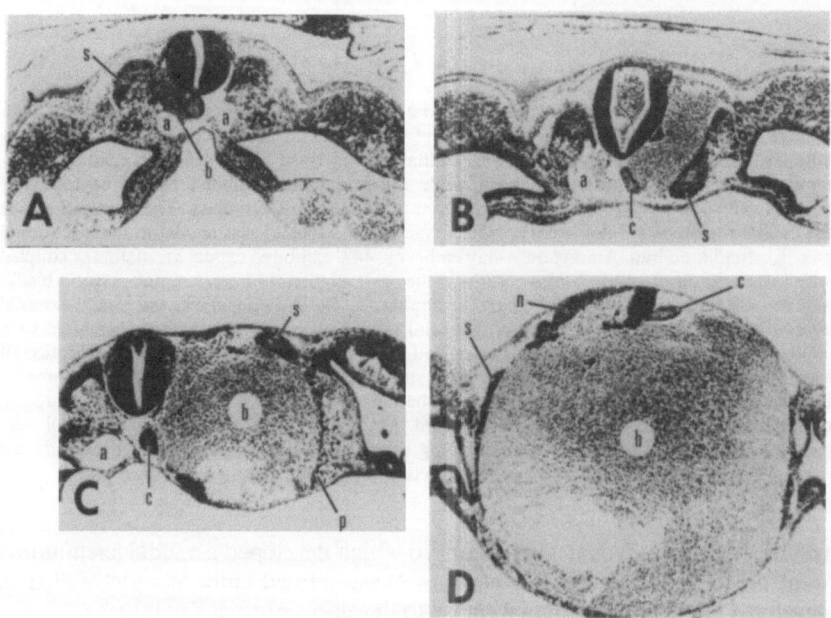

Figure 11.4 Sections from four embryos treated with trypan blue at 48 h of development and preserved 24 h later. Abbreviations: a = aorta; b = blood cells; c = notochord; p = pyknotic nuclei; n = neural tube; s = somite. × 36 (continued on p. 187)

logical study: (1) rupture of the caudal dorsal aortá (Figure 11.4A); (2) extravasation of blood into the surrounding tissue spaces (Figure 11.4B and 11.4C); (3) damage to the neural tube, notochord, somites and other caudal structures due to the expanding haematoma (Figure 11.4B, 11.4C and 11.4D); (4) necrosis, as evidence by pyknotic nuclei; (5) resorption of dead and dying tissues; (6) ultimately, rumplessness.

This represents the first conclusive evidence that rumplessness is caused by caudal haematomas. Furthermore, it seems likely that the haematoma

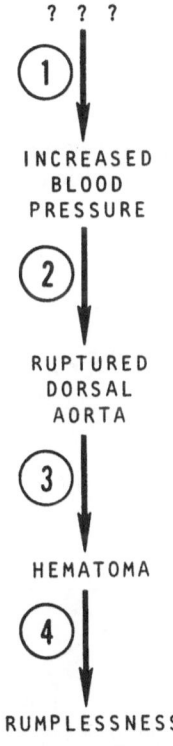

Figure 11.5   Sequential steps in the formation of rumplessness

Figure 11.4 (continued from p. 186)

A. Section through the cephàlic part of a caudal haematoma. Note that blood cells from the broken aorta are wedged between a somite and the neural tube.

B. This section is closer to the centre of a haematoma than that shown in A. The notochord and somite in the vicinity of the haematoma are misshapen, presumably due to the pressure exerted on these structures by the haematoma. The somite is markedly shifted in position and blood cells are present in the neural canal.

C. Still closer to the centre, the limits of the haematoma are well defined by a border of pyknotic nuclei. Distortion and shift in position of a somite, the notochord and other structures is evident.

D. Section through the centre of a haematoma. Massive damage is evident. Somites are represented by amorphous strands of cells located at the lateral margin of the haematoma. The neural tube is split and lies flattened against the overlying ectoderm. The notochord is also flattened. Pyknotic nuclei are prevalent, suggesting that massive cell death was taking place.

(From Kaplan and Grabowski[45], by kind permission of the Wistar Institute Press)

results from a burst dorsal aorta, and we have established several steps in the sequence leading to this specific defect (steps 3 and 4, Figure 11.5). The next obvious question concerns the preceding step, the cause for the bursting of these caudal blood vessels.

Physiological analysis

One hypothesis concerning the rupture might be that the teratogen, trypan blue, selectively attacks the dorsal aorta in the caudal region. However, when teratogens exhibit the ability to malform a specific area, the specificity resides not in the agent itself, but rather in the reacting system – the embryo. That is, we should look for some property of the developing organ system to under-stand why it is easily malformed.

The working hypothesis that was developed concerned a possible increase in blood pressure. There was an indication that the dorsal aortae were some-

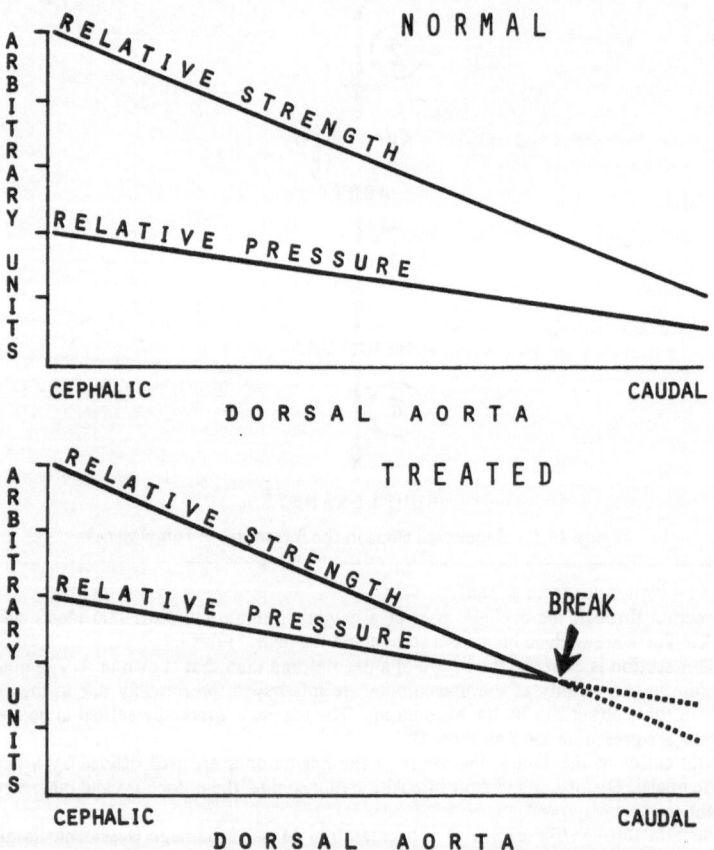

Figure 11.6 Hypothetical relationship between relative strength of the dorsal aorta and blood pressure in the normal (upper) and trypan blue treated (lower) embryo. Strength and pressure are plotted in arbitrary units along the embryonic axis, from head to tail. See text for discussion

what distended in the treated embryos, when compared with controls. Might this be an indication that there was an increase in blood pressure? If there was an increase in blood pressure, why would this lead to a *selective* break in the aortae in the caudal region?

To understand this, one must recall that development of the embryo takes place in a cephalic to caudal sequence. For example, in human embryos, the digits of the hand are carved out and separate a few days before the toes[46]. Since caudal structures are 'younger', it is suggested that they are more fragile than structures in the cephalic region, at least during the period of early (primary) morphogenesis. This hypothesis can be depicted graphically. As shown for a normal embryo in the upper part of Figure 11.6, relative strength of the aorta would decrease along the embryonic axis, from head to tail, due to the fragility just mentioned. Concomitantly, relative blood pressure would also decrease, as we move farther from the pressure source – the heart. In a treated embryo, if the blood pressure were to increase beyond the normal level (shown in the lower graph), the relative pressure would exceed the relative strength *at some point in the caudal region of the embryo*. Thus, the conditions for a selective caudal rupture would be met.

Critical period study

We tested the blood pressure hypothesis, but only after some preliminary experiments which were necessary to further define our system. Previously injections had been made at 48 h of incubation, and we knew that caudal haematomas developed in a large proportion of the embryos treated at this time. Perhaps at 48 h all the embryos were not at the same stage of maturity. Furthermore, any given embryo might or might not develop a haematoma. To prepare for measuring blood pressure in embryos *before* haematomas were

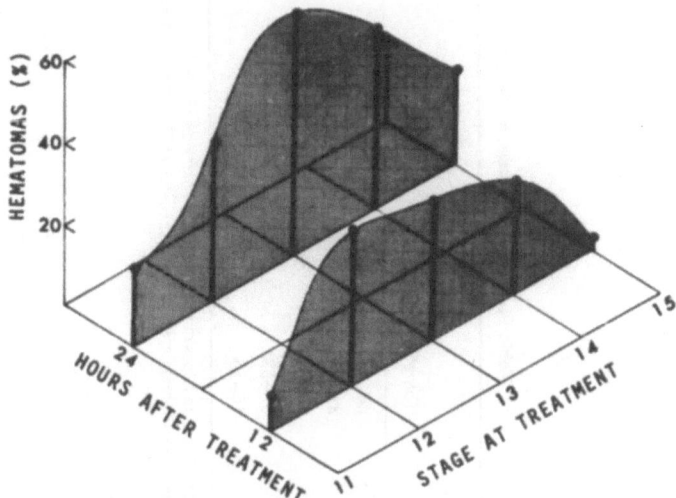

Figure 11.7 Effect of developmental stage, at the time of treatment with trypan blue, on the development of caudal haematomas. Treatment at stage 13 and 14 yields the greatest number of haematomas 24 h later

Table 11.3   Effect of stage at the time of injection on the development of caudal haematomas and on mortality

Observed stage at injection	Hours*	No.	Mean stage	% Haematomas	% Dead	Mean stage	% Haematomas†	% Dead†
11	40–45	32	12.8	9	50	14.6	19	75
12	45–49	37	14.4	38	30	16.4	43	38
13	48–52	44	16.1	34	7	17.5	59	16
14	50–53	36	16.8	28	6	18.1	44	14
15	50–55	30	17.1	3	10	18.4	23	10

* Published variation in hours of incubation for a particular stage[27]
† Percentages in this column are cumulative, i.e. 19% at 24 h includes the 9% seen at 12 h

visible (a time when the blood pressure should be high, if the hypothesis were correct, since the increased pressure was postulated to be the cause of the haematoma), we had to be sure that we were treating the embryos during the 'critical period' and causing a maximum number of embryos to have high blood pressure. Since all treated embryos do not develop haematomas, a sharper method for assessing the stage of maturity was necessary if we were to detect pressure changes in a population containing both embryos which would develop haematomas and embryos which would not. To do this, we treated a large population of embryos with trypan blue and recorded the Hamburger–Hamilton stage[27] of each embryo when treated. Each embryo was then examined 12 h after injection, and then again at 24 h after injection, if still living, for signs of caudal haematomas. The results are shown in Table 11.3 and Figures 11.7 and 11.8. Clearly, by treating at Hamburger–Hamilton

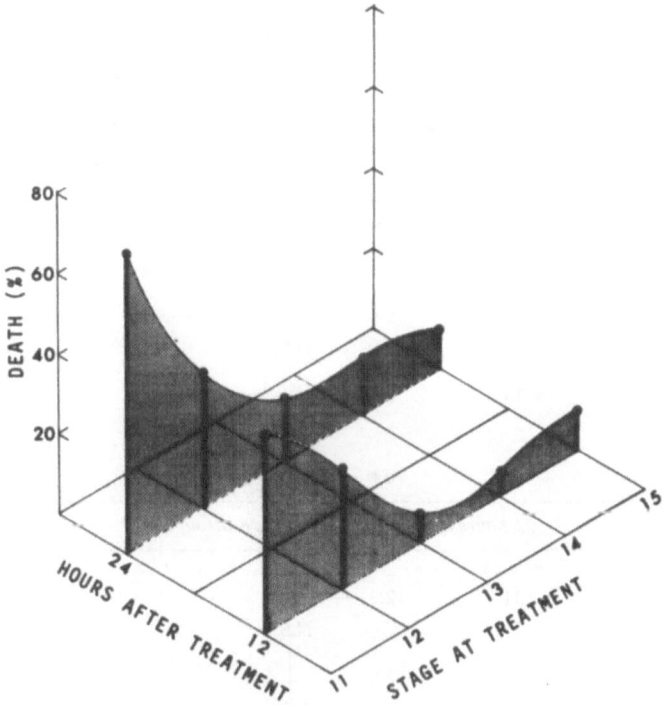

Figure 11.8 Effect of developmental stage, at the time of treatment with trypan blue, on the incidence of death. Treatment at stage 13 and 14 yields a reasonably low incidence of embryonic death 12 or 24 h later. Compare with Figure 11.7

stage 13 or 14, we could expect about half of the embryos to develop a caudal haematoma and relatively few embryos to die. The optimum procedure, then, was to choose only stage 13 and 14 embryos for injection, and this was done for all subsequent studies of blood pressure[47].

Blood pressure study

The apparatus used to measure mean ventricular blood pressure (MVBP) has been fully described elsewhere[48]. In short, it consisted of a saline manometer system, including a micrometer syringe, a saline reservoir, a manometer calibrated in millimetres and a glass cannula connected by flexible tubing. In practice, the cannula was inserted into the embryonic heart, and successful cannulation was signalled when erythrocytes began to enter the cannula tip. At that time, the micrometer syringe was turned until the stream of erythrocytes ceased entering the cannula tip and oscillated at the tip with systole and diastole. The MVBP was then recorded as the height (in mm) above heart level of the saline column in the manometer.

The results of such measurements are shown in Table 11.4, for embryos treated at stage 13 or 14 and measured at stage 17 (upper part of table) or 18–20 (lower part of table).

There was a 17% increase in blood pressure in the stage 17 experimental embryos, when compared with either class of controls, and this difference was statistically highly significant. Thus, treatment really did increase the pressure.

Table 11.4 Blood pressure in trypan blue treated and control embryos, at stages 17 and 18 through 20

Group	No.	MVBP†	p-value‡
*12–16 hours after injection (stage 17)**			
Unopened controls	14	8.6	N.S. (Unopened vs Saline)
Saline injected controls	14	8.6	<0.01 (Unopened vs Trypan blue)
Trypan blue injected§	14	10.4	<0.01 (Saline vs Trypan blue)
*24 hours after injection (stages 18 through 20)**			
Unopened controls	31	12.3	N.S. (Unopened vs Saline)
Saline injected controls	22	12.0	<0.05 (Saline vs Trypan blue no haematoma); N.S.; <0.001
Trypan blue injected (no haematoma)	14	13.5	<0.001
Trypan blue injected (haematoma)	21	7.6	<0.001

* All embryos were injected at stage 13 or 14
† MVBP = mean ventricular blood pressure in mm of saline
‡ NS = not significant ($p > 0.05$)
§ Embryos exhibiting caudal haematomas at stage 17 were removed from the study. See text for discussion

One must realize also that the magnitude of this difference might actually be much higher than 17% in embryos which would have gone on to develop a haematoma, since at least half of the measured population consisted of embryos which would not develop a haematoma (assuming such embryos also would have a normal or slightly elevated blood pressure).

The stage 17 embryos which exhibited caudal haematomas were excluded from the study, since the object at this stage was to measure pressure before the haematoma formed. The data in Table 11.3 and Figure 11.7 indicate that a proportion (16–25%) of the stage 17 embryos would have developed a caudal haematoma, had it been possible to permit them to continue development beyond the time when blood pressure measurements were made.

Among the stage 18–20 embryos (lower half of Table 11.4), dye treated embryos with haematomas showed a large and highly significant *decrease* in blood pressure, when compared with treated embryos without haematomas or either class of controls. A *decrease* in pressure among these embryos was expected, since there was a large pooling of blood in the haematoma, leading to a decrease in circulating blood volume.

Treated embryos with no haematomas exhibited an increased blood pressure, but this was statistically significant only when compared with saline injected controls. Perhaps the embryos with no haematoma did not develop a high enough blood pressure to result in vessel rupture at earlier stages.

Therefore, it seems reasonable to add yet an earlier step (2) to Figure 11.5, at least tentatively. The next question, as yet unanswered (step 1), concerns the cause for the increased blood pressure. Is it an increase in circulating fluids caused by an inrush of fluids past a leaky membrane? Is it an increase in cardiac output? Such questions must await further experimentation.

DISCUSSION

Caudal haematomas

A major unresolved question concerning caudal haematomas is whether the haematoma forms directly from burst vessels or whether it forms secondarily from clear blisters. Several investigators have suggested that clear blisters form first, and that these are later converted to haematomas by rupture of nearby vessels and bleeding into the clear fluid of the blister[41,43]. We have never observed a case, among the thousands of chick embryos that we have treated with trypan blue, in which a clear blister preceded the formation of a haematoma, although we have looked for this phenomenon. On the other hand, perhaps a very slow increase in blood pressure after treatment could lead to leakage of fluid from caudal blood vessels into surrounding tissues spaces, creating a clear blister. Later rupture of the caudal vessels would then produce a haematoma. It may be that other investigators, using different species, different teratogens or even a different strain of chick embryo, have created conditions in which clear blister formation and haematoma formation can be seen as separate events. Perhaps, in our hands, the abnormal pressure rise is more rapid, leading to the direct formation of a haematoma by a swift rupture of the aorta.

Relevance of the model

Animal model systems in experimental teratology teach us much about both normal and abnormal development, but often the question is raised concerning the direct relevance of such studies to problems of human development. Why should the chick embryo be studied as a model of human caudal dysplasia syndrome? Why not study a mammalian model?

Aside from questions of cost or the convenience of surgical approach to the embryo itself, there appears to be a very real reason why species other than the chick are less suitable as models of caudal dysplasia syndrome in man, and this reason stems from similarities between the chick and man in the normal development of the caudal region. Briefly, this basis is as follows.

In all species, the *caudal* neural tube arises by the hollowing out of a solid rod of neural tissue. In the rat, mouse, pig and opossum, this hollowing out proceeds evenly, from the pre-existing central canal toward the end of the embryonic axis. In man and the chick, however, the extension of the central canal proceeds by isolated cavitation of the rod of neural tissue, which results in irregular pockets. These pockets later fuse to form a continuous cavity, but there still exists, for a time, a branched and irregular lumen[49]. Furthermore, in tailed animals, the site of closure of the posterior neuropore is at the tip of the tail. In man and chicks (both of which have lost their tails during the course of evolution) the posterior neuropore closes before formation of the tail bud.

Perhaps these parallels in normal development between man and the chick are the basis of the peculiar susceptibility of both species to malformations of the caudal axis. Hughes and Freeman[49] have pointed out that caudal defects in man and the chick are relatively common, but relatively rare in tailed mammals. Perhaps this also is the basis for such enigmas as the finding that the alkaloid cyclopamine, which causes only cyclopia in sheep[50,51] induces caudal abnormalities in the chick[52].

It is suggested, then, that the peculiar susceptibility of the caudal portion of the chick to the influence of teratogenic agents may well reflect a similar situation in man. Finally, parallels in the development of this region between humans and the chick seem to indicate that the chick is, indeed, the best available animal model for human caudal dysplasia syndrome.

A note on approach

Contemporary experimental teratology is a field rightfully concerned with the elucidation of mechanisms. The work presented in this chapter has shown that a fruitful approach to the study of mechanisms is the intensive study of a single malformation, organized in a retrograde progression beginning at the morphological level. This style of approach cannot help but lead the investigator to problems at the level of embryonic physiology and, ultimately, to the biochemistry of development. The investigator must be willing and able to be led to these levels, for the embryo does not confine its workings to just those processes studied by an anatomist or a physiologist or a biochemist. The late A. F. W. Hughes pointed out[53] that, 'Until the artificial barriers, largely imposed by disciplinary organization are surmounted, Stockard's second principle [that the same abnormalities may be induced in the embryos of various species by a great number of different experimental treatments[54]] will continue to present a virtual impasse, blocking attempts to lift the study of abnormal development out of its present morass of empirical observation'. For this reason, the modern experimental teratologist can often feel like a 'jack-of-all-trades', borrowing a technique from one discipline or another to answer the question currently under study in the laboratory.

There is a benefit to be gleaned from studies of abnormal development which experimental teratologists often ignore. Because relatively little is still known of the biochemistry and physiology of normal development, and because normal control embryos are almost always studied along with those experimentally treated, studies in experimental teratology often yield valuable information concerning *normal* development.

A case in point was our recent discovery of the way in which the embryonic heart rate is controlled, prior to neural control[48]. This discovery was made while we were studying blood pressure in trypan blue treated *and in control* embryos. It is this author's contention that such studies, which help us also to understand the normal embryo, are doing double duty in the search for knowledge concerning development.

ACKNOWLEDGMENTS

I thank my students, Drs Richard Schmidt, Gary Kolesari, Gregory Rajala and Don Hilbelink, for keeping me intellectually young through their discussions, questions and arguments and for bringing to bear their varied points of view on the central problem represented in this work. I thank my wife Shari for her constant support; but for her none of this would have come to pass. My thanks to Dr Frank D. Anderson, Mrs Shari A. Kaplan, Dr Sally Y. Long and Ms Melissa G. Sherman for their helpful suggestions with the manuscript. The continued financial support of the Wisconsin Heart Association and PHS GRS 5 SO1 FR-5434 is gratefully acknowledged.

References

1. Price, D. L., Dooling, E. C. and Richardson, E. P., Jr. (1970). Caudal dysplasia (caudal regression syndrome). *Arch. Neurol.*, 23, 212
2. Gellis, S. S. and Feingold, M. (1968). Caudal dysplasia syndrome (caudal regression syndrome). *Am. J. Dis. Child.*, 116, 407
3. Kucera, J. and Lenz, W. (1967). Caudale Regression mit Oesophagusatresie und Nierenagenesie – ein Syndrom. *Z. Kinderheilkd.*, 98, 326
4. Hohl, A. F. (1952). *Zur Pathologie des Beckens: 1. Das Schräg-Ovale Becken.* pp. 61–63 (Leipzig: Wilhelm Engelmann)
5. Wertheim, C. C. (1857). Vollständiger Mangel des Kreuz- und Steissbeins bei einem Neugebornen. *Monatsschr. Geburtsk. Frauenkr.*, 9, 127
6. Blumel, J., Evans, E. B. and Eggers, G. W. N. (1959). Partial and complete agenesis or malformation of the sacrum with associated anomalies. *J. Bone Jt. Surg.*, 41-A, 497
7. Rusnak, S. L. and Driscoll, S. G. (1965). Congenital spinal anomalies in infants of diabetic mothers. *Pediatrics*, 35, 989
8. White, R. I. and Klauber, G. T. (1976). Sacral agenesis. Analysis of 22 cases. *Urology*, 8, 521

9. Smith, D. W. (1976). *Recognizable Patterns of Human Malformation.* 2nd Ed. (Philadelphia: W. B. Saunders) (*Major Problems in Clinical Pediatrics,* Vol. 7, A. J. Schaffer, ed.)

10. Miller, H. C. (1956). Offspring of diabetic and prediabetic mothers. *Adv. Pediatr.*, **8,** 137

11. Driscoll, S. G., Benirschke, K. and Curtis, G. W. (1960). Neonatal deaths among infants of diabetic mothers: Postmortem findings in ninety-five infants. *Am. J. Dis. Child.,* **100,** 818

12. Pedersen, L. M., Tygstrup, I. and Pederson, J. (1964). Congenital malformations in newborn infants of diabetic women: Correlation with maternal diabetic vascular complications. *Lancet,* **1,** 1124

13. Passarge, E. and Lenz, W. (1966). Syndrome of caudal regression in infants of diabetic mothers: observations of further cases. *Pediatrics,* **37,** 672

14. Kucera, J. (1971). Rate and type of congenital anomalies among offspring of diabetic women. *J. Reprod. Med.,* **7,** 61

15. Karlsson, K. and Kjellmer, I. (1972). The outcome of diabetic pregnancies in relation to the mother's blood sugar level. *Am. J. Obstet. Gynecol.,* **112,** 213

16. Cohen, A. M. and Schenker, J. G. (1972). The effect of insulin treatment on fetal mortality and congenital malformations in diabetic pregnant women. *Adv. Exp. Med. Biol.,* **27,** 377 (In: M. A. Klingberg, A. Abramovici and J. Chemke (eds.)). *Drugs and Fetal Development.* (New York: Plenum Press)

17. Ismajovich, B., Mashiack, S., Zakut, H. and Serr, D. M. (1972). The effects of insulin on fetal development in 'gestational diabetes'. *Adv. Exp. Med. Biol.,* **27,** 383 (In: M. A. Klingberg, A. Abramovici and J. Chemke (eds.)). *Drugs and Fetal Development.* (New York: Plenum Press)

18. Hughes, A. F., Freeman, R. B. and Fadem, T. (1974). The teratogenic effects of sugars on the chick embryo. *J. Embryol. Exp. Morphol.,* **32,** 661

19. Davenport, C. B. (1909). *Inheritance of Characteristics in Domestic Fowl.* Carnegie Institution of Washington Publication 121, 100 pp. (Philadelphia: J. B. Lippincott)

20. Du Toit, P. J. (1913). Untersuchungen über das Synsacrum und den Schwanz von *Gallus domesticus* nebst Beobachtungen über Schwanzlosigkeit bei Kaulhühnern. *Jena. Z. Naturwiss.,* **49,** 1

21. Landauer, W. and Baumann, L. (1943). Rumplessness of chicken embryos produced by mechanical shaking of eggs prior to incubation. *J. Exp. Zool.,* **93,** 51

22. Danforth, C. H. (1932–33). Artificial and hereditary suppression of sacral vertebrae in the fowl. *Proc. Soc. Exp. Biol. Med.,* **30,** 143

23. Dunn, L. C. (1925). The inheritance of rumplessness in the domestic fowl. *J. Hered.,* **16,** 127

24. Landauer, W. (1928). The morphology of intermediate rumplessness in the fowl. *J. Hered.,* **19,** 453

25. Zwilling, E. (1945). The embryogeny of a recessive rumpless condition of chickens. *J. Exp. Zool.,* **99,** 79

26. Ancel, P. and Lallemand, S. (1941). Sur l'arrêt de développement du bourgeon caudal obtenu experimentalement chez l'embryon de poulet. *Arch. Phys. Biol.,* **15,** 43 (Suppl.), 27

27. Hamburger, V. and Hamilton, H. L. (1951). A series of normal stages in the development of the chick embryo. *J. Morphol.,* **88,** 49

28. Landauer, W. (1945). Rumplessness of chicken embryos produced by the injection of insulin and other chemicals. *J. Exp. Zool.,* **98,** 65

29. Landauer, W. and Bliss, C. I. (1946). Insulin-induced rumplessness of chickens. III. The relationship of dosage and of developmental stage at time of injection to response. *J. Exp. Zool.,* **102,** 1

30. Romanoff, A. L. (1972). *Pathogenesis of the Avian Embryo.* (New York: Wiley-Interscience)

31. Beaudoin, A. R. and Wilson, J. G. (1958). Teratogenic effect of trypan blue on the developing chick. *Proc. Soc. Exp. Biol. Med.,* **97,** 85

32. Beaudoin, A. R. (1961). Teratogenic activity of several closely related disazo dyes on the developing chick embryo. *J. Embryol. Exp. Morphol.,* **9,** 14

33. Gillman, J., Gilbert, C., Gillman, T. and Spence, I. (1948). A preliminary report on hydrocephalus, spina bifida and other congenital anomalies in the rat produced by trypan

blue. The significance of these results in the interpretation of congenital malformations following maternal rubella. *S. Afr. J. Med. Sci.*, 13, 47

34. Hamburgh, M. (1952). Malformations in mouse embryos induced by trypan blue. *Nature (Lond.)*, 169, 27

35. Hamburgh, M. (1954). The embryology of trypan blue induced abnormalities in mice. *Anat. Rec.*, 119, 409

36. Waddington, C. H. and Carter, T. C. (1952). Malformations in mouse embryos induced by trypan blue. *Nature (Lond.)*, 169, 27

37. Gunberg, D. L. (1954). Spina bifida and herniation of hindbrain in the offspring of trypan blue injected rats. *Anat. Rec.*, 118, 387 (Abstract)

38. Wilson, J. G., Beaudoin, A. R. and Free, H. J. (1959). Studies on the mechanism of teratogenic action of trypan blue. *Anat. Rec.*, 133, 115

39. Ancel, P. (1950). *La chimiotératogenèse: réalisation des monstruosités par des substances chimiques chez les vertébrés.* (Paris: G. Doin)

40. Mulherkar, L. (1960). The effects of trypan blue on chick embryos cultured *in vitro*. *J. Embryol. Exp. Morphol.*, 8, 1

41. Stephan, F. and Sutter, B. (1961). Réaction de l'embryon de poulet au bleu trypan. *J. Embryol. Exp. Morphol.*, 9, 410

42. Ferm, V. H. and Carpenter, S. J. (1967). Developmental malformations resulting from the administration of lead salts. *Exp. Mol. Pathol.*, 7, 208

43. Carpenter, S. J. and Ferm, V. H. (1977). Embryopathic effects of lead in the hamster. A morphologic analysis. *Lab. Invest.*, 37, 369

44. Kolesari, G. L. and Kaplan, S. (1979). Amphetamines: Their ability to reduce embryonic size and produce caudal hematomas during early chick morphogenesis. *Teratology*, (in press)

45. Kaplan, S. and Grabowski, C. T. (1967). Analysis of trypan blue-induced rumplessness in chick embryos. *J. Exp. Zool.*, 165, 325

46. Moore, K. L. (1977). *The Developing Human.* 2nd Ed. (Philadelphia: W. B. Saunders)

47. Rajala, G. M. and Kaplan, S. (1979). Evidence that trypan blue induced caudal hematomas are caused by abnormally elevated blood pressure in the chick embryo. (In preparation)

48. Rajala, G. M., Kalbfleisch, J. H. and Kaplan, S. (1976). Evidence that blood pressure controls heart rate in the chick embryo prior to neural control. *J. Embryol. Exp. Morphol.*, 36, 685

49. Hughes, A. F. and Freeman, R. B. (1974). Comparative remarks on the development of the tail cord among higher vertebrates. *J. Embryol. Exp. Morphol.*, 32, 355

50. Keeler, R. F. and Binns, W. (1968). Teratogenic compounds of *Veratrum californicum* (Durand). V. Comparison of cyclopian effects of steroidal alkaloids from the plant and structurally related compounds from other sources. *Teratology*, 1, 5

51. Keeler, R. F. (1969). Teratogenic compounds of *Veratrum californicum* (Durand). VI. The structure of cyclopamine. *Phytochem. Newsl.*, 8, 223

52. Freeman, R. B. and Hughes, A. F. W. (1973). Multiple lumina in the neuraxis of the cyclopamine-treated chick. *Teratology*, 7, A15 (Abstract)

53. Hughes, A. F. W. (1976). Developmental biology and the study of malformations. *Biol. Rev.*, 51, 143

54. Stockard, C. R. (1921). Developmental rate and structural expression. *Am. J. Anat.*, 28, 115

12
Reversed susceptibility of lung to teratogenic insult

E. M. JOHNSON

INTRODUCTION

The adverse effects of diverse agents on embryonic development are widely documented and the object of increasing governmental regulation and public awareness. A major impediment to protection of the unborn human is our lack of adequate fundamental knowledge of teratogenic mechanisms. Until we know how teratogens divert embryonic tissues from normal sequences of differentiation and morphogenesis into abnormal patterns, prevention of congenital defects will elude us. Experimentation on mechanisms is extremely time consuming and lacking in obvious and immediate applicability toward improved protective measures. Development of usable data will come from continued studies of younger scientists already at work in laboratories with their students[1]. In spite of our limited knowledge, senior teratologists and pharmacologists have done an admirable job in guiding development of the current guidelines attempting protection of human pregnancies from teratogens. The best of efforts today cannot, however, preclude another thalidomide-type tragedy tomorrow. The reasons for this lamentable state are largely known and surely are beyond the topic of this writing[2].

Much research and the current guidelines have been predicated largely on the premise that the embryonic period of development is the time of gestation most susceptible to teratogenic insult. We have come to think of embryogenesis as a time of minimal repair. Once a developmental sequence has been altered it cannot revert to a normal developmental sequence.

The fetal stages, and especially the later fetal stages, were considered to be largely resistant to abnormal development. It has come to be accepted by some that once an organ is formed, one cannot then apply an agent and cause it to develop abnormally. A notable and important exception to this concept is later development and histogenesis of a functioning central nervous system. However, functional teratology must not come to be considered solely relevant to the CNS; while examining postnatal animals for effects of prenatal insult close attention must be applied to other organ systems as well.

GOALS

Evidence will be presented here to demonstrate that an embryonic organ system undergoing rapid organogenesis can be altered at a very fundamental level and still recover to become apparently normal by term. Contrasted with this will be data illustrating marked vulnerability of the same organ system to several teratogens during late fetal stages. A reversal of developmental stages susceptible to teratogenesis.

NORMAL LUNG DEVELOPMENT

The rat lung develops as a bifurcating single midline evagination from the floor of the primitive foregut. Expansion of the endodermally-derived epithelium causes it to encroach rapidly into, and become surrounded by, an equally proliferative mesoderm[3-5]. These form a reciprocal inductive–responsive system. Each component, endoderm and mesoderm, depends upon normal development of the other to ensure its own normal growth[4, 6, 7]. The initial endodermal evagination begins, in the rat embryo, during day 11 of gestation. (Day of finding spermatozoa in a smear of vaginal contents is considered day zero of gestation). On day 12 the lungs are developing rapidly in the rat embryo[8]. The laryngotracheal groove is already becoming moulded and the paired lung buds are surrounded by an abundant mesenchyme (Figures 12.1 and 12.2). One day later (Figures 12.3 and 12.4) the multiple generations of bronchiolar bifurcation begin to give the lung an increasingly vesicular appearance. This embryonic period extends through day 16 (Figures 12.5 and 12.6) and is superseded during days 17 and 18 (Figures 12.7 and 12.8) by a pseudoglandular stage, so termed because in histologic section the bronchials are rarely observed interconnecting and thereby superficially resemble an exocrine gland. The bronchiolar lumina lengthen and dilate somewhat during days 19 and 20 (Figures 12.9 and 12.10) when the lung is termed canalicular. The term fetal lung (day 21) is at a terminal sac stage (Figures 12.11 and 12.12). These terminal dilations are slightly similar in appearance to the alveoli which apparently develop entirely during postnatal life in the rat. Human lung development proceeds through the same stages but on a different time schedule: embryonic – up to six weeks of gestational age, pseudoglandular – week 7 through 17, canalicular – week 18 through 23, terminal sac – week 24 through term, and alveolar – term through the first postnatal decade[9, 10].

ALTERED LUNG DEVELOPMENT

Effect of folic acid deficiency on embryonic lung

Incorporation of the potent teratogen 9-methyl pteroylglutamic acid (9me-PGA) into a diet compounded of purified foodstuff is a potent teratogenic regimen[11]. It was fed to pregnant Long-Evans rats on days 11, 12 and 13 of pregnancy. A minimal 15 g per day food consumption was required for a female to continue in the experiment.

Biochemical development in these lungs was monitored by assay of

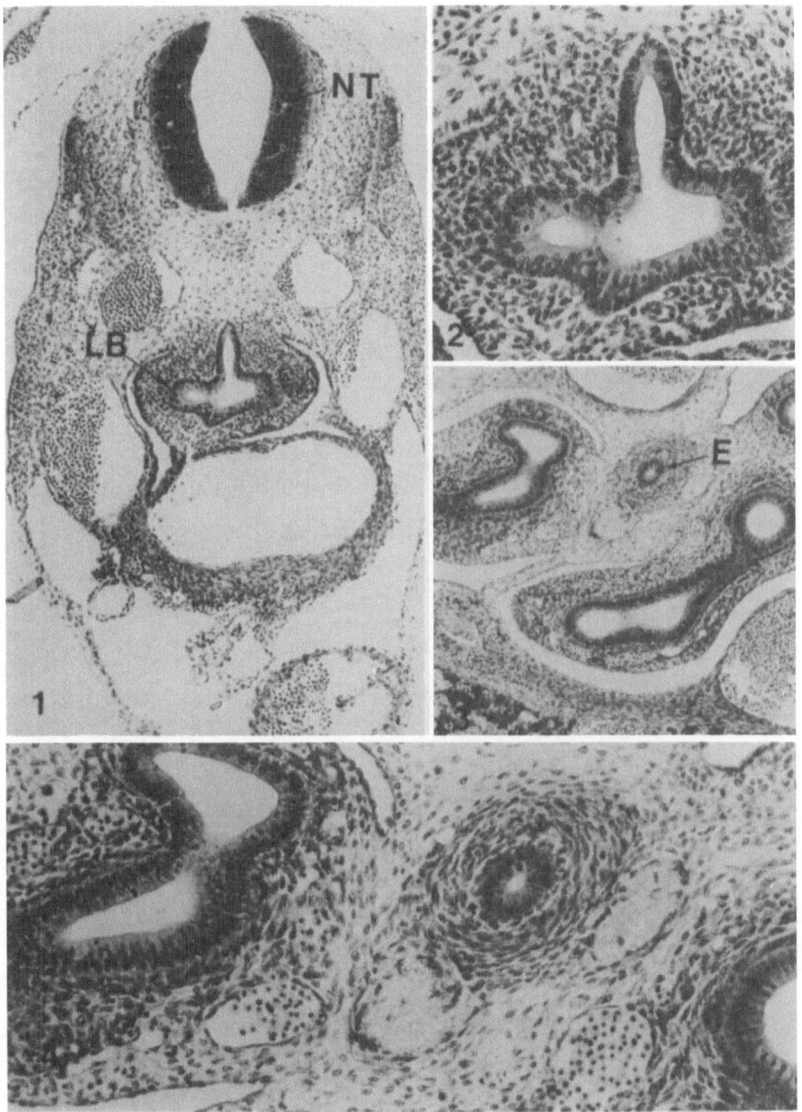

Figure 12.1   Cross section of a day 12 rat embryo through the laryngotracheal groove
(NT = neural tube, LB = lung bud). H & E, × 56

Figure 12.2   Enlargement of Figure 12.1. Note mitotic figures in both epithelium and mesoderm. H & E, × 188

Figure 12.3   Cross-section through thorax of a day 13 rat embryo showing branching of
conducting bronchioles (E = oesophagus). H & E, × 56

Figure 12.4   Enlargement of Figure 12.3. Note the tall columnal epithelium with large ovoid
nuclei. H & E, × 188

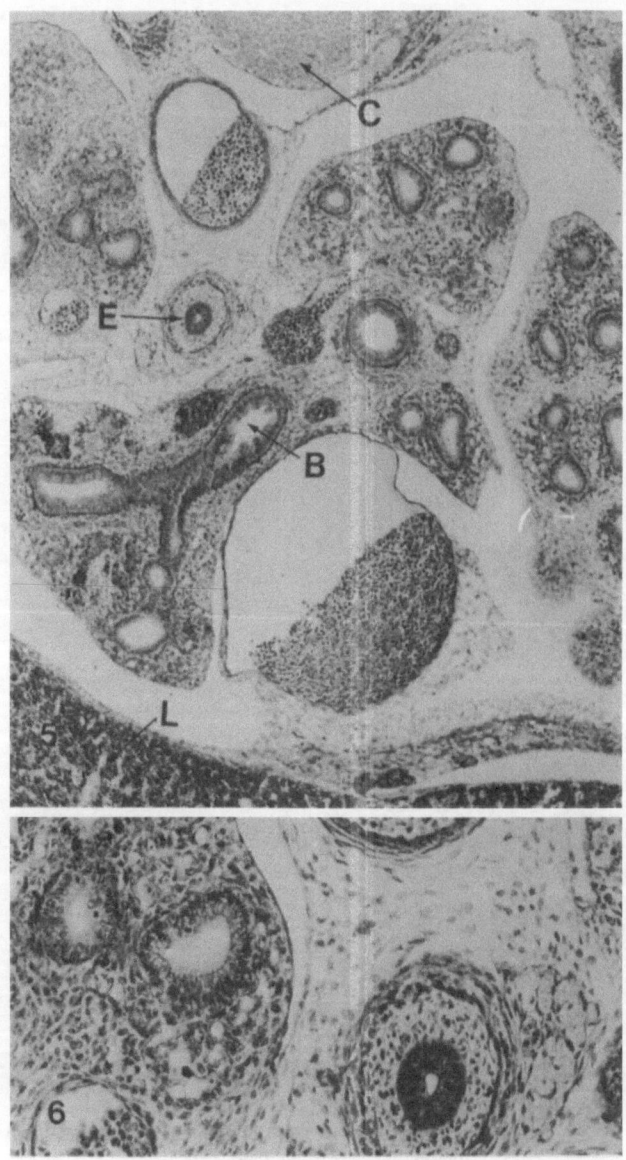

Figure 12.5   Cross section of day 16 fetus showing relationships of lung. Note progression of branching from day 13 (Figure 12.3). C = centrum of vertebra, E = oesophagus, B = bronchiole, L = liver. H & E, × 56

Figure 12.6   Enlargement of Figure 12.5. Note that the bronchiolar epithelium is now less tall and that the mesoderm is still dense. H & E, × 188

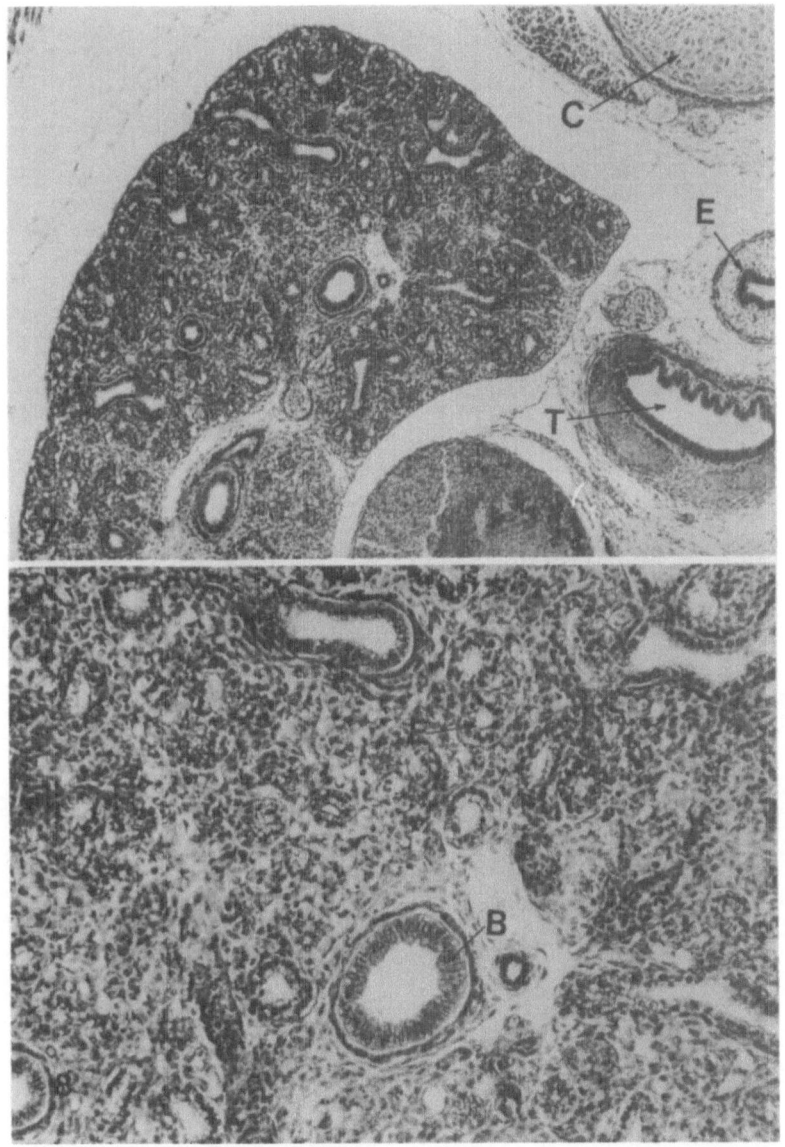

Figure 12.7 Cross section through thorax of a day 18 fetus. C=centrum of a vertebra, E=oesophagus, T=trachea. × 19

Figure 12.8 Enlargement of Figure 12.7. Note that the conducting bronchial epithelium is more cuboidal at the level indicated and that the more terminal epithelium appears glandular. E=epithelial tube, B=bronchiole. H & E, × 188

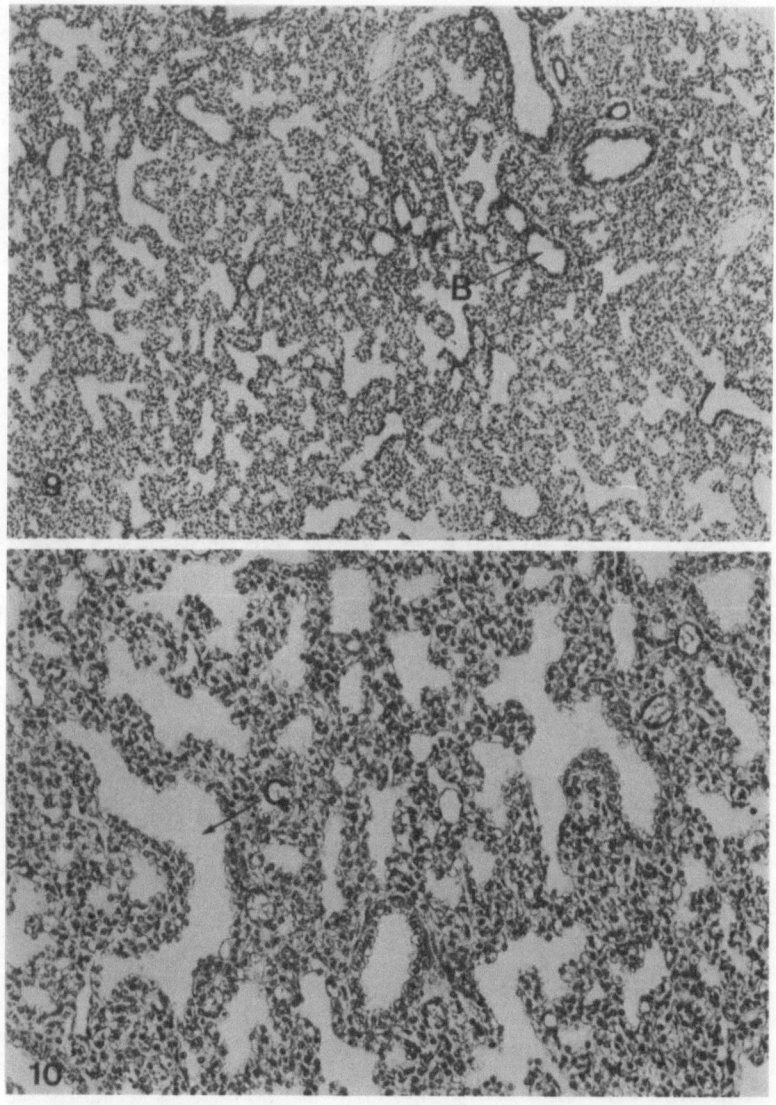

Figure 12.9 Section of day 20 fetal lung. B = conducting bronchiole. H & E, × 38

Figure 12.10 Enlargement of Figure 12.9. Note (c) the canalicular appearance caused by dilation and growth of the epithelial organizations in Figure 12.8. This canalicular structure is lined by cuboidal epithelium which thins toward the upper left to a squamous cell type. This represents a future respiratory bronchiole. H & E, × 150

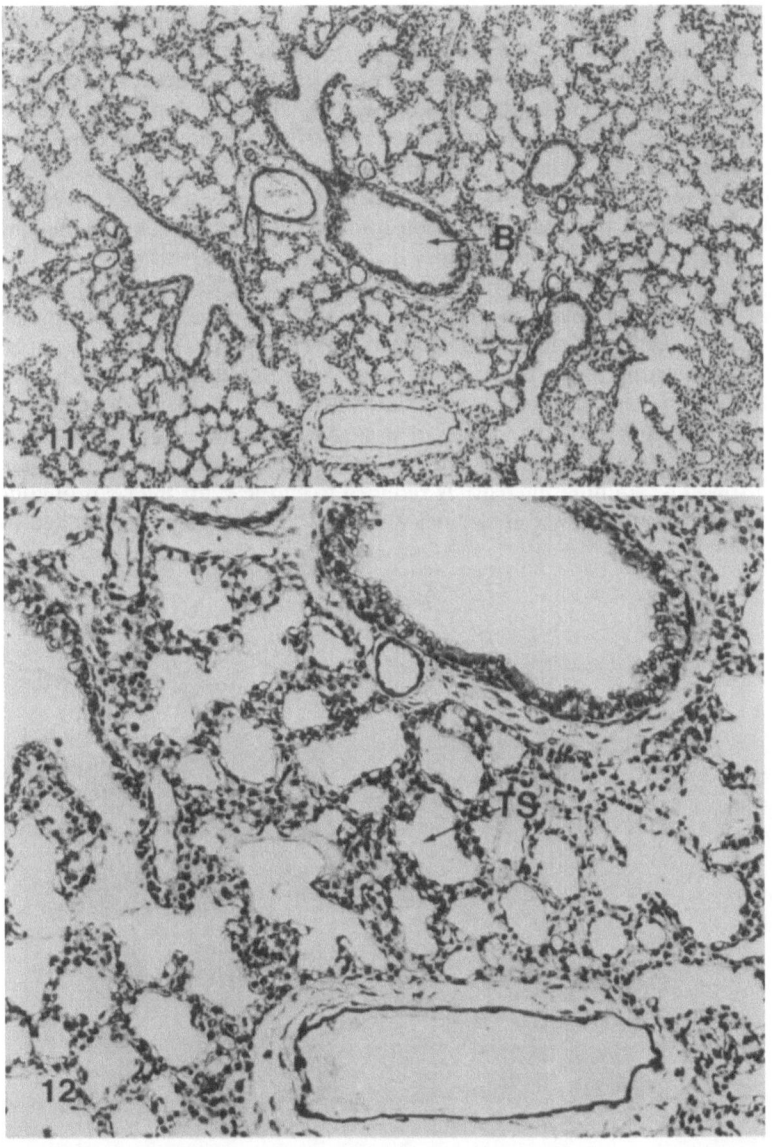

Figure 12.11   Section of day 21 fetal lung. Note that the terminal airways are dilated into saccular configurations. B = bronchiole. H & E, × 38

Figure 12.12   Enlargement of Figure 12.11 to show the terminal sac epithelium and the markedly thinned septal walls. TS = terminal sac, H & E, × 188

multiple molecular forms of enzymes in homogenates of fetal lungs taken on gestational days 16, 18 and 20. These enzyme molecules present changing zymogram repertoires in an ontogenic sequence[12, 13] leading to the adult pattern as development continues. Electrophoretic resolution followed by specific histochemical staining provides information of basic cellular metabolic abilities closer to differential gene expression than assay of, for example, complex structural proteins or cell organelles. The adult pattern is attained by qualitative and quantitative additions to and/or deletions from repertoires of molecular forms present at an earlier stage. Electrophoretic resolution was obtained in 7% polyacrylamide gels within 45 min at 3 mA direct current per gel. Details of the standard electrophoretic technique and histochemical visualization of specific isoenzymes have been previously described[14, 15].

Fetal lungs harvested from PGA-deficient dams are markedly altered in their biochemical development. Normally the lung contains all five molecules of lactate dehydrogenase (LDH) by day 16 (Figure 12.13) with the more slowly migrating forms reacting more intensly than the faster molecules. This pattern holds true through day 20 but the staining intensities become nearly uniform toward term. The lungs of teratogen-treated pups exhibit only three of the LDH molecules on day 16 and these are less well represented than the controls. The staining intensity is reduced from the control level probably due to the presence of fewer molecules since the amount of protein added to the

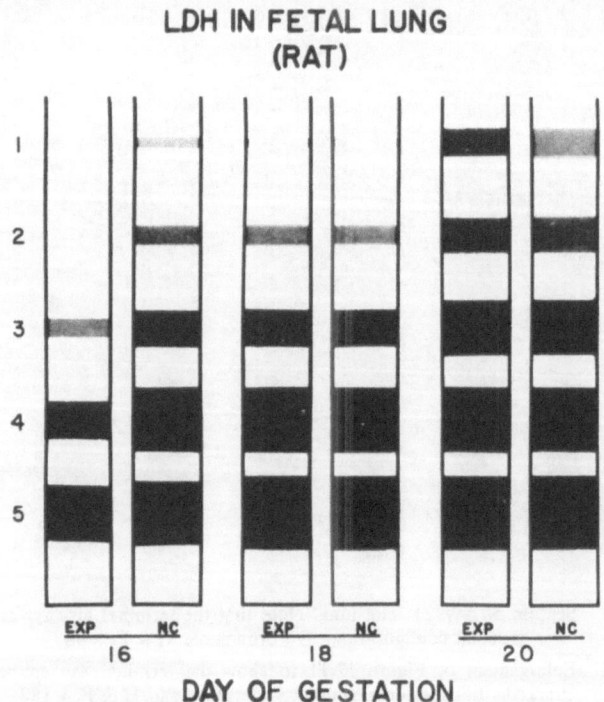

LDH IN FETAL LUNG
(RAT)

DAY OF GESTATION

Figure 12.13 Diagrammatic representation of LDH zymograms. Migration was from the bottom upward. The maximum number of molecular forms resolved is 5

gels was the same as the controls. The treated lungs appear to have caught up or to have undergone repair by days 18 and 20 as the zymograms are identical to those of normal controls.

Examination of the molecular repertories of glucose-6-phosphate dehydrogenase (G-6-PD) shows a slightly different type of response to the teratogen (Figure 12.14). Both molecular forms of G-6-PD undergo quantitative differentiation between days 16 and 20 in both drug-treated and untreated lungs. Both molecules are at the limits of electrophoretic/histochemical resolution in treated lungs on day 16 but appear to be at normal levels by day 20. Quantitatively, the enzyme activity is markedly reduced in the experimental group on day 16 but appears normal by day 20.

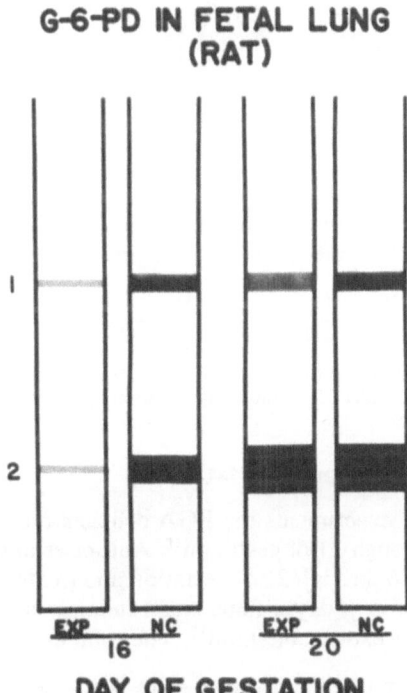

G-6-PD IN FETAL LUNG (RAT)

DAY OF GESTATION

Figure 12.14   Diagrammatic representation of G-6-PD zymograms. Migration as in Figure 12.13. Two different molecular forms of this enzyme are present

The maternal folic acid deficiency did not appear to alter the formation or expression of the two malate dehydrogenese (MDH) molecules (Figure 12.15).

Alterations in the qualitative and quantitative development of enzyme repertoires have proved to be reliable indicators of incipient teratogenesis in other systems[16, 17]. Such was not the case for the fetal lung. By day 20 the lungs from PGA-deficient mothers were both biochemically and histologically normal by the parameters evaluated. More importantly the treated pups inflated their lungs, breathed normally at delivery, and exhibited normal postnatal survival rates.

MDH IN FETAL LUNG
(RAT)

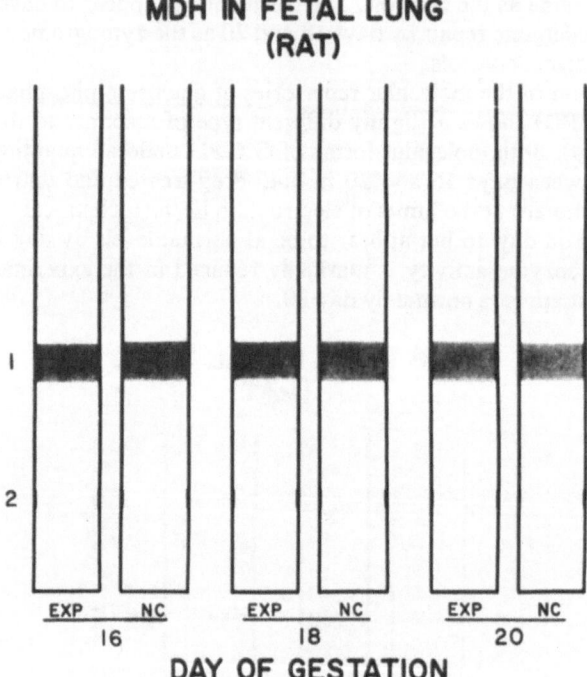

EXP   NC   EXP   NC   EXP   NC

16   18   20

DAY OF GESTATION

Figure 12.15   Diagrammatic representation of MDH zymograms. Migration as in Figure 12.13. Two molecular forms are present

Effect of folic acid deficiency on fetal lung

In a second series of experiments the PGA-deficient diet was fed to pregnant rats from day 14 through 21 of gestation[18]. Autopsies of normal control pups were made on days 16 through 21 of gestation and treated pups were obtained on day 21. Each autopsy day a fetus was randomly selected from the litter and assayed for total extractable lipid[19]. The total extractable lipid ($CHCl_3$:

Table 12.1   Total extractable lipid in lungs

Treatment	Gestational age (days)	Number of fetuses (determinations)	Mg lipid/g wet weight lung
NC*	16	3	17.3
NC	17	2	11.6
NC	18	4	7.4
NC	19	5	8.7
NC	20	4	9.1
NC	21	13	12.3
FAD†	21	3	17.3

* Normal control
† Folic acid-deficient

MeOH,2:1) normally declines by day 18 and then slowly increases through term (Table 12.1). However, when the pups were from folic acid-deficient dams, the extractable lipid was increased above the control level on day 21 which is one day before expected delivery. As the primary surface-active agent in lung is considered to be the lipid phosphatidylcholine one might believe these lungs were prepared to inflate and stay inflated. Such, however, is not the case, as can be observed in Table 12.2. These fetal and newborn

Table 12.2   Assay of surface activity from lung (bubble stability)

	Gestational age (days)	Height of bubble column* (mm)	
		5% Homogenate	Lipid extract
NC	21	8.0	5.5
FAD	21	5.5	—
Actinomycin D†	21	—	2.8
NC	DD‡	—	4.5
Actinomycin D	DD	—	2.0
NC	ID§	—	5.5
Actinomycin D	ID	—	3.8

* 30 min after formation
† Actinomycin D (0.1, 0.05 and 0.05 mg/kg) days 15, 15.5 and 16
‡ Day of delivery (live pups)
§ One day after delivery

lungs were homogenized, filtered and diluted with physiological saline to a 5% dilution. This diluted homogenate was shaken for 15 seconds by a vortex agitator and allowed to stand for 30 min. In a normal fetus the height of the stable bubble column was 8 mm. This crude assay for surface activity gave a stable column of only 5.5 mm from the lungs of folic acid-deficient pups.

Other agents affecting fetal lung development

A possible reduction in surface-active lipid was not unique for folic acid deficiency but was caused by actinomycin also. Another method of measuring surface activity is by resuspension of a lipid extract in saline and determining the stability of bubbles after shaking. This was done with spontaneously delivered and day-old pups of untreated pregnancies and with pups delivered of dams injected with actinomycin D on days 15, 15.5 and 16 of pregnancy. The total height of the stable bubble columns is reduced from that of controls (Table 12.2). This could have resulted from the absence of proteins in the chloroform methanol extract. Bubble stability was reduced by actinomycin D treatment to 44% of the control level on the day of delivery and to 69% of control values in day-old treated pups. Litters of pups were allowed to be born subsequent to prenatal actinomycin D insult during the fetal period and 7% of them were non-viable and had collapsed lungs at autopsy.

The results of the preceding experiments revealed that the fetal lung is remarkably susceptible to abnormal development and led to further exploration of adverse structural and functional reactions by fetuses which would be manifest at term and during neonatal life. Of the several agents explored, vitamin A proved to be an excellent tool, and vitamin A acetate (160 000 USP units in corn oil) was administered once per day on days 15 through 19 of gestation. All the pups were externally normal at term but only 82% were born alive (Table 12.3). On each of the three ensuing days the number of live pups was reduced and only 2% survived through weaning. The sequence of death was similar for pups regardless of the postnatal day on which they died. Each in turn came into respiratory distress. The forelimbs and neck would be extended and the mouth held open. The pups would then become progressively cyanotic and slower in respiratory rate prior to death. The adverse effects of excess vitamin A acetate appeared directly related to pup survival as a clear dose–response existed (Table 12.3).

Table 12.3   Effects of maternal excess vitamin A on pup survival

Dose	No. of females	% of live pups		
		At delivery	At 4 days	At 21 days
NC*	18	99	98	98
VC†	9	98	98	97
80K‡	4	100	100	100
160K	10	82	4	2
320K	3	0	0	0

* Pups of untreated dams
† Pups of vehicle-treated (corn oil) dams
‡ Pups of dams treated with 80 000, 160 000, and 320 000 USP units vitamin A acetate per day from day 15 through day 19 of pregnancy

A clinically applied test for surfactant levels characteristic or indicative of lung maturity is the lecithin sphingomyelin ratio of lipids extracted (chloroform/methanol) from amniotic fluid. However thin layer chromatography of fluid lipids extracted from all vitamin A-treated pups was identical to normal controls. Clement's foam test was applied and again showed no differences due to the teratogen (Table 12.4) at either the 1 : 1 or 1 : 2 dilutions. The same was true of normal and vitamin A-treated pregnancies when amniotic fluid

Table 12.4   Clements' foam test for surfactant

Group	Day 18		Days 19 and 20		Day 21	
	1:1	1:2	1:1	1:2	1:1	1:2
NC fluid*	—	—¶	+**	—	+	+
VC fluid†	—	—	+	—	+	+
80K Vitamin A‡	—	—	+	—	+	+
160K Vitamin A§	—	—	+	—	+	+
320K Vitamin A‖	—	—	all pups dead		all pups dead	

*   Amniotic fluid from normal control pregnancies
†   Amniotic fluid from corn oil (vehicle) treated pregnancies
‡, §, ‖ Amniotic fluid from pregnant dams treated with 80 000, 160 000, or 320 000 USP units vitamin A acetate per day from day 15 through 19 of pregnancy
¶   No stable foam
**   Stable foam

lipids were extracted[20], chromatographed ($CHCl_3 : H_3COH$: Acetic acid, $50 : 25 : 8$, v/v), scraped and the phosphotidylcholine and sphingomyelin spots (stained with iodine vapours) analysed quantitatively for phosphorus[21, 22].

The lungs and other vital organs of live-born pups were studied histologically in haematoxylin and eosin stained sections in an attempt to determine the cause of neonatal death.

At the 80 000 USP units dosage, no differences from normal control, newborn (first day) were observed in histology or cytology of lung epithelium, or blood vessels. No gross or histological abnormalities or other differences from controls were found in other organ systems either (adrenal, kidney, heart, spleen, thymus, stomach, ileum, transverse colon, gonad, liver, pancreas, haematocrit, or blood cell morphology) in this treatment group.

The survival table shows that a majority of the pups of the 160 000 units group survived into the first postpartum day. Two classes of pups existed: those dead at delivery and those killed during their first 24 h for histological evaluation. Lungs removed from stillborn pups were red in colour and sank in the fixative. Their histological picture was of non-inflation and somewhat similar to normal day 21 *fetuses* with exception of the septal wall thickness. Septal walls normally are 2–3 cells thick on the day prior to birth while those of stillborns from these vitamin A-treated mothers were 3–4 cells thick.

The lungs removed from live pups were pink in colour prior to fixation though not as pale as normal controls. They tended to float in the fixative. Marked peripheral atelectasis was found in several cases. In other instances, atelectasis alternated with marked overexpansion of terminal passageways[23, 24]. Frequently, dilated terminal bronchioles were observed and in several sections some degree of emphysema was present. No changes were detected in the blood vessels or connective tissue in these haematoxylin and eosin stained paraffin sections with the exception of some reduction in connective tissue fibre staining. All vital organs appeared normal.

All pups delivered of females treated with 320 000 units vitamin A from days 15–19 were born dead and the fetuses showed moderate stages of maceration.

SUMMARY

The fetal rat lung is an organ undergoing marked histological development late in gestation as preparation for its transition from a liquid to an air environment. These developmental sequences are easily altered by exogenous agents administered to the dam. Unfortunately such effects tend to be overlooked by most investigators. Studies designed to detect teratogenic effects tend to exclude evaluation of stillborns and concentrate instead on gross structural examinations. Careful microscopic examination is necessary for detection of the more subtle effects characteristic of the fetus and neonate. Additionally, the functional parameters served by the microscopic structures require different methods of evaluation for detection of effects manifest primarily through effects on function.

Maternal excess vitamin A does not appear to alter surfactant production or release, and organs other than the lung appear unaltered on routine histological examination. It is beyond the point of fetal functional susceptibility being made in this report to explain methods for evaluating neonatal Q_{O_2} lung compliance or circulation and blood gases. Obviously such studies are necessary. The broader question remains. To what extent can teratogens alter other fetal structural/functional relationships and to what extent may human neonatal disease be the resultant of prenatal insult during the fetal stages of development?

References

1. Kretchmer, N. (1978). Perspectives in teratologic research. *Teratology*, 17, 203
2. Fraser, F. C. (1978). Prevention of birth defects. *Teratology*, 17, 193
3. Rudnick, D. (1933). Developmental capacities of the chick lung in chorioallantoic grafts. *J. Exp. Zool.*, 66, 125
4. Wessels, N. K. (1970). Mammalian lung development. Interactions in formation and morphogenesis of tracheal buds. *J. Exp. Zool.*, 175, 455
5. Alescio, T. (1973). Effect of a proline analogue acetidine-2-carboxylic acid on the morphogenesis *in vitro* of mouse embryonic lung. *J. Embryol. Exp. Morphol.*, 29, 439
6. Taderera, J. V. (1967). Control of lung differentiation *in vitro*. *Dev. Biol.*, 16, 489
7. Alescio, T. and Cassini, A. (1962). Induction *in vitro* of tracheal buds by pulmonary mesenchyme grafted on tracheal epithelium. *J. Exp. Zool.*, 150, 83
8. Johnson, E. M. (1959). Effect of transitory maternal pteroylglutamic acid deficiency on embryonic and placental development. Dissertation, Univ. of California, Berkeley
9. Dunnill, M. S. (1962). Postnatal growth of the lung. *Thorax*, 17, 329
10. Davies, G. and Reid, L. (1970). Growth of the alveoli and pulmonary arteries in childhood. *Thorax*, 25, 669
11. Johnson, E. M., Nelson, M. M. and Monie, I. W. (1963). Effects of transitory pteroyl-glutamic acid (PGA) deficiency on embryonic and placental development in the rat. *Anat. Rec.*, 146, 215
12. Markert, C. L. and Moller, F. (1959). Multiple form of enzymes: tissue, ontogenetic and species specific patterns. *Proc. Nat. Acad. Sci. (USA)*, 45, 753
13. Johnson, E. M., Takano, K. and Nishimura, H. (1972). Molecular differentiation of several enzymes in early human embryos. *Teratology*, 5, 89
14. Johnson, E. M. and Spinuzzi, R. (1966). Enzymic differentiation of the rat yolk-sac placenta as affected by a teratogenic agent. *J. Embryol. Exp. Morphol.*, 16, 271
15. Johnson, E. M. and Spinuzzi, R. (1968). Differentiation of alkaline phosphatase and glucose-6-phosphate dehydrogenase in the rat yolk-sac. *J. Embryol. Exp. Morphol.*, 19, 137
16. Johnson, E. M. (1965). Electrophoretic analysis of abnormal development. *Proc. Soc. Exp. Biol. Med.*, 118, 9
17. Johnson, E. M. (1975). Organ and tissue specificity of response to teratogenic insult. In: D. Neubert and H. J. Merker (eds.). *New Approaches to the Evaluation of Abnormal Embryonic Development*, pp. 573–590. (Stuttgart: Georg Thieme Verlag)
18. Armenti, V. T., Masters, E., Cliver, A. and Johnson, E. M. (1977). Effects of teratogenic insult on fetal lung maturation. *Teratology*, 15, 30A
19. Folch, J., Lees, M. and Sloane Stanely, G. H. (1957). A simple method for the isolation and purification of total lipids from animal tissues. *J. Biol. Chem.*, 226, 497
20. Bligh, E. G. and Dyer, W. J. (1959). A rapid method of total lipid extraction and purification. *Can. J. Biochem. Physiol.*, 37, 911
21. Rouser, G., Siakotos, A. N. and Fleischer, S. (1966). Quantitative analysis of phospholipids by thin-layer chromatography and phosphorous analysis of spots. *Lipids*, 1, 85
22. Chalvardjian, A. and Rudnicki, E. (1970). Determination of lipid phosphorus in the nanomolar range. *Anal. Biochem.*, 36, 225
23. Johnson, E. M. and Armenti, V. T. (1978). Postnatal effects of prenatal insult on lung development in the rat. *Anat. Rec.*, 190, 432 (Abstract)
24. Johnson, E. M., Armenti, V. T. and Newman, L. M. (1978). Postnatal effects of prenatal hypervitaminosis A in the rat. *Teratology*, 17, 39A (Abstract)

13
New approaches to the study of malformations resulting from maternal diabetes

ELIZABETH M. DEUCHAR

INTRODUCTION

Despite the success of present-day treatments for diabetes mellitus in man, this disease still poses major problems to obstetricians. It has long been known that if a woman has diabetes, even mildly, during pregnancy there is an increased risk that her baby may either die during the perinatal period[1] or be born with a serious congenital abnormality[2]. Now that improved management of the perinatal period has greatly reduced the numbers of deaths at that stage, the problem of the congenital abnormalities is thrown into sharper relief. It has recently been re-emphasized by Beard and Hoet[3] that the incidence of abnormalities is still three times as high in children of diabetic mothers than in the rest of the population, in both Britain and Belgium. Data collected in Holland[4] and in the United States[5, 6] confirm these estimates. There is therefore an urgent need to develop new approaches in research, aimed at discovering the causes of these abnormalities and how they may be prevented.

Several new approaches have, in fact, been used in clinical research on diabetic pregnancy over the past ten to twenty years. It has become possible to follow and to control the concentrations of a number of metabolites in the blood of diabetic patients during pregnancy (see, for example, Freinkel[7]). Besides this, amniocentesis and other monitoring techniques can now be used to diagnose some kinds of abnormalities and distress in the fetus[8]. As Mintz and Chez[9] have pointed out, a limited degree of 'fetal therapeutics' is already possible. This consists mainly of careful and precise insulin dosage to the pregnant diabetic woman, to avoid surges of hyperglycaemia after meals (which would result in a corresponding, though less marked fetal hyper-glycaemia[10]) and also to avoid the more serious condition of hypoglycaemia, severe attacks of which could cause death of the fetus.

So far, however, no therapeutic measures have been developed to prevent the kinds of congenital malformation that arise early in embryonic development, during the first phases of organogenesis, which in humans take place from 3–10 weeks after fertilization of the ovum. Clinical investigations rarely cover this period, since human pregnancies are seldom confirmed before the tenth week. Yet, if one is to discover the causes of malformations such as anencephaly, spina bifida, cardiac anomalies and hindgut defects, which are among the most common as well as the most serious of the malformations arising in children of diabetic as well as normal mothers, attention must be concentrated on these first few weeks of embryonic development. During this period the embryo is, of course, too small to be accessible for observation *in vivo*, even by the most sophisticated monitoring techniques. For the groundwork on such early embryonic stages, therefore, we have to turn to work on other mammalian species, where the female can be sacrificed and the embryos removed for examination.

Studies of malformations in animals have until recently been limited to observation of a few arbitrarily chosen stages, sacrificing both mother and offspring at these stages. In our own work on Wistar rats[11], embryos and fetuses were examined at mid-gestation and at 20 days. It was found that maternal diabetes induced either with alloxan or streptozotocin resulted in increased incidences of nervous system and heart abnormalities in 11–13 day embryos (Table 13.1). At 20 days the commonest abnormality was incom-

Table 13.1 Incidence of abnormalities in early embryos from diabetic rats, compared with controls

	Embryos resorbed	Total live embryos	Abnormalities of CNS	Abnormalities of heart
Alloxan-induced diabetics	34	120	7	2
Controls	10	136	3	0
Streptozotocin-induced diabetics	38	190	19	10
Controls	8	176	5	3

Table 13.2 Abnormalities in 20-day fetuses of diabetic and control rats compared

	Resorptions	Live fetuses	Small	Viscera everted	Sacral defects
Alloxan-diabetic rats	4	333	12	0	27
Controls	0	141	2	0	0
Streptozotocin-diabetic rats	27	145	35	3	40
Controls	0	156	1	0	0
Insulin-treated Streptozotocin-diabetic rats	0	180	0	0	0

plete ossification of sacral vertebrae (Table 13.2). It was therefore of particular interest to study in detail the phases of development immediately preceding those of the observations, i.e. organogenesis from 9 to 11 days, and sacral osteogenesis from 17 to 20 days. Such studies are greatly facilitated by *in vitro* culture methods.

A NEW APPROACH

In 1966 New[12] described a method for culturing early post-implantation rat embryos *in vitro* for short periods. This has made it possible to observe development in detail during phases of organogenesis that are crucial for an understanding of the aetiology of certain major malformations. The embryo is cultured in rat serum, with its placental membranes essentially intact. Thus, the conditions are similar to those *in vivo*, in that maternal serum is in contact with embryonic trophoblast. The placenta in the rat resembles that of the human in being haemochorial, with the uterine capillaries and connective tissue eroded by trophoblast. This culture system therefore provides a useful model for investigating the effects of factors in the maternal serum on the early, organ-forming stages of embryonic development. Possible differences in the serum due to maternal diabetes can also be tested in this system.

In our laboratory we have started to use this new approach in the study of embryonic abnormalities caused by maternal diabetes. The work is only in its beginnings, but has already yielded some interesting and unexpected results, which it is hoped may give useful indications for future 'embryonic therapeutics' to prevent such abnormalities from occurring.

THE CULTURE METHOD

The simplest method for culturing 9- and 10-day rat embryos for short periods is that described by New[12]. The apparatus is illustrated in Figure 13.1. Each embryonic vesicle, bounded by yolk sac and trophoblast, is dissected out of the uterus in sterile conditions in Tyrode saline and, after deflection of the outer elastic boundary known as Reichert's membrane, is placed in a watchglass containing rat serum. As little as 0.3 ml of serum per embryo has been found adequate to support the growth of 10-day rat embryos for 24 h[13]. The watchglasses, each in a moistened petri dish, are stacked under a beaker inverted over liquid paraffin and are then gassed with an oxygen-rich mixture. The whole apparatus is then placed in an incubator at 37°C. Embryos may be grown successfully in this system for up to 30 h without renewal of the medium. If longer-term cultures are desired, a roller-bottle method[14] can be used, with 1 ml of serum per embryo. Even longer-term cultures, for 4–5 days, may be achieved using a continuously circulating medium[15]; in this system Cockroft[16] has succeeded in culturing rat embryos up to early fetal stages equivalent to 14.5 days' gestation.

The external morphology of rat embryos ranging from 10 to 12 days' gestation age is shown in Figures 13.2–13.4. All the embryos depicted have been grown *in vitro*. New, Coppola and Cockroft[17, 18] have made detailed comparisons of the growth of embryos *in vitro* and *in vivo*, in assessing the

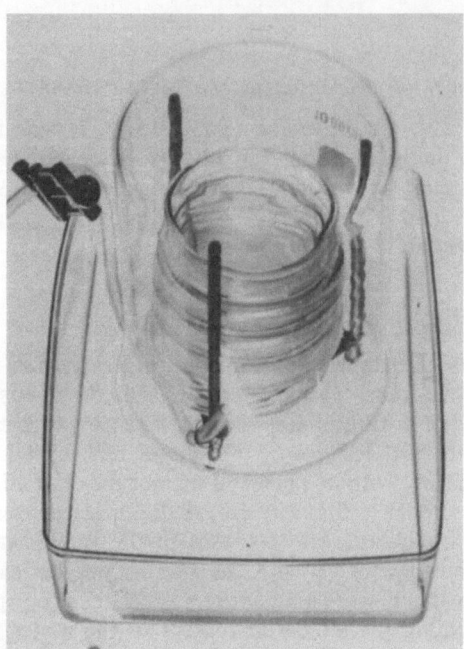

Figure 13.1 Simplest apparatus for cultures of rat embryos (cf. New, 1966). A watch-glass containing one embryo in rat serum is placed in a petri dish moistened with saline. The dishes are stacked on a triangular metal stand under an inverted beaker, in a dish containing liquid paraffin. Gas is led in via plastic tubing which is then clamped during the incubation period. ($\times \frac{1}{3}$).

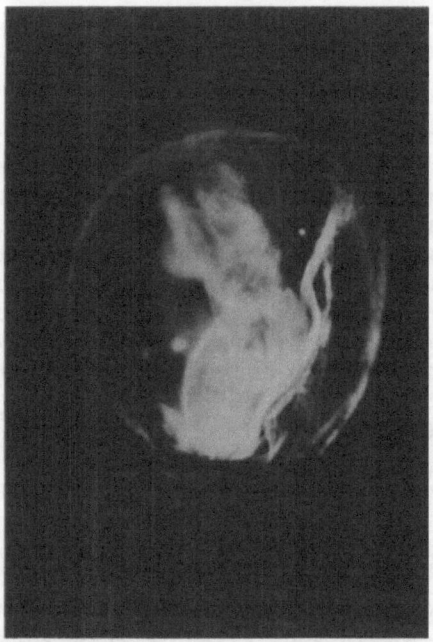

Figure 13.2 Ten-day rat embryo in its membranes. The neural folds (nf) are still open in the brain region and the embryo is concave dorsally ($\times$ 36)

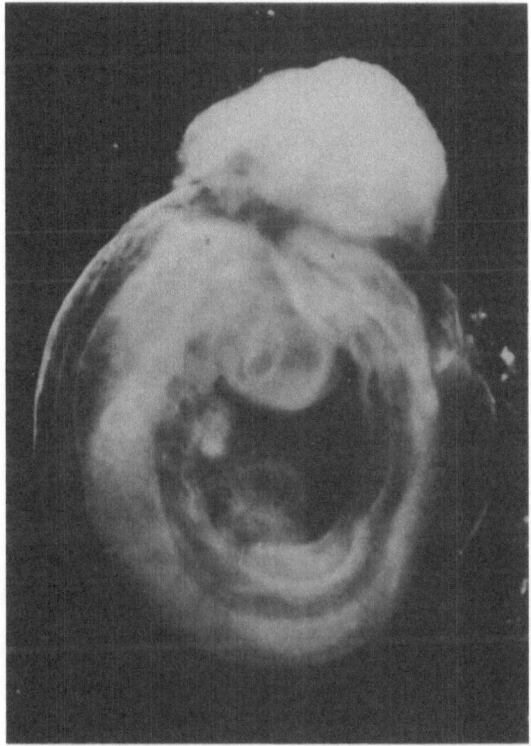

Figure 13.3   Eleven-day embryo, well developed and fully rotated after 24 h in culture. Brain (br), eye (ey), and ear (ea) rudiments are visible and the heart (h) is relatively large. Somites (s) also are shown. The trophoblast (t) forms a knob of tissue at the apex of the vesicle formed by the yolk sac. The embryo is now convex dorsally. (× 36)

success of their culture methods. It is to be noted that the main organ systems – central nervous system, vascular system, gut, segmental muscles and limb rudiments – are formed in the period from 9 to 12 days after mating.

USE OF THE CULTURE METHOD TO STUDY THE EFFECTS OF MATERNAL DIABETES ON ORGANOGENESIS IN RAT EMBRYOS

So far we have used only New's original watchglass culture method, since this requires minimal volumes of serum and is adquate for embryos of 9, 10 and 11 days' gestation age, i.e. for the preliminary and most vulnerable phases of organogenesis. To start with, we have looked for answers to the following questions:

(i) Are embryos of diabetic females any less viable at these early stages than embryos of non-diabetic females?

(ii) Does organogenesis more often proceed abnormally in embryos of diabetics than in those of non-diabetics?

217

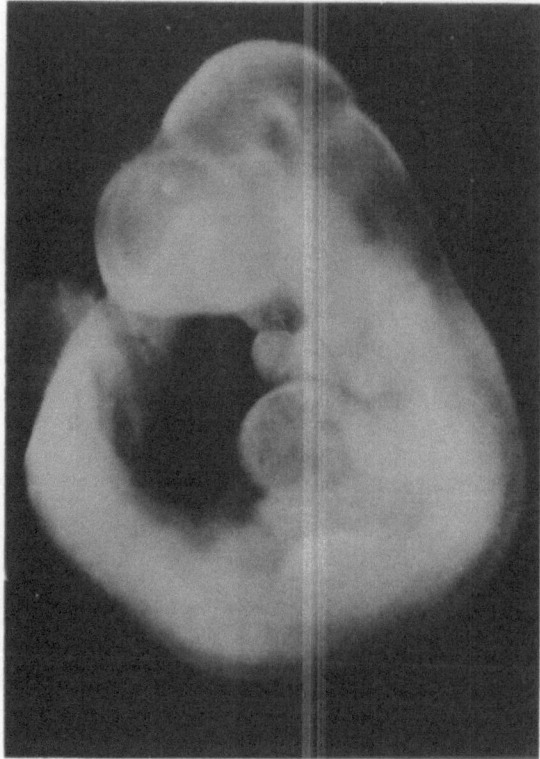

Figure 13.4 Twelve-day embryo from culture, with membranes removed. Fore-, mid- and hindbrain regions are shown (fb, mb, hb) and also the limb buds (lb). (×36)

(iii) Are there differences in growth rate between embryos of diabetics and non-diabetics?

(iv) Does the serum of diabetic females have any adverse effects on development?

The answers are all as yet incomplete, but they have shown some of the potentialities of this new approach and have given useful pointers for future research. We will take each question in turn below.

In all this work the female rats were made diabetic by an intravenous injection of 45 mg/kg body weight of streptozotocin (Upjohn Ltd.) on the day after mating. Controls were injected with just the citrate buffer used to dissolve the streptozotocin. Blood glucose levels in diabetic females ranged from 175–400 mg/100 ml, while in controls the levels were from 90–130 mg/100 ml.

Are embryos of diabetic females any less viable at early organogenesis stages than embryos of normal females?

To answer this question, embryos of as nearly as possible identical size and stage of development were dissected from the uteri of normal and diabetic rats at 10 days' gestation and each embryo was cultured in 0.3 ml of normal serum. They were examined alive after 24 h culture. The most successful cultures, classed as 'Good' in Table 13.3, were embryos with well-expanded membranes, which had grown to at least twice their original size and had rotated fully (cf. Figure 13.3) so that the dorsal surface was now convex. They showed a rapidly beating heart, and red blood cells circulating through the vessels of the body and the yolk sac. Embryos that were somewhat smaller than these, with a slower heartbeat and few or no red blood cells, and which were not fully turned, but were nevertheless healthy and normal for the stage they had reached, were classed as 'Fair'. Any stunted or abnormal embryos were classed as 'Poor'. The very few that died were also recorded; these had become white and opaque, usually with shrunken membranes and, of course, no heartbeat.

Table 13.3   External morphology of cultured embryos from diabetic and control rats

	Total embryos cultured	Good	Fair	Poor	Dead
Diabetic rats	60	33	17	5	5
Control rats	59	45	3	4	7

It can be seen from Table 13.3 that a high proportion (76%) of the embryos from normal rats developed well and could be classed as 'Good', whereas a lower proportion (55%) were 'Good' among the embryos from diabetic rats. Conversely, more of the embryos from diabetic rats were only 'Fair' in their development. The numbers of 'Poor' and dead embryos in each group were too small to show significant differences.

From these preliminary criteria, then, it appears that embryos from diabetic Wistar rats are somewhat less viable than those from normal rats, when tested at this stage of organogenesis under identical conditions *in vitro*.

Does early organogenesis more often proceed abnormally in embryos of diabetics than in those of controls?

This was answered by a histological study of embryos cultured for 24 h as described above. Our earlier findings (cf. Table 13.1) already indicated that *in vivo*, organogenesis was abnormal more often in embryos of diabetic animals than in controls. But using the culture method, the question could be framed in more precise terms. One could test whether, *given the same chances*, i.e. when exposed to identical environmental conditions and when nourished by a normal animal's serum, the embryos from diabetic rats had any tendency to develop less normally than embryos from non-diabetic rats.

The central nervous system, the heart, the gut, the liver, the somites and the limb-buds were examined in histological sections of embryos from culture. Other than occasional retardation of development, very few abnormalities were seen, and these were all in either the brain or the heart; distortions of the brain vesicles, or reduction in size of the heart chambers. The total results are summarized in Table 13.4. It can be seen that not quite so many of the embryos from diabetic mothers were completely normal histologically as among those from controls. Conversely, significantly more of the embryos from diabetics showed retarded development of their organs than did embryos from controls. There were too few other abnormalities to show significant differences.

Table 13.4 Histological findings in cultured embryos from diabetic and control rats

	Total embryos examined	Normal	Abnormal CNS or heart	Retarded development	Cell death
Diabetic rats	52	21	4	10	20
Control	50	28	2	3	19

Although this is only a small number of data, it does suggest that diabetes in these female rats increases the risk that embryonic development may become retarded in the 10–11 day period of gestation.

Are there differences in growth rate between embryos of diabetic and non-diabetic rats?

To provide a preliminary estimate of growth rate in culture, embryos of equal size were chosen to explant at 10 days' gestation, and their final lengths, as well as the diameter of their membranes, were measured at the end of the culture period. The results are included in Table 13.5. There were no significant differences in size between embryos of diabetic rats and those from controls, or in their membranes. This is specially interesting in view of findings by other workers that fetuses of diabetic rats are smaller than normal[19, 20]. Evidently the size differences seen in late fetuses do not owe their origin to any differences in rate of increase in size that can be detected in these early stages of development. As will be discussed later, however, factors in a diabetic mother's serum may influence the size of early embryos when it is used as culture medium for them.

Growth can be more accurately defined in terms of either increases in cell number or increases in total protein. So the best estimates of growth rate are obtained by measurements of total DNA and protein in the embryos. Data of this kind are now being collected.

As stressed earlier, in the culture system embryos from diabetic and from non-diabetic animals are compared in exactly equivalent environments. So, if there are indeed consistent differences between them under these standard conditions, the differences must be due to influences already imparted from the mother to her embryos before the 10-day stage at which they were explanted. Our data so far indicate, therefore, that the diabetic rat must have some deleterious influence on her embryos during the period between day 0 and day 10 of gestation. This influence, whatever it is, causes them to be less viable and more frequently retarded than the embryos of normal rats, when

Table 13.5 Comparison of development of embryos in normal and diabetic serum

	Externals				Histology				Size		
	% Good	% Fair	% Poor	% Dead	% Normal	% Abnormal	% Retarded	% with Cell death	Diameter of membranes	Crown-rump length	Tail length
(a) Embryos from control rats											
Normal serum	76	5	7	12	56	4	6	19	3.01 ± 0.07	2.18 ± 0.06	1.84 ± 0.06
Diabetic serum	83	7	7	3	78	11	18	1	3.30 ± 0.08	2.45 ± 0.09	2.15 ± 0.07
(b) Embryos from diabetic rats											
Normal serum	55	28	8	8	40	8	19	38	3.00 ± 0.08	2.08 ± 0.06	1.91 ± 0.05
Diabetic serum	58	25	11	6	50	9	29	18	3.18 ± 0.07	2.26 ± 0.09	2.07 ± 0.07

growing in culture during these crucial phases of early organogenesis. This emphasizes the importance of maternal metabolism in the very early stages after fertilization – stages almost never studied in human diabetic pregnancies. As a follow-up to the present findings, biochemical data are now being collected on samples of blood from both diabetic and normal female rats during this early period. It is particularly pertinent now to consider what are the effects of a diabetic female's serum when used as culture medium for 10–11 day embryos.

Does the serum of a diabetic rat have deleterious effects on development of embryos *in vitro*?

To answer this question, individual 10-day embryos from the same litters were cultured in watchglasses in either normal female rats' serum or diabetic female rats' serum. After 24 h their development was compared by the criteria used above. The results are summarized in Table 13.5. It can be seen that there were no differences in the frequencies of 'Good' or 'Fair' embryos judged from external appearance, and no differences in the numbers that were normal or abnormal histologically, when development in the two types of serum was compared. Surprisingly, however, diabetic serum had two *favourable* effects: fewer embryos showed cell death in diabetic serum than in normal serum, and the size of embryos at the end of culture in diabetic serum was significantly larger than in normal serum. These comparisons held both for embryos from non-diabetic rats (Table 13.5a) and for those from diabetic rats (Table 13.5b).

It may be the higher glucose content of diabetic serum that produces these unexpectedly favourable effects on cultured embryos. We have since found that embryos consume *more* glucose when cultured in diabetic serum than when in normal serum (0.58 mg/embryo/24 h and 0.49 mg/embryo/24 h, respectively), if larger volumes, 1 ml per watchglass, are used. So it seems likely that the smaller volume of 0.3 ml of normal serum provided less than ideal quantities of glucose for the embryo's energy requirements. New and Cockroft (personal communication) have found that addition of glucose to culture sera improves the growth of embryos. Excess glucose is to be avoided, however, since it causes malformations[21].

CONCLUSIONS AND FUTURE PROSPECTS

Although all the findings reported here are only preliminary, they show some of the possible applications of this 'embryo culture approach' in investigating maternal influences during early organogenesis. This is not to say that equally important factors operating in late organogenesis, at fetal stages, may not also be investigated by means of cultures *in vitro*, of the particular organs concerned. A line of approach now being pursued in our laboratory, as a follow-up from our findings of delayed sacral ossification in fetuses of diabetic rats (cf. Table 13.2), is to culture the sacral vertebral rudiments *in vitro* and to study their ossification under carefully controlled conditions.

Just as work on fetal physiology in the 1940s to 1960s gained greatly from studies on externalized animal fetuses maintained for short periods in the laboratory[22], so the current clinical studies on maternal metabolism and its effects on the fetus in diabetes could gain from use of the rat embryo as a model, cultured for short periods *in vitro*, with its haemochorial placenta intact[23]. Cultures of individual organs have and will continue to be used too for certain specific problems. Cockroft's method[16] of maintaining rat embryos up to early fetal stages in a continuously circulating medium offers even wider possibilities, for it should enable the effects of fluctuations in the concentrations of maternal metabolites to be studied. Ultimately, the aim of these researches must be to find ways of controlling the composition of the embryo's environment so that congenital abnormalities are prevented.

ACKNOWLEDGEMENTS

I am grateful to the British Diabetic Association for supporting this research, to Mrs C. Jeynes for skilled technical assistance and to Mr M. Alexander for the photographs.

References

1. Miller, H. C. (1946). The effects of diabetic and prediabetic pregnancies on the fetus and newborn infant. *J. Pediatr.*, 29, 455
2. Day, R. E. and Insley, J. (1976). Maternal diabetes and congenital malformations: survey of 205 cases. *Arch. Dis. Child.*, 51, 935
3. Beard, R. W. and Hoet, J. J. (1979). Introduction. In: *Ciba Foundation Symposium No. 63, Pregnancy Metabolism, Diabetes and the Fetus.* (Amsterdam and New York: Elsevier/ Excerpta Medica/North-Holland)
4. Bergstein, N. A. M. (1979). The influence of preconceptional glucose values on the outcome of pregnancy. (*Ibid.*)
5. Naeye, R. L. (1979). The outcome of diabetic pregnancies: a prospective study. *Ibid.*
6. Bennett, P. H., Webner, C. and Miller, M. (1979). Congenital anomalies in the prediabetic and diabetic pregnancy. (*Ibid.*)
7. Freinkel, N. (1979). Physiological Changes in CHO and lipid metabolism. In: Sutherland, H. W. and Stowers, J. (eds.). *2nd Aberdeen International Colloquium on Carbohydrate Metabolism in Pregnancy and the Newborn.* (New York: Springer-Verlag)
8. Milunsky, A. (1973). *The Prenatal Diagnosis of Hereditary Disorders.* (Springfield: C. C. Thomas)
9. Mintz, D. H. and Chez, R. A. (1976). Effects of diabetes mellitus on fetal growth and development. In: S. S. Fajans (ed.). *Diabetes Mellitus.* Fogarty International Center Series on Preventive Medicine, Vol. 4. (Maryland: Nat. Insts. Health)
10. de Gasparo, M., Pictet, R. L., Rall, L. B. and Rutter, J. (1975). Control of insulin secretion in the developing pancreatic rudiment. *Dev. Biol.*, 47, 106
11. Deuchar, E. M. (1977). Embryonic malformations in rats, resulting from maternal diabetes: a preliminary study. *J. Embryol. Exp. Morphol.*, 41, 93
12. New, D. A. T. (1966). Development of rat embryos cultured in blood sera. *J. Reprod. Fertil.*, 12, 509
13. Payne, G. S. and Deuchar, E. M. (1972). An *in vitro* study of functions of embryonic membranes in the rat. *J. Embryol. Exp. Morphol.*, 27, 533
14. New, D. A. T., Coppola, P. T. and Terry, S. (1973). Culture of explanted rat embryos in rotating tubes. *J. Reprod. Fertil.*, 35, 135
15. New, D. A. T. (1967). Development of explanted rat embryos in circulating medium. *J. Embryol. Exp. Morphol.*, 17, 513
16. Cockcroft, D. (1973). Development in culture of rat foetuses explanted at 12.5 and 13.5 days of gestation. *J. Embryol. Exp. Morphol.*, 29, 473

17. New, D. A. T., Coppola, P. T. and Cockroft, D. L. (1976). Comparison of growth *in vitro* and *in vivo* of post-implantation rat embryos. *J. Embryol. Exp. Morphol.*, 36, 133
18. Cockroft, D. L. (1976). Comparison of *in vitro* and *in vivo* development of rat fetuses. *Dev. Biol.*, 48, 163
19. Golob, E. K., Rishi, S., Becker, K. L. and Moore, C. (1970). Streptozotocin diabetes in pregnant and non-pregnant rats. *Metabolism*, 19, 1014
20. Van Assche, F. A. (1975). The fetal endocrine pancreas. In: H. Sutherland and J. Stowers, (eds.) p. 68, *1st Aberdeen International Congress on Carbohydrate Metabolism*. (Edinburgh & London: Churchill Livingstone)
21. Cockroft, D. L. and Coppola, P. T. (1977). Teratogenic effects of excess glucose on headfold rat embryos in culture. *Teratology*, 16, 141
22. Cross, K. W. and Dawes, G. S. (eds.) (1966). The foetus and the new-born: recent research. *Br. Med. Bull.*, 22, 1
23. New, D. A. T. (1978). Whole-embryo culture and the study of mammalian embryos during organogenesis. *Biol. Rev.*, 53, 81

14
Prolonged pregnancy and intra-uterine development

J. A. THLIVERIS

INTRODUCTION

Pregnancy in man is considered prolonged or post-term when it exceeds 294 days (the average duration of normal pregnancy being 280 days, calculated from the first day of the last menstrual period). Approximately 7% of all pregnancies extend beyond 42 weeks and 5% beyond 43 weeks[1,2]. While many post-term infants do not appear adversely affected by prolonged gestation, a significant number display a clinical syndrome commonly referred to as 'postmaturity'. This syndrome is characterized by loss of the vernix caseosa, meconium staining, and a thin physical appearance associated with loose, wrinkled, often parchment-like skin[3,4]. In prolonged pregnancy both normal and postmature fetuses are frequently of higher body weight, compared with normal term fetuses[5]. On the other hand, the mean birth weight of the post-mature fetuses is 200–300 g less than that of the unaffected post-term fetuses[5]. Moreover, the perinatal mortality rate is highest in those post-term fetuses displaying the clinical features associated with postmaturity.

Congenital malformations such as anencephaly, hydrocephaly and osteogenic imperfecta have been correlated with an increased incidence of both prolonged pregnancy and fetal postmaturity[1]. This correlation in cases of anencephaly and hydrocephaly is believed to be related to an altered state in the fetal pituitary–adrenal axis. Several studies have provided evidence that with total or subtotal absence of the fetal pituitary gland, the fetal adrenal becomes hypoplastic. The hypoplastic adrenal in turn is incapable of secreting adequate levels of corticosteroids, which together with maternal oxytocin and possibly prostaglandins, are believed to play an important role in the initiation of labour at term[6-9]. Hence, with hypoplasia of the fetal adrenal gland, pregnancy becomes extended beyond term. Furthermore, according to Anderson et al.[6], the smaller the fetal adrenal gland the greater the length of gestation. On the other hand, when ACTH is given to an anencephalic fetus, delivery occurs prior to term[8]. In contrast to the above studies the findings of Honnebier and Swaab[10] failed to support the concept of the role the fetal adrenal gland has in the initiation of parturition. It was noted that by far the majority of anencephalic fetuses were delivered prior to term whereas only a few extended past term.

One of the main difficulties encountered in studying the clinical problem of postmaturity is the fact that 85% of postmature babies die *in utero*[4], and as a result there is little information as to how an extended pregnancy affects the fetus. An approach to this problem, however, has been to use animal models to investigate postmaturity as it relates to aetiology, fetal development, carbohydrate metabolism, parturition delay, placental insufficiency and alterations in a number of fetal organs.

ANIMAL MODELS AND OBSERVATIONS

Rabbit

Barcroft and Young[11] studied the effects of prolonged pregnancy on oxygen levels in the fetal rabbit brain. Gestation was prolonged in this species beyond term (31 days) to day 35 by a single injection of 100 IU chorionic gonadotrophin and 5 mg progesterone on day 25 of gestation. The results of this study revealed that oxygen saturation of the brain decreased from 50% at term to 17% by day 35. Oxygen saturation in the placenta decreased as well, from 30% at term to 20% by day 35. In addition, fetal weights increased *in utero* from an average of 45 g at term to 80 g on day 35 of gestation. Placental weights did not differ statistically, being on the average 4.7 g at term and during prolonged gestation. Barcroft and Young[11] concluded from their findings that fetal death was most likely caused by fetal anoxia.

Roux *et al.*[12] expanded the above study to include glycolysis and the oxidative cycle in the liver, heart, and in the brain of postmature, newborn and neonatal rabbits. Their method of prolonging gestation was by injecting intraperitoneally 40 units per kilogram of Antuitrin-S (human chorionic gonadotrophin) on day 24 of gestation. Their results showed a 100% increase in oxygen consumption, pyruvate utilization and conversion of pyruvate to carbon dioxide in livers of postmature rabbits compared with the livers of newborn animals. Moreover, glycogen content in postmature livers was markedly less than that found in newborn animals.

When brain tissue was incubated, the oxygen consumption and conversion of pyruvate to lipid in postmature brains was less (18%) than that of the five-day old brain. Comparison of newborn and postmature brains did not reveal significant differences. In contrast, conversion of pyruvate to carbon dioxide was similar in postmature and five-day old brains but 46% higher than that in the newborn brains.

Heart tissue utilized pyruvate and oxygen and converted pyruvate to carbon dioxide at similar rates in all three groups of animals studied. Postmature and newborn hearts contained two to three times more glycogen than that of five-day old hearts. Furthermore, glycogen utilization by the postmature and newborn heart was 200% greater than that of five-day old animals. It was also noted that while the five-day old heart contained higher levels of endogenous lactate, the postmature and newborn hearts produced two to three times more lactate.

In addition to the biochemical data, the authors also noted that experimental postmaturity in the rabbit was accompanied by the same physical features observed by Clifford[4] in the human postmature fetus, and that intrauterine mortality occurred in 60% of the postmature fetuses.

Roux et al.[12] concluded that experimentally induced postmaturity in the rabbit reproduced the clinical condition seen in human postmaturity and serves as a reliable model in assessing the effect of chronic hypoxia on fetal development in vivo.

Sheep

In the sheep, the association between prolonged gestation and congenital malformations was noted in offspring of ewes injesting Veratrum californicum, commonly referred to as skunk cabbage or wild corn. The chemical compounds in V. californicum causing the malformations have been identified by Keeler[13, 14] as cyclopamine (11-deoxyjervine) and cycloposine (3-glucosyl-11-deoxyjervine). Binns et al.[15] reported the presence of cyclopia in the affected newborn lambs as well as fusion of the cerebral hemispheres, reduction in numbers of gyri and sulci, a single fused optic nerve, dilation of the lateral ventricles and agenesis of the pituitary gland. In a more recent study by Van Kampen and Ellis[16], fetal pituitary aplasia, hypoplasia of the thyroid and adrenal glands were noted, in addition to the central nervous system malformations and cyclopia described by Binns et al.[15].

The finding of pituitary agenesis, adrenal hypoplasia and prolonged pregnancy lends additional support to the role of the fetal hypophyseal–adrenal axis in the initiation of parturition in this species[17, 18].

Rat

The laboratory rat has also been used in investigating the effects of prolonged gestation on prenatal growth and development. In a study by Barrow and Rowland[19], gestation was prolonged in order to examine the effect of two to three additional days of intra-uterine growth on abnormalities induced by teratogenic agents.

Two groups of animals served as controls in this investigation. The first group received daily intramuscular injections of 5–10 mg progesterone from days 21 to 26 of pregnancy. The second group consisted of animals in which a ligature was placed around the uterus, proximal to the junction of the horns, on day 14 of gestation. Females of both control groups were killed daily from days 21 to 26 of pregnancy.

Experimental animals were also divided into two groups: one treated only with teratogens, and the other with progesterone plus teratogens. The teratogenic agents were administered as follows: (a) 80 mg per kg chlorocyclizine on days 11–16 and 13–16; 100 mg per kg chlorocyclizine on days 11–16; (b) 100000 or 200000 IU of vitamin A palmitate on days 11–16; (c) 200 mg of B-aminopropionitrile (BAPN) on days 11–16. Animals in the first group were killed on day 21; those in the second group were killed either on day 24 (progesterone plus BAPN or Vitamin A palmitate) or on days 22–24 (progesterone plus chlorcyclizine).

Congenital abnormalities were detected in the offspring of animals from both experimental groups. The following malformations were induced by chlorcyclizine: hydronephrosis, oedema, lack of calcification of the vertebral bodies and undescended testes. All of these abnormalities were accentuated during prolonged pregnancy with one exception; there was no change in the relative position of the undescended testes. It was also noted that when chlorcyclizine was administered in doses of 80 mg per kg on days 13–16, approximately 50% of the fetuses exhibited calcified centra on day 21 and nearly all on day 24. On the other hand, when the same dosage was given on days 11–16, calcification of the centra was absent at day 21 but present in 80% of the fetuses on day 24. Furthermore, when the dosage was increased to 100 mg per kg, calcification was inhibited in a majority of the day 24 fetuses.

In regards to intra-uterine fetal mortality, 50% of the fetuses died by day 24 and 100% by day 26 of gestation, irrespective of whether progesterone administration or uterine ligation was used to prolong gestation. It was also noted that intra-uterine growth continued beyond term with fetal weight increasing up to day 24. Two additional parameters were used to assess continued growth *in utero*: (a) skeletal maturation, which continued past term, and (b) skin pigmentation which normally appears one day after birth was present on days 23 and 24.

Vitamin A palmitate also produced malformations which were present in both experimental groups. Dosages of 100000 units induced limb deformities, varying degrees of mandibular anomalies as well as ectopic calcification adjacent to the facial bones. These features seen at day 21 were more pronounced during prolonged gestation, with severe prognathism being the most prominent. When the dosage of vitamin A was increased to 200000 units, a generalized decrease in calcification occurred in addition to severe deformities of the limbs and face.

B-aminopropionitrile caused marked kyphosis of the vertebral column and contractures of the distal forelimbs; both anomalies persisted during prolonged gestation.

Persaud[20] used an inhibitor of prostaglandin synthesis to study not only the duration of pregnancy but also its effects on prenatal development. Indomethacin was given orally to pregnant rats in concentrations of 25 mg per 2000 ml water daily from the first day of gestation until parturition occurred in the nontreated control animals (day 21.5). Various organs from the treated mothers were examined for evidence of damage resulting from treatment by the drug; however, no changes were encountered. It was noted that in the drug-treated group, gestation was prolonged beyond term, the longest gestational period observed being 24.6 days. In post-dated animals the uterine tubes contained blood which was attributed to separation of the placentae from the wall of the uterus. The average litter size in these animals was reduced, whereas the occurrence of resorption sites was increased significantly compared with that observed in control animals. Moreover, there was a marked increase in the number of dead fetuses (Table 14.1) as well as fetuses exhibiting a variety of malformations (Table 14.2). Also, mean fetal body weights, crown rump lengths and weights of the placentae were significantly higher in the drug treated animals (Table 14.1).

Table 14.1 Indomethacin treatment during pregnancy in rats

	Treatment	
	Indomethacin	Control
No. of pregnant animals	11	6
Implantations	133	82
Gestational duration (in days)*	22.8 ± 0.2†	21.5 ± 0
Resorptions (%)	13 (9.8)†	1 (1.2)
No. of offspring recovered	120	81
Dead (%)	6 (5)†	1 (1.2)
Alive (%)	114 (95)	80 (98.8)
Abnormal (%)	43 (35.8)†	0 (0)
Mean fetal weight (g)*	6.7 ± 0.1†	6.4 ± 0.1
Mean fetal CR length (mm)*	46.0 ± 0.4†	44.0 ± 0.4
Mean placental weight (mg)*	715 ± 0.2†	673 ± 0.4

* Mean ± S.E.
† $p < 0.01$ (Significance of difference from control was evaluated by Student's 't' test)
From Persaud[20]

Table 14.2 Summary of indomethacin-induced fetal abnormalities

Treatment	Fetuses abnormal/ fetuses recovered	Fetal abnormalities
Control	0/81	None
Indomethacin	43/120	Stunting
		Exencephalus (10)
		Hydrocephalus (2)
		Microphthalmia (4)
		Hydronephrosis (25)
		Abnormal position of kidneys (2)
		Poor ossification of skull bones
		Missing metacarpals
		Missing phalanges
		Wavy ribs
		Necrotic liver (4)
		Cerebral (1), renal (10)
		and intraperitoneal haemorrhage (5)

From Persaud[20]

These results demonstrated that indomethacin not only prolonged gestation but also induced severe fetal malformations. Prolongation of pregnancy was undoubtedly related to the inhibition of synthesis and release of prostaglandins which are now known to be involved in the initiation of parturition in several species including man[21–23].

Studies in our laboratory have focussed not only on the effects of prolonged gestation on various organs of the fetus but also on the effects of exogenous progesterone on the fetus prior to term[24–28]. Prolonged gestation was induced by subcutaneous injection of 5 mg progesterone daily from day 20 through 24 of gestation. In order to evaluate the effect of exogenous progesterone on fetal organs and the placenta, another group of animals received 5 mg progesterone daily from day 14 through day 21 of gestation. Fetal tissues and placentae from this group were removed at term (day 22) and compared with fetal tissues and placentae obtained from normal term females.

Observations on the external features of the fetuses revealed no differences between those from 'preterm progesterone' treated and nontreated term animals. Neither were any changes detected on day 23; however, all viable day 24 and day 25 fetuses exhibited features typical of the postmaturity syndrome. Fetal malformations were not detected in any of the offspring.

The fetal organs studied to date, in addition to the placenta, have been the adrenal, thyroid and parathyroid glands, as well as the endocrine pancreas and liver. Electron microscopy revealed that morphological changes occurred only on days 24 and 25; fetal tissues and placentae from preterm progesterone treated and day 23 animals were similar to term controls.

The adrenal cortex revealed cellular hypertrophy, numerous profiles of dilated smooth endoplasmic reticulum, enlarged mitochondria and reduction in lipid content. Cells of the medulla also contained enlarged mitochondria along with a depletion of catecholamine storage granules. Similar results have been reported in adult animals subjected to stress[29, 30].

Fetal hepatocytes on days 24 and 25 contained less glycogen than at term (Figures 14.1 and 14.2). There was also a marked increase in the quantity of smooth endoplasmic reticulum, an organelle containing glucose-6-phosphatase which is essential in glycogenolysis[31]. Similar features are present in hepatocytes of newborn rats[32] and fasting adults[33].

Alpha and beta cells of the pancreas at term contained numerous glucagon and insulin storage granules respectively. Both cells types are functional

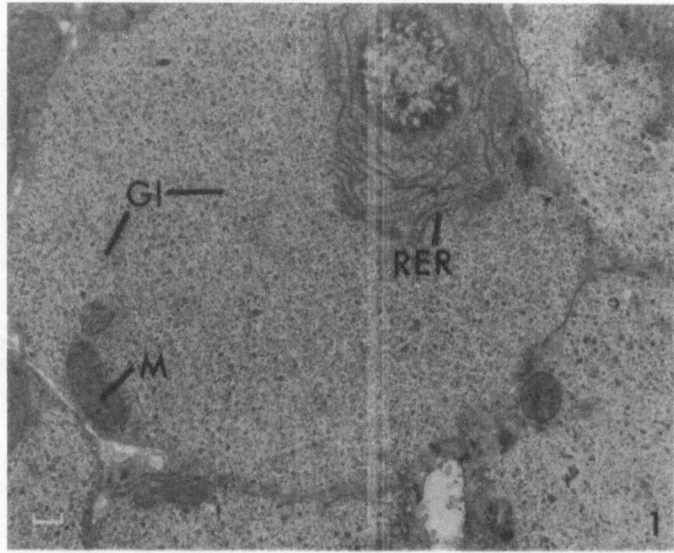

Figure 14.1   Fetal hepatocytes at term (day 22) are characterized by the presence of copious amounts of glycogen (Gl). Rough surfaced endoplasmic reticulum (RER), mitochondria (M)
× 4160

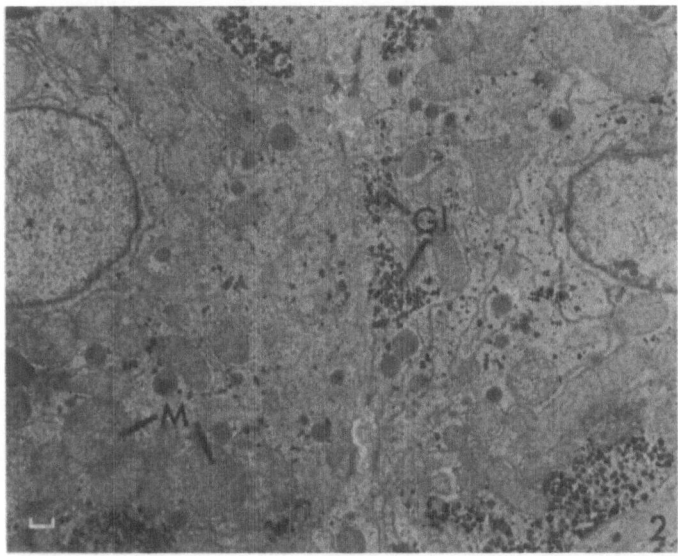

Figure 14.2 Fetal hepatocytes on day 25 of gestation exhibit marked reduction in glycogen (Gl). Mitochondria (M) × 2880

during gestation and respond to fluctuations in glucose levels[34]. During prolonged gestation there occurred an accumulation of beta cell granules whereas the alpha cells revealed a decrease in granule numbers as well as increased Golgi complex activity. These morphological features coupled with those of the liver are indicative of decreased peripheral blood glucose levels, i.e. hypoglycaemia, which was substantiated by biochemical data (Table 14.3).

Table 14.3 Fetal rat blood glucose levels

Group	Number of fetuses	Mean ± SEM (mg%)
1	9	58.8 ± 7.77
2	10	65.6 ± 2.26
3	8	54.6 ± 1.96
4	10	37.1 ± 5.17
5	6	38.0 ± 6.50

Groups: 1 = Preterm progesterone treated; 2 = Term (Day 22); 3 = Prolonged gestation (Day 23); 4 = Prolonged gestation (Day 24); 5 = Prolonged gestation (Day 25)

From Thliveris, J. A.[28]

As is the case for the fetal adrenal[35] and endocrine pancreas[34, 36], the thyroid and parathyroid glands of the fetus are known to be autonomous with respect to maternal hormonal influence[37–39]. Moreover, both organs respond to fluctuations in calcium levels[40–42].

In contrast to the above findings, no morphological changes were encountered in either the fetal thyroid or the parathyroid glands, suggesting that

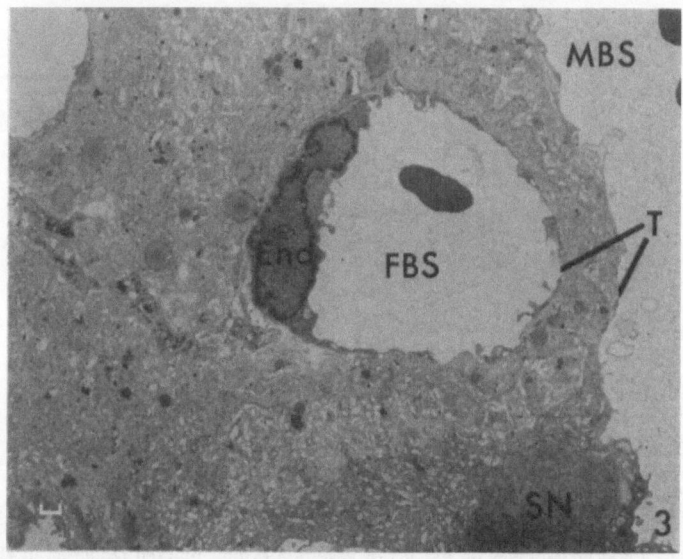

Figure 14.3   The placental labyrinth at term (day 22) consists of maternal (MBS) and fetal (FBS) blood spaces separated by trophoblast tissue (T) and fetal endothelium (End.). Syncytio-trophoblast nucleus (SN) ×3040

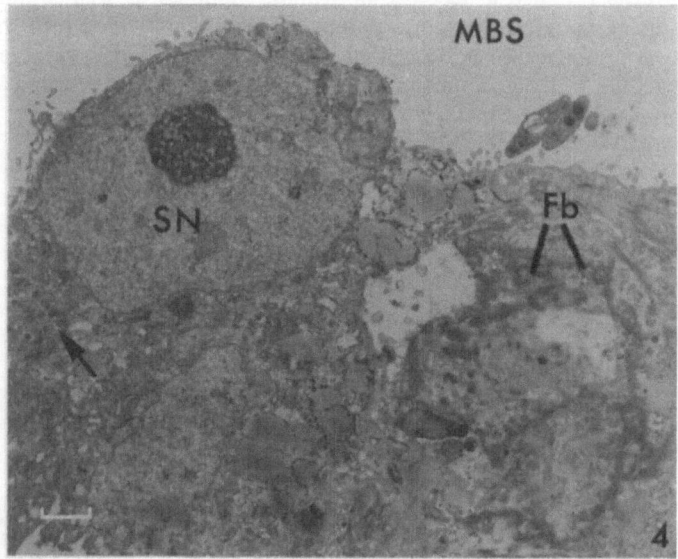

Figure 14.4   The placental labyrinth on day 25 of gestation is characterized by disruption of the trophoblast elements (arrow) and presence of fibrinoid material (Fb). Maternal blood space (MBS), syncytiotrophoblast nucleus (SN) × 6880

fetal calcium levels are probably within normal limits during prolonged gestation.

Morphological changes were observed in the placenta on days 24 and 25 of gestation. When compared with placentae at term, the layers of trophoblast became discontinuous with loss of cytoplasmic organelles. Fibrinoid material was also detected throughout the trophoblast layers (Figures 14.3 and 14.4). The changes observed in the placenta are indicative of 'placental insufficiency' which has been suggested by several authors to be the probable cause of fetal distress encountered in the postmature fetuses (see Thliveris[28] for review of the literature).

CONCLUDING REMARKS

Postmaturity in man ranks second only to prematurity as a cause of perinatal mortality[4]. This is based on the premise that as gestation extends beyond term, the greater the likelihood for development of an adverse intra-uterine environment which is reflected by changes in fetal homeostasis. Postmature fetuses do in fact suffer in this condition as is seen by the presence of lung complications, hypovolaemia, metabolic acidosis, dehydration and symptoms of cerebral hypoxia[43]. Moreover, there is also intra-uterine weight loss in these fetuses[44,45], as well as an increased tendency towards neonatal hypoglycaemia[46]. There is concern for the well-being of these infants because prolonged episodes of hypoxia and or hypoglycaemia may lead to severe complications such as brain damage[47,48].

The various changes in fetal homeostasis leading to fetal distress and in some instances fetal death, have been thought to result from placental insufficiency. The precise cause or causes of placental insufficiency has remained an unsolved problem. Several lines of evidence have suggested that loss of placental integrity and deposition of fibrinoid material may be due to alteration in placental haemodynamics[49,50], or result from an auto-immune phenomenon[51,52]. On the other hand there is evidence contrary to the auto-immune theory in that the placental trophoblast lacks sufficient antigenicity to warrant an immunological reaction resulting in rejection[53,54].

Regardless of the precise aetiology of placental insufficiency, it is apparent that lengthy duration of this condition is detrimental to the well-being of the fetus. The aforementioned experimental models of prolonged gestation have been of value for investigating the adverse influence of an altered maternal condition on feto-placental development. It is anticipated that further studies of this problem will provide much needed information in assessing not only fetal growth and development but also prognosis for survival.

References

1. Clifford, S. H. (1957). Postmaturity. *Adv. Pediatr.*, 9, 13
2. Perkins, R. P. (1974). Antenatal assessment of fetal maturity, *Obstet. Gynecol. Surv.*, 29, 369
3. Kortenover, M. E. (1950). Pregnancy of long duration and postmature infant. *Obstet. Gynecol. Surv.*, 5, 812
4. Clifford, S. H. (1954). Postmaturity with placental dysfunction. *J. Paediatr.*, 44, 1

5. Sjostedt, S., Engleson, G. and Rooth, G. (1958). Dysmaturity. *Arch. Dis. Child.*, 33, 123
6. Anderson, A. B. M., Laurence, K. M. and Turnbull, A. C. (1969). The relationship in anencephaly between the size of the adrenal cortex and the length of gestation. *J. Obstet. Gynaecol. Br. Commonw.*, 76, 196
7. Anderson, A. B. M., Laurence, K. M., Davis, K., Campbell, H. and Turnbull, A. C. (1971). Fetal adrenal weight and the cause of premature delivery in human pregnancy. *J. Obstet. Gynecol. Br. Commonw.*, 78, 481
8. Price, H. V., Cone, B. A. and Keogh, M. (1971). Length of gestation in congenital hyperplasia. *J. Obstet. Gynaecol. Br. Commonw.*, 78, 430
9. Murphy, B. E. P. (1973). Does the human fetal adrenal play a role in parturition? *Am. J. Obstet. Gynecol.*, 115, 521
10. Honnebier, W. J. and Swaab, D. F. (1973). The influence of anencephaly upon intrauterine growth of fetus and placenta and upon gestation length. *J. Obstet. Gynaecol. Br. Commonw.*, 80, 577
11. Barcroft, J. and Young, M. I. (1945). Internal oxygen environment of the brains of postmature rabbit embryos. *J. Exp. Biol.*, 21, 70
12. Roux, J. F., Romney, S. L. and Dinnerstein, M. A. (1964). Environmental and aging effects of postmaturity on fetal development and carbohydrate metabolism. *Am. J. Obstet. Gynecol.*, 90, 546
13. Keeler, R. F. (1969). Teratogenic compounds of *Veratrum californicum* (Durand). VI. The structure of cyclopamine. *Phytochemistry*, 8, 223
14. Keeler, R. F. (1969). Teratogenic compounds of *Veratrum californicum* (Durand). VII. The structure of the glycosidic cycloposine. *Steroids*, 13, 579
15. Binns, W., Thacker, E. J., James, L. F. and Huffman, W. T. (1959). A congenital cyclopian-type malformation in lambs. *J. Am. Vet. Med. Assoc.*, 134, 180
16. Van Kampen, K. R. and Ellis, L. C. (1972). Prolonged gestation in ewes ingesting *Veratrum californicum*: morphological changes and steroid biosynthesis in the endocrine organs of cyclopic lambs. *J. Endocrinol.*, 52, 549
17. Liggens, G. C. (1969). Premature delivery of foetal lambs infused with glucocorticoids. *J. Endocrinol.*, 45, 515
18. Liggens, G. C., Kennedy, P. C. and Holm, L. W. (1971). Failure of initiation of parturition after electrocoagulation of the pituitary of the fetal lamb. *Am. J. Obstet. Gynecol.*, 98, 1080
19. Barrow, M. V. and Rowland, C. A. (1969). Prolonged gestation in the rat and its usefulness in experimental teratology. *Teratology*, 2, 125
20. Persaud, T. V. N. (1974). Inhibitors of prostaglandin synthesis during pregnancy. 2. The effects of indomethacin in pregnant rats. *Anat. Anz.*, 136, 354
21. Liggens, G. C. (1973). The physiological role of prostaglandins in parturition. *J. Reprod. Fertil.*, 18 (Suppl.), 143
22. Eliasson, R. (1973). Prostaglandins and reproduction: a general survey. *J. Reprod. Fertil.*, 18 (Suppl.), 127
23. Labhsetwar, A. P. (1974). Prostaglandins and the reproductive cycle. *Fed. Proc.*, 33, 61
24. Thliveris, J. A. and Connell, R. S. (1973). Ultrastructure of the fetal rat adrenal gland at full-term and during prolonged gestation. *Anat. Rec.*, 175, 607
25. Thliveris, J. A. (1974). Ultrastructure of fetal liver at term and during prolonged gestation in the rat. *Am. J. Obstet. Gynecol.*, 118, 864
26. Thliveris, J. A. (1975). Fine structure of foetal rat pancreatic alpha and beta cells at term and during prolonged gestation. *Virch. Arch. B*, 19, 157
27. Thliveris, J. A. (1976). Fine structure of the fetal rat thyroid and parathyroid glands at term and during prolonged gestation. *Cell Tiss. Res.*, 166, 421
28. Thliveris, J. A. (1976). Fine structure of the placental labyrinth in the rat at term and during prolonged gestation. *Virch. Arch. B*, 21, 169
29. Swartz, W., Merker, H. J. and Suchowski, G. (1962). Elektronenmikroskopische Untersuchungen über die Wirkungen von ACTH und Stress auf die Nebennierenrinde der Ratte. *Virch. Arch. Pathol. Anat.*, 335, 165
30. D'Anzi, F. A. (1969). Morphological and biochemical observations on the cathecholamine storing vesicles of rat adrenomedullary cells during insulin induced hypoglycemia. *Am. J. Anat.*, 125, 381

31. Leskes, A., Siekevitz, P. and Palade, G. E. (1971). Differentiation of endoplasmic reticulum in hepatocytes. I. Glucose-6-phosphatase distribution *in situ. J. Cell Biol.*, 49, 264
32. Dallner, G., Siekevitz, P. and Palade, G. E. (1966). Biogenesis of endoplasmic reticulum membranes. II. Synthesis of constitutive microsomal enzymes in developing rat hepatocyte. *J. Cell Biol.*, 30, 97
33. Cardell, R. R. (1971). Action of metabolic hormones on the fine structure of rat liver cells. I. Effects of fasting on the ultrastructure of hepatocytes. *Am. J. Anat.*, 131, 21
34. Girard, J. R., Kervan, A., Soufflet, E. and Assan, R. (1974). Factors affecting the secretion of insulin and glucagon by rat fetus. *Diabetes*, 23, 310
35. Jost, A., Jacquot, R. and Cohen, A. (1962). The pituitary control of the fetal adrenal cortex. In: T. Symington, J. K. Grant and A. R. Currie (eds.). *The Human Adrenal Cortex*, pp. 569–579 (Baltimore: Williams and Wilkins)
36. Goodner, C. J. and Freinkel, N. (1959). The effect of pregnancy on insulin metabolism in the rat. *Clin. Res. Proc.*, 7, 247
37. Garel, J. M., Milhaud, G. and Sizoneko, P. (1969). Thyrocalcitonine et barrière placentaire chez le rat. *C.R. Acad. Sci. (D)*, 269, 1785
38. Garel, J. M. and Dumont, C. (1972). Distribution and inactivation of labelled parathyroid hormone in rat fetus. *Horm. Metab. Res.*, 4, 217
39. Fisher, D. A., Hobel, C. J., Garze, R. and Pierce, C. A. (1970). Thyroid function in the preterm fetus. *Pediatrics*, 46, 208
40. Cameron, D. A. (1968). Fine structure and function of thyroid 'C' cells and parathyroid cells. In: R. U. Talmage, and C. F. Belanger (eds.). *Parathyroid Hormone and Thyrocalcitonin (Calcitonin)*; Proc. 3rd Parathyroid Conf., pp. 437–439 (Amsterdam: Excerpta Medica)
41. Garel, J. M. (1971). Fetal calcemia and fetal parathyroids. *Isr. J. Med. Sci.*, 7, 349
42. Roth, S., Feinblatt, J. D. and Raise, L. G. (1974). Effect of calcium on embryonic rat thyroid C cells *in vitro. Am. J. Pathol.*, 75, 27
43. McKay, R. J. and Smith, C. A. (1964). Postmaturity and placental dysfunction. In: W. E. Nelson (ed.), *Textbook of Pediatrics*, 8th Ed. pp. 358–359. (Philadelphia: W. B. Saunders)
44. McClure Browne, J. C. (1963). Postmaturity. *Am. J. Obstet. Gynecol.*, 85, 573
45. Gruenwald, P. (1964). The fetus in prolonged pregnancy. *Am. J. Obstet. Gynecol.*, 89, 503
46. Khattab, A. K. and Forfur, J. O. (1971). The interrelationship between calcium, phosphorus, and glucose levels in mother and infant in conditions commonly associated with placental insufficiency. *Biol. Neonate*, 18, 1
47. Haworth, J. C., Coodin, F. J., Finkel, K. C. and Weidman, M. L. (1963). Hypoglycemia associated with symptoms in the newborn period. *Can. Med. Assoc. J.*, 88, 23
48. Anderson, J. M., Milner, R. D. G. and Strich, S. J. (1966). Pathological changes in the nervous system in severe neonatal hypoglycemia. *Lancet*, ii, 372
49. Fox, H. (1967). Perivillous fibrin deposition in the human placenta. *Am. J. Obstet. Gynecol.*, 98, 245
50. Maclennen, A. H. and Sharp, F. (1972). The ultrastructure of human trophoblast in spontaneous and induced hypoxia using a system of organ culture. *J. Obstet. Gynaecol. Br. Commonw.*, 79, 113
51. Burstein, R., Frankel, S., Soule, S. D. and Blumenthal, H. T. (1973). Aging of the placenta: Autoimmune theory of senescence. *Am. J. Obstet. Gynecol.*, 116, 271
52. Gille, J., Borner, P., Reinecke, J., Krause, P. and Deicher, H. (1974). Über die Fibrinoid Ablagerugen in den Endozotten der menschlichen Plazenta. *Arch. Gynäk.*, 217, 271
53. Douglas, G. W. (1965). The immunologic role of the placenta. *Obstet. Gynecol. Surv.*, 20, 442
54. Simmons, R. L., Cruse, V. and McKay D. G. (1967). The immunologic problem of pregnancy. II. Ultrastructure of isogenic and allogeneic trophoblast transplants. *Am. J. Obstet. Gynecol.*, 97, 218

Index